Cargo Work

T0300033

An essential reference written for the marine industry and seafarers around the world, *Cargo Work* has been fully revised and expanded to cover the key classes of cargo, regarding the handling, stowage and carriage of all major commodities by marine transports.

The book provides a general guide to the movement of a wide range of cargoes safely, under the latest international regulations affecting all cargo work, equipment and operational systems. The work includes marine movements in both the passenger and offshore environments as well as the unit load systems of containerisation, Roll on, Roll off practice, hazardous goods and project cargoes.

- Covers the latest International Maritime Organization (IMO) Codes, plus key elements of the International Port and Ship Security Code (ISPS)
- Includes a new chapter on Heavy Lift Practice and Project Cargo
- Updated throughout with colour diagrams and photographs

Cargo Work 9th edition is a crucial reference for both maritime students and serving crew.

D.J. House has now written and published 19 marine titles, many of which are in multiple editions. After commencing his seagoing career in 1962, he was initially engaged on general cargo vessels. He later experienced worldwide trade with cargo passenger, container, Ro-Ro, reefer ships and bulk cargoes. He left the sea in 1978 with a Master Mariner's qualification and commenced teaching at the Fleetwood Nautical College. He retired in 2012 after 33 years of teaching in nautical education. He continues to write and research maritime aspects for future works.

Other Works Published by D.J. House

Seamanship Techniques (4th Edition) (2014), Routledge.
ISBN 9780415829526 (hbk), 9780415810050 (pbk), 9780203796702 (ebk)

Seamanship Techniques Volume III 'The Command Companion' (2000), Butterworth/Heinemann.
ISBN 0750644435

Marine Survival (3rd Edition) (2011), Witherby Publishing Group.
ISBN: 9781856093552 (hbk), 9781856094856 (ebk)

Navigation for Masters (4th Edition) (2012), Witherby Publishing Group.
ISBN 9781856094030, (ebk) 9781856095099

An Introduction to Helicopter Operations at Sea – A Guide for Industry (2nd Edition) (1998), Witherby Publishing Group.
ISBN 1856091686

Anchor Practice – A Guide for Industry (2001), Witherby Publishing Group.
ISBN 1856092127

Marine Ferry Transports – An Operator's Guide (2002), Witherby Publishing Group.
ISBN 1856092313

Dry Docking and Shipboard Maintenance (2003), Witherby Publishing Group.
ISBN 1856092453

Heavy Lift and Rigging (2005), Brown, Son & Ferguson.
ISBN 0851747205

Seamanship Examiner (2005), Elsevier.
ISBN 075066701X

The Ship Handling (2007), Elsevier.
ISBN 9780750685306

Elements of Modern Ship Construction (2010), Brown, Son & Ferguson.
ISBN 9780851748146

The Ice Navigation Manual (2010), Witherby Seamanship International.
ISBN 9789053315989 (joint authorship)

Marine Emergencies (2014), Routledge.
ISBN 9781138020450 (pbk), 9781315770697 (ebk)

Also:

Marine Technology Reference Book (Safety Chapter) (1990), Nina Morgan (ed.), Butterworths.
ISBN 0408027843

Cargo Work
For Maritime Operations

NINTH EDITION

D.J. House

Based on the original series by Kemp & Young

Routledge
Taylor & Francis Group

LONDON AND NEW YORK

Designed cover image: © Getty Images/narvikk

Ninth edition published 2024
by Routledge
4 Park Square, Milton Park, Abingdon, Oxon, OX14 4RN

and by Routledge
605 Third Avenue, New York, NY 10158

Routledge is an imprint of the Taylor & Francis Group, an informa business

© 2024 D.J. House

First edition published by Stanford Maritime Ltd 1960
Eighth edition published by Routledge 2015

British Library Cataloguing-in-Publication Data
A catalogue record for this book is available from the British Library

Library of Congress Cataloging-in-Publication Data
Names: House, D. J., author.
Title: Cargo work : for maritime operations / David J House.
Description: Ninth edition. | Abingdon, Oxon ; New York, NY : Routledge, 2024. | "Based on the original series by Kemp & Young."—Title page. | Includes bibliographical references and index.
Identifiers: LCCN 2023033413 (print) | LCCN 2023033414 (ebook) | ISBN 9781032526522 (hardback) | ISBN 9781032331843 (paperback) | ISBN 9781003407706 (ebook)
Subjects: LCSH: Cargo handling. | Ships—Cargo. | Stowage.
Classification: LCC VK235 .H68 2024 (print) | LCC VK235 (ebook) | DDC 623.88/81—dc23/eng/20230922
LC record available at https://lccn.loc.gov/2023033413
LC ebook record available at https://lccn.loc.gov/2023033414

ISBN: 978-1-032-52652-2 (hbk)
ISBN: 978-1-032-33184-3 (pbk)
ISBN: 978-1-003-40770-6 (ebk)

DOI: 10.4324/9781003407706

Typeset in Berling and Futura
by Apex CoVantage, LLC

Disclaimer: The author's views and expressions do not necessarily reflect the views and insight of Statutory Authorities. Data used has been sourced from worldwide information and has not been authenticated by any other authority. The author does not claim 100% accuracy nor accepts responsibility for opinion and does not constitute guidelines or recommendations for any course of action taken by the reader.

Contents

Preface to 9th Edition

Cargo Operations of the maritime sector continue to evolve with containerisation playing a greater and leading role across all aspects of commodity movement.

The Container ships have become much larger, carrying increased tonnage in excess of 23,000 teu. The subsequently increased draughts of these vessels have forced many of the major ports to increase their own water depths, to be able to continue alongside working by the Container Gantries.

Door-to-door deliveries have become the accepted norm, involving container units via Ro-Ro activity and trans-shipments to all regions of the globe. With such changes, freight rates have increased, generating economic problems for shippers and manufacturers alike. All have experienced changes with the Global Pandemic and the legislation changes with the BREXIT event of 2020. Cargo movement suffered from extensive administration and customs changes. With Harbour blockages alongside berths, increasing fuel and labour costs, shipping overheads have been pushed to increasing limits.

> The saving grace of shipping, is the fact that we all rely on the food chain being sustained. Extensive trade in foodstuffs, especially to Island communities, is an essential element in sustaining the populace. Alongside the working of bulk products like sugar, wine and cereals, etc., operators are engaged in oil, gas and ore movements to fuel modern day living.

Changes are taking place to increase efficiency, right across the shipping fraternity to meet the growing needs to combat climate change. Cargo movement will be under the spotlight, as fossil fuels find their appeal waning with an increase in cleaner energies. Hydrogen gas imports can expect to increase within the tanker sector as offshore oil and gas platforms experience decommissioning, as and when they become no longer operationally economic.

The heavy lift sector has expanded with the project cargoes overlapping into the offshore geographic regions. The environmental changes continue to occur, like the expansion of the Panama Canal which will only enhance trade. All the continents are developing energy resources alongside their transport infrastructures and will rely ultimately on a world of efficient shipping.

Preface to 1st Edition

This book has been written to serve as an introduction to cargo work both for those newly at sea and those whose experience of cargo has been confined to one or two trades.

We are mindful that practical experience is one of the most important factors contributing to the successful carriage of cargo and we would emphasise that this book will supplement, not supplant, the knowledge gained through experience.

Particular emphasis has been placed throughout on the safety of the ship and we are indebted to the controller of her Majesty's Stationary Office for permission to reproduce the extracts from the Ministry of Transport 'M' Notices, the Statutory Instruments concerning grain and timber and the extracts from the Dangerous Goods Rules. These latter extracts have also been approved by the Marine Safety Division of the Ministry of Transport.

We acknowledge with thanks the assistance given by those who have helped to prepare the text and who have contributed drawings.

J. F. Kemp
Peter Young

About the Author

David House commenced his seagoing career in the 'General Cargo' ships of the early 1960s. During his seagoing career he experienced container movements, reefer cargoes, heavy lifts, hazardous goods, livestock, bulk commodities, Roll on, Roll off cargoes as well as periods with passengers. His seagoing activities took him into the Atlantic, the Pacific and the Indian oceans, as well as the Mediterranean and Baltic Sea ports.

His experiences included the carriage of many specific cargoes such as bulk sugar, coal, tallow and grain. Other cargo parcels including heavy lifts, palletised commodities, cased glass, hides, chilled fruit, frozen meats, together with all kinds of steelwork and forestry products. Additionally, special cargoes of mails, bank notes, spirits, munitions and stamps were all transported in one form or another.

His later time was involved with containers from Europe to North America, short sea ferry routes and coastal movements to and from the United Kingdom and European ports. He gained extensive Roll on, Roll off experience on the Irish Sea routes prior to taking up a lecturing post at the Fleetwood Nautical College.

He has recently retired from 34 years of teaching seamanship, navigation and most other marine disciplines. His research and continued writing activities have included nineteen published marine textbooks which are read widely around the world's maritime nations.

Acknowledgements

Allseas Group S.A. Switzerland
B&V Industrietechnik GmbH
BigLift Shipping B.V. Amsterdam, Netherlands
Boskills-Dockwise Netherlands
British Nuclear Fuels (Shipping)
British Standards Institution
Brown, Son and Ferguson, Ltd. (marine publishers)
Bruntons (Musselburgh) Ltd.
Dubai Dry Docks UAE
International Maritime Organization (publications)
James Fisher Shipping Company
Lisnave Estaleiros Navais, S.A. Setúbal, Portugal
Loveridge Ltd.
MacGregor International Organisation
Maritime and Coastguard Agency
Mitsubishi Heavy Industries Ltd., Shimonoseki Shipyard & Machinery Works
Motor Ship (published by IPC Industrial Press Ltd.)
Overseas Containers Ltd.
P&O European (Irish Sea) Ferries
Scheuerle Fahrzeugfabrik GmbH
Seaform Design (Isle of Man)
Smit International
TTS – Mongstad AS Marine Cargo Gear
Witherby Publishing Group Ltd.

ADDITIONAL PHOTOGRAPHY

Capt. K. B. Millar, Master Mariner, Lecturer, Senior Nautical Studies,
Millar Marine Services (Deceased)
Capt. J. G. Swindlehurst (MN) Master Mariner
Capt. A. Malpass (MN) Master Mariner

Capt. D. MacNamee (MN) Master Mariner, FNI
Mr M. Gooderman, Master Mariner, B.A. Lecturer Nautical Studies
Mr G. Edwards, Chief Engineer (MN) (rtd.)
Mr P. Brooks, Chief Officer (MN)
Mr J. Leyland, Nautical Lecturer
Mr M. Ashcroft, Nautical Lecturer
Mr S. Trivedi, Chief Officer (MN)
Miss Martel Fursden, Deck Officer (MN)
IT Consultant: Mr C. D. House
Additional research: Mr A. P. G. House

The author would like to thank all the companies and the many individuals who have contributed and helped in the formulation of this illustrated text, the assistance has been greatly appreciated.

List of Abbreviations used in the context of cargo work and this text

°A	Degrees Absolute
AAA	Association of Average Adjusters
ABS	American Bureau of Shipping
AC	Alternating Current
AHV	Anchor Handling Vessel
AIS	Automatic Identification System
AMD	Advanced Multi-Hull Design
B	Representative of the ship's Centre of Buoyancy
B/A	Breathing Apparatus
BACAT	Barge Catamaran
BCH	Bulk Chemical Code (now IBC Code)
BLU (code)	Code of Practice for Loading and Unloading of Bulk Cargoes
BNFL	British Nuclear Fuels Ltd.
BOG	Boil Off Gas
BOHS	British Occupational Hygiene Society
BP	British Petroleum
BS (i)	Breaking Strength
BS (ii)	British Standard
BS (iii)	Broken Stowage
BSI	British Standards Institute
BST	British Summer Time
BT	Ballast Tank
BV	Bureau Veritas
BWM	Ballast Water Management (systems)
°C	Centigrade
CAS	Condition Assessment Scheme
CBM	Conventional Buoy Mooring
CBT	Clean Ballast Tank
CCTV	Close Circuit Television
CEU	Car Equivalent Unit
CG	Coast Guard
CGLC	Compressed Gas Liquid Carriers
cms	centimetres

Ch/Off (C/O)	Chief Officer
CL	Centre Line
CNG	Compressed Natural Gas
CoF (i)	Certificate of Fitness
CoF (ii)	Centre of Flotation
COSHH	Control of Substances Hazardous to Health
C of G	Centre of Gravity
CoT	Change of Trim
COW	Crude Oil Washing
CO_2	Carbon Dioxide
C/P	Charter party
CSC	Certificate for Safety of Container
CSM	Cargo Securing Manual
CSO	Company Security Officer
CSS	Cargo Stowage and Securing (IMO Code of Safe Practice of)
CSWP	Code of Safe Working Practice
CTU	Cargo Transport Unit
cu	Cubic
D	Density
D (dia)	Diameter
DC	Direct Current
DGN	Dangerous Goods Notice
DNV	Det Norske Veritas
DOC	Document of Compliance
DP	Dynamic Positioning
DSV	Diving Support Vessel
DWA	Dock Water Allowance
Dwt	Deadweight tonnage.
EC	European Community
ECPD	Export Container Packing Document
EDI	Electronic Data Interchange
EEBD	Emergency Escape Breathing Device
EFSWR	Extra Flexible Steel Wire Rope
EMS	Emergency Response Guide
EMSA	European Maritime Safety Agency
ETA (i)	Estimated Time of Arrival
ETA (ii)	Emergency Towing Arrangement
ETV	Emergency Towing Vessel
EU	European Union
°F	Fahrenheit
F (i)	Fresh
FloFlo	Float On, Float Off
FMEA	Failure Mode & Effect Analysis
F.O.	Fuel Oil
F.Pk	Fore Peak (tank)
FPSO	Floating Production Storage Offloading System

FSE	Free Surface Effect
FSM	Free Surface Moment
FSRU	Floating Storage and Re-gasification Unit
FSU	Floating Storage Unit
FSWR	Flexible Steel Wire Rope
ft	Feet
FW	Fresh Water
FWA	Fresh Water Allowance
G	Representative of Ship's Centre of Gravity
G/A	General Average
gals	Gallons
GG_1	Representation of the movement of the ship's C of G when moving a weight aboard the vessel
GHz	Gigahertz
GL	Germanischer Lloyd (Classification Society)
GM	Metacentric Height
GMT	Greenwich Mean Time
GP	Greenhouse Potential
GPS	Global Positioning System
grt	Gross Registered Tonnage
GZ	Representative of Ship's Righting Lever
H_2	Hydrogen (gas)
HCFC	Hydro chlorofluorocarbons
HDFD	Heavy Duty, Floating Derrick
HFC	Hydrofluorocarbon
HMPE	High Molecular Weight Polyethylene
HMSO	Her Majesty's Stationery Office
HNS	Hazardous and Noxious Substance
HP (i)	High Pressure
HP (ii)	Horse Power
HSC	High Speed Craft
HSE	Health and Safety Executive
HSMS	Hull Stress Monitoring System
HSSC	Harmonised System of Survey and Certification
IACS	International Association of Classification Societies
IALA	International Association of Lighthouse Authorities
IAPPC	International Air Pollution Prevention Certificate
IBC	International Bulk Chemical (code)
ICS	International Chamber of Shipping
IG	Inert Gas
IGC	International Code of the Construction and Equipment of Ships Carrying Liquefied Gases in Bulk (2016 Edition)
IGS	Inert Gas System
ILO	International Labour Organization
IMDG	International Maritime Dangerous Goods (code)
IMFO	International Maritime Fumigation Organisation

IMO	International Maritime Organization
IMSBC	International Maritime Solid Bulk Cargoes (code)
IOPP	International Oil Pollution Prevention (certificate)
ISGOTT	International Safety Guide for Oil Tankers & Terminals
ISM	International Safety Management
ISPS	International Ship and Port facility Security (code)
ISSC	International Ship Security Certificate
ITU	Inter-modal Transport Unit
ISU	International Salvage Union
K	Representative of the ship's Keel
kgs	Kilograms
KM	Representative of the distance from the ship's keel to the Metacentre
kN	Kilo-Newtons
kPa	Kilopascal
kts	Knots
kW	Kilowatt
LASH	Lighter Aboard Ship (system)
LAT	Lowest Astronomical Tide
LBP	Length Between Perpendiculars
lbs	Pounds
LCB	Longitudinal Centre of Buoyancy
LCF	Longitudinal Centre of Floatation
LCG	Longitudinal Centre of Gravity
LEL	Lower Explosive Limit
LFL	Lower Flammable Limit
L/H	Lower Hold
LNG	Liquefied Natural Gas
LOA	Length Over All
LOLER	Lifting Equipment Regulations-2006 Inspection and Testing.
Lo-Lo	Load On, Load Off
LP	Low Pressure
LPG (i)	Liquid Propane Gas
LPG (ii)	Liquid Petroleum Gas
LR	Lloyds Register
LRS	Lloyds Register of Shipping
LSA	Life Saving Appliances
LW	Low Water
m	Metres
M	Representative of the ships Metacentre
MA	Mechanical Advantage
MAIB	Marine Accident Investigation Branch
MAREP	Marine Reporting System
MARPOL	Maritime Pollution (Convention)
MARV	Maximum Allowable Relief Valve
MBL	Minimum Breaking Load

MCA	Maritime and Coastguard Agency
MCTC	Moment to Change Trim per Centimetre
MEPC	Marine Environment Protection Committee
MFAG	Medical First Aid Guide (for use with accidents involving dangerous goods)
MGN	Marine Guidance Notice
MIN	Marine Information Notice
MLC	Maritime Labour Convention
mm	Millimetres
MN	Mercantile Marine (Merchant Navy)
MMS	Minerals Management Service
MMSI	Maritime Mobile Service Identity Number
MODU	Mobile Offshore Drilling Unit
MPCU	Marine Pollution Control Unit
mrad	Metre radians
MSC (i)	Maritime Safety Committee (of IMO)
MSC (ii)	Mediterranean Shipping Company
MSD	Material Safety Data Sheets
MSL	Maximum Securing Load
MSN	Merchant Shipping Notice
mt	Metric tons
MTC	Moment to Change Trim 1 Centimetre
MTSA	Maritime Transport Security Act (US)
MV	Motor Vessel
MW	Mega Watt
NLS	Noxious Liquid Substances (NOx)
NMVOC	Non Methane Volatile Organic Compound
NOS	Not Otherwise Specified
NPSH	Net Positive Suction Head
NRV	Non Return Valve
OBO	Oil, Bulk, Ore (carrier)
OCIMF	Oil Companies International Marine Forum
ODP	Ozone Depletion Potential
OLB	Official Log Book
OOW	Officer of the Watch
OPIC	Oil Pollution Insurance Certificate
ORB	Oil Record Book
P	Port
P/A	Public Address System
PCC	Pure Car Carrier
PCTC	Pure Car & Truck Carrier
PEL	Permissible Exposure Limit
PES	Periodic Examination Scheme
PFSP	Port Facility Security Plan
PHA	Preliminary Hazard Analysis

P&I	Protection and Indemnity Association
P/L	Protective Location
PMA	Permanent Means of Access
PNG	Pressurised Natural Gas
ppm	Parts per million
PRS	Polish Register of Shipping (Class Society)
PSC	Port State Control
psi	Pounds per square inch
PUWER	Provision and Use of Work Equipment Regulations 2006
P/V	Pressure Vacuum
R	Representative of electrical Resistance
RD	Relative Density
RINA	Registro Italiano Navale (Italian Classification Society)
RMC	Refrigerated Machinery Certificate
RMS	Royal Mail Ship
RoPax	Roll On, Roll Off + Passengers
Ro-Ro	Roll On, Roll Off
rpm	Revolutions per minute
RVP	Reid Vapour Pressure
S (stb'd)(i)	Starboard
S (ii)	Summer
SAR	Search and Rescue
SatComs	Satellite Communications
SBM	Single Buoy Mooring
SBT	Segregated Ballast Tank
SCBA	Self-Contained Breathing Apparatus
SeaBee	Sea Barge
SECU	StoraEnso Cargo Unit
S/F	Stowage Factor
SG	Specific Gravity
s.h.p.	Shaft Horse Power
S.I.	Statutory Instrument
SL	Summer Loadline
SMC	Safety Management Certificate
SMS	Safety Management System
SOLAS	Safety of Life at Sea (Convention)
SOPEP	Ship's Oil Pollution Emergency Plan
SO_x	Oxides of Sulphur
SPG	Self-Supporting Prismatic-shape Gas tank
SRV	Shuttle and Re-gasification Vessel system
SSO	Ship Security Officer
SSP	Ship Security Plan
STEL	Short Term Exposure Limit
SW	Salt Water
SWL	Safe Working Load

SWR	Steel Wire Rope
T (i)	Tropical
T (ii)	Tons (tonnes)
Tan	Tangent
T/D	Tween Deck
TEU (t.e.u.)	Twenty-foot Equivalent Unit
TF	Tropical Fresh
Tk	Tank
TLVs	Threshold Limit Values
TMD	True Mean Draught
TML	Twin Marine Lifter
TPC	Tonnes per centimetre
TWA	Time Weighted Average
U	Union Purchase – safe working load
UEL	Upper Explosive Limit
UFL	Upper Flammable Limit
UHP	Ultra-High Pressure
UK	United Kingdom
UKC	Under Keel Clearance
ULCC	Ultra Large Crude Carrier
ULCV	Ultra Large Container Vessel (over 14,500 teu)
UN	United Nations
US	United States
USA	United States of America
USCG	United States Coast Guard
U–SWL	Union Rig – Safe Working Load
VCM	Vinyl Chloride Monomer
VDR	Voyage Data Recorder
VDU	Visual Display Unit
VLCC	Very Large Crude Carrier
VOCs	Volatile Organic Compounds
VR	Velocity Ratio
W (i)	Winter (Loadline)
W (ii)	Representative of the ship's displacement
WBT	Water Ballast Tank
W/L	Water Line
WNA	Winter North Atlantic
wp	Waterplane (area)
wps	Wires per strand
YAR	York Antwerp Rules (2016)

Conversion and Measurement Tables

IMPERIAL/METRIC

1 inch = 2.5400 centimetres 1 centimetre = 0.3937 inches.
1 foot = 0.3048 metres 1 metre = 3.2808 feet.
1 square inch = 6.4516 square cms 1 sq.cm = 0.1550 sq. inches.
1 square foot = 0.09293 square metres 1 sq. metre = 10.7639 sq. feet.
1 cubic inch = 16.3871 cubic cms 1 cu.cm. = 0.0610 cu. mts.
1 cubic foot = 0.02832 cubic metres 1 cu.metre = 35.3146 sq. feet.

xxii   Conversion and Measurement Tables

MEASUREMENT

Metres To Feet

Cms.	Feet	Metres	Feet
1	0.03	14	45.93
2	0.06	15	49.21
3	0.09	16	52.49
4	0.13	17	55.77
5	0.16	18	59.06
6	0.19	19	62.34
7	0.22	20	65.62
8	0.26	21	68.90
9	0.30	22	72.18
10	0.33	23	75.46
20	0.66	24	78.74
30	0.98	25	82.02
40	1.31	26	85.30
50	1.64	27	88.58
60	1.97	28	91.86
70	2.30	29	95.15
80	2.62	30	98.43
90	2.95	40	131.23
		50	164.04
Metres	Feet	60	196.85
1	3.28	70	229.66
2	6.56	80	262.47
3	9.84	90	295.28
4	13.12	100	328.08
5	16.40	200	656.17
6	19.69	300	984.25
7	22.97	400	1312.33
8	26.25	500	1640.42
9	29.53	600	1968.50
10	32.81	700	2296.58
11	36.09	800	2624.66
12	39.37	900	2952.74
13	42.65	1000	3280.83

Feet To Metres

Inches	Metres	Feet	Metres
1	0.03	80	24.38
2	0.05	90	27.43
3	0.08	100	30.48
4	0.10	150	45.72
5	0.13	200	60.96
6	0.15	250	76.20
7	0.18	300	91.44
8	0.20	350	106.68
9	0.23	400	121.92
10	0.25	450	137.16
11	0.28	500	152.40
12	0.30	550	167.64
		600	182.88
Feet	Metres	650	198.12
1	0.30	700	213.36
2	0.61	750	228.60
3	0.91	800	243.84
4	1.22	850	259.08
5	1.52	900	274.32
6	1.83	950	289.56
7	2.13	1000	304.80
8	2.44	1100	335.28
9	2.74	1200	365.76
10	3.05	1300	396.24
20	6.10	1400	426.72
30	9.14	1500	457.20
40	12.19	2000	609.60
50	15.24	3000	914.40
60	18.29	4000	1219.20
70	21.34	5000	1524.00

TONNAGE AND FLUID MEASUREMENT

	US Gallons	Imperial Gallons	Capacity Cubic Feet
1 Gallon (imp)	x 1.2	x 1	x 0.1604
1 Gallon (US)	x 1.0	x 0 .8333	x 0. 1337
1 Cubic Foot	x 7.48	x 0.2344	x 1.0
1 Litre	x 0.2642	x 0.22	x 0.0353
1 ton Fresh Water	x 269	x 224	x 35.84
1 ton Salt Water	x 262.418	x 218.536	x 35

Weight	Short Ton	Long Ton	Metric Tonne
Long ton (imp)	x 1.12	x 1.0	x 1.01605
Short ton (USA)	x 1.0	x 0.89286	x 0.90718
Metric ton	x 1.10231	x 0.98421	x 1.0

Grain	Bushel (imp)	Bushel (US)	Cu. Feet.
1 Bushel (imp)	x 1.0	x 1.0316	x 1.2837
1 Bushel (USA)	x 0.9694	x 1.0	x 1.2445
1 Cubic Foot	x 0.789	x 0.8035	x 1.0

MISCELLANEOUS

Pounds 1 lb = 0.45359 kgs 1 kg = 2.20462 lbs.
1 ft^3/ ton = 0.16 Imperial Gallons per ton.
1 tonne/m^3 = 0.02787 ton/ft^3.
1 m^3/tonne = 35.8816 ft^3/ton.

MASS/WEIGHT

1 pound (lb) = 0.4536 kilograms (kgs).
1 ton = 1016.05 kgs = 2240 lbs.

1 tonne = 1000 kgs = 0.985 ton.
1 kg = 2.205 lbs.
1 tonne = 2205 lbs.
1 pint (pt) = 0.568 Litre (lt).
1 quart = 1.13648 lts.
1 gallon (gal) = 4.5469 lts.
1 gallon US = 3.7853 lts.
1 lt. = 1.7599 pts.
1 lt. = 0.220 gals.
10 lts = 2.19981 gals.

MISCELLANEOUS

1 Fathom = 1.829 metres (m) = 6.0 feet (ft).
1 Nautical mile = 10 cables = 6080 ft.
1 Cable = 600 ft.
1 Shackle (Anchor Cable) = 27.5 m = 90 ft. or 15 fathoms.
1 tonne of Sea Water = 1m³ (approx).
1 Cubic Metre = 35.314 ft³.
1 Horse Power = 746 Watts (W).
1000 Watts = 1 kilowatt.
1 atmosphere = 14.7 lbs/inch².
1 Barrel (imperial) = 36,000 imp/gallons.
1 Barrel (US) = 42,000 US/gallons.

CHAPTER 1

General Principles of the Handling, Lifting, Stowage and Carriage of Cargoes

INTRODUCTION

The transport of cargo dates back through the centuries to the Egyptians, the Phoenicians, ancient Greeks and early Chinese, long before the Europeans ventured beyond the shores of the Atlantic. Strong evidence exists that the Chinese treasure ships traded for spices, and charted the Americas, Antarctica, Australia, the Pacific and Indian Oceans, before Columbus supposedly discovered America.*

The stones for the Pyramids of Egypt had to be brought up the River Nile or across the Mediterranean and this would reflect means of lifting heavy weights, and transporting the same was a known science before the birth of Christ. Marco Polo reported 200,000 vessel movements a year were plying the Yangtze River of China in 1271 and it must be assumed that commerce was very much alive, with a variety of merchandise being transported over water.

Produce from the world's markets has grown considerably alongside technology. Bigger and better ships feed the world populations and the methods of faster and safer methods of transport have evolved.

The various cargoes or merchandise which are carried may be broadly divided into the following six types:

1   Bulk Solids
2   Bulk Liquids
3   Containerised Units
4   Refrigerated Frozen/Chilled
5   General, which includes virtually everything not in (1), (2), (3) or (4) above (including Heavy Lifts and Hazardous Goods)
6   Roll-On, Roll-Off cargoes (may include livestock). (Passenger transports are not considered as cargo, but require trade requisites.)

DOI: 10.4324/9781003407706-1

Bulk cargoes can be loaded and discharged quickly and efficiently but we have yet to see 10,000 tons of grain being loaded into a jumbo jet. Ships remain the most efficient means of transport for all cargo parcels of any respectable weight or size.

This book investigates the business of how cargo is loaded, how it is stowed and subsequently shipped to its destination. Later chapters will deal with specifics on the commodities, but the methods of handling prior to starting the voyage and the practical stowage of goods should be considered an essential element of the foundation to successful trade.

*Gavin Menzies, *1421: The Year China Discovered the World* (London: Bantam, 2002).

DEFINITIONS AND CARGO TERMINOLOGY

Air draught – The vertical distance from the surface of the water to the highest point of the ship's mast or aerial.

Angle of repose – Is that natural angle between the cone slope and the horizontal plane when bulk cargo is emptied onto this plane, in ideal conditions.

Bale space capacity – The cubic capacity of a cargo space when the breadth is measured from the inside of the cargo battens (spar ceiling) and the measured depth is from the wood tank top ceiling to the underside of the deck beams – the length being measured from the inside of the fore and aft bulkhead stiffeners.

Bill of lading – A document issued by a carrier to a shipper recording the receipt by the carrier of a consignment of goods for carriage to a stated destination. It incorporates the carrier's conditions of carriage.

Break bulk – (verb) Meaning to commence discharging cargo.

Boom – An American term used to describe a derrick.

Broken stowage – The space between packages which remains unfilled. The percentage that has to be allowed varies with the type of cargo and with the shape of the ship's hold. It is greatest when large cases are stowed in an end hold or at the turn of a bilge.

Cargo information – Appropriate information relevant to the cargo and its stowage and securing, which should specify in particular the precautions necessary for the safe carriage of that cargo by sea.

Cargo manifest – A ship's list of all cargo parcels being carried on board. It is an essential document for use with customs when the vessel is entering inwards to a port or harbour and forms part of the Master's declaration. A Chief Officer would use the manifest to identify any Hazardous/dangerous goods, heavy lift cargo and valuable parcels.

Cargo plan – A ship's plan which shows the distribution of all cargo parcels stowed on board the vessel for the voyage. Each entry onto the plan would detail the quantity, the weight, and the port of discharge. The plan is constructed by the ship's cargo officer and would effectively show special loads such as heavy lifts, hazardous cargoes and valuable cargo, in addition to all other commodities being shipped.

Cargo runner – General term used to describe the cargo lifting wire used on a derrick. It may be found rove as a 'single whip' or doubled up into a 'gun tackle' (two single blocks) or set into a multi-sheave lifting purchase. It is part of the derrick's 'running rigging' passing over at least two sheaves set in the head block and the heel block, prior to being led to the barrel of the winch. Normal size is usually 24 mm and its construction is FSWR of 6 x 24 w.p.s.

Cargo securing manual – A manual pertinent to an individual ship, which will show the lashing points and details of the securing of relevant cargoes carried by the vessel.
   It is a ship's reference which specifies the on-board securing arrangements for cargo units including vehicles and containers and other entities. The securing examples are based on the transverse, longitudinal and vertical forces which may arise during adverse weather conditions at sea. The manual is drawn up to the standard contained in MSC circular of the organisation, MSC/Circ 745.

Cargo ship – Any ship which is not a 'Passenger ship', troop ship, pleasure vessel or fishing boat.

Cargo spaces (e.g. Cargo hold) – All enclosed spaces which are appropriate for the transport of cargo to be discharged from the ship. Space available for cargo may be expressed by either the vessel's deadweight or her cubic capacity in either bale or grain space terms.

Cargo unit – Includes a cargo transport unit and means wheeled cargo, vehicles, containers, flat pallet, portable tank packaged unit or any other cargo and loading equipment or any part thereof, which belongs to the ship and which is not fixed to the ship.

Centre of buoyancy – The geometric centre of the underwater volume. The point through which all the forces due to buoyancy are considered to act.

Centre of gravity (C of G) – The point through which all the forces due to gravity are considered to act. Each cargo load will have its own C of G.

Cradle – A lifting base manufactured in wood or steel or a combination of both employed to accept and support a heavy or awkward shaped load (e.g. a boat). It would normally be engaged with a bridle and/or heavy duty lifting slings shackled to each corner to provide stability to the load movement.

Crutch – A term used to describe the stowage support for a derrick/boom or crane jib. They are usually fitted with a securing band or lashing arrangement to prevent derrick movement when at sea.

Dangerous goods – Goods defined as such in the Merchant Shipping (Dangerous Goods and Marine Pollutants) Regulations 1990.

Deadweight – The difference in tonnes between the displacement of a ship at the summer load water line in water of specific gravity of 1,025, and the lightweight of the ship.

Deadweight cargo – Cargo on which freight is usually charged on its weight. While no hard-and-fast rules are in force, cargo stowing at less than 1.2m²/tonne (40 cu. ft./ton) is likely to be rated as deadweight cargo.

Deadweight scale – One of the ship's stability documents. It is frequently displayed on a convenient bulkhead, usually in the Chief Mate's Office. Once loading or discharge has taken place, the ship's draughts are noted and compared to the scale. This allows the estimated cargo tonnage transferred to be quickly established by the cargo officer.

Double gear – An expression used when winches are engaged in making a heavy lift. The topping lift and lifting purchase of the derrick rigging are locked in 'Double Gear' to slow the lifting operation down to a manageable safe speed.

Double up – A term used with derricks which allows a load greater than the SWL of the runner wire, but less than the SWL of the derrick to be lifted safely. It is achieved by means of a longer wire being used in conjunction with a floating block. This effectively provides a double wire support for the load turning a single whip runner wire into a gun tackle.

Dunnage – An expression used to describe timber boards which can be laid singularly or in double pattern under cargo parcels to keep the surface of the cargo off the steel deck plate. Its purpose is to provide air space around the cargo and so prevent 'cargo sweat'. Heavy lift cargoes would normally employ heavy timber bearers to spread the load and dunnage would normally be used for lighter load cargoes.

Flemish eye – A name given to a reduced eye made of three strands (not six), spliced into the end of a cargo runner which is secured to the barrel of a winch. (Alternative names are Spanish Eye, or Reduced Eye.)

Flemish hook – A large hook, often used in conjunction with the lower purchase block in the rigging of a heavy lift derrick. The hook can be opened to accommodate the load slings and then bolt locked.

Floodable length – The maximum length of a compartment which can be flooded so as to bring a damaged vessel to float at a waterline which is tangential to the margin line. Note: In determining this length, account must be taken of the permeability of the compartment.

Free surface – A term extensively used aboard ships when assessing the stability of the vessel, especially if the ship is expected to experience an angle of heel. Any slack tanks or slack fluids aboard the vessel, while the ship heels over, could generate sur face movement of the fluid. Free surface could be detrimental to the ships positive stability and can be removed by pressing up tanks to a full capacity or pumping the tank empty.

Freight – The term used to express the monetary charge which is levied for the carriage of the cargo.

Gooseneck – The bearing and swivel fitment found at the heel of a derrick which allows the derrick to slew from port to starboard, and luff up and down when in operation.

Grain capacity – The cubic capacity of a cargo space when the length, breadth and depth are measured from the inside of the ship's shell plating, all allowances being made for the volume occupied by frames and beams.

Gross tonnage (GT) – Defined as that measurement of the total internal volume of the ship.
  The GT is found from the formula GT = K1 V
  When K1 = 0.2 + 0.02 $\log_{10}$ V and V = Total volume of all enclosed spaces (measured in Cubic meters).

Hallen universal swinging derrick – A single swinging derrick with lifting capacity up to about 100 tons SWL. The original design employed a 'D' frame to segregate the leads of the combined slewing and topping lift guys. The more modern design incorporates 'out-riggers' for the same purpose.

Heavy lift – A generic term that denotes a heavy load within a single lift, pertinent to the size of vessel that is engaged in the transport of the same. Most vessels would consider any load over 10 tonnes as falling into the category of being a heavy lift.

Homogeneous – A term used to describe a full cargo whose density doesn't change along its overall length. e.g. A floating log, suffers little or no longitudinal stress because the log is 'Homogeneous'.

Hounds band – A lugged steel band that is strapped around a 'Mast'. It is used to shackle on shrouds and stays. It is also employed to secure 'Preventer backstays' when a heavy derrick is being deployed in order to provide additional strength to the mast structure when making the heavy lift.

International Maritime Solid Bulk Cargoes, Code (IMSBC code) – A replacement code (for what was known as the BC code) which became effective from 1 January 2011 under the SOLAS convention. The code addresses the risks associated with the carriage of solid bulk cargoes inclusive of reduced stability which could lead to capsize, cargo liquefaction, fire, explosion and chemical hazards.

Jumbo derrick – Colloquial term used to describe a conventional 'heavy lift derrick'.

Lateral drag – A term used to describe the movement of a suspended load from a derrick or crane when loading or discharging, where the load is caused to move in a horizontal direction as opposed to the expected vertical movement. It can occur on the point of landing the load during a ship to shore transfer. Movement being caused by the ships roll or upright turn of the vessel after the ship has been heeled. The sudden ship movement can cause the plumb line of the load to experience a sideways motion as the load is dragged by the angled lifting purchase.

Lead block – A single sheave block secured in such a position as to change the direction of the weight bearing wire or rope. Its use generally pulls the load to an off centre stowage position.

Limit switch – A crane or winch feature acts as a cut out if working movement extends outside its operational working limits.

Load density plan – A ship's plan which indicates the deck load capacity of cargo space areas of the ship. The ship's Chief Officer would consult this plan to ensure that the space is not being overloaded by very dense, heavy cargoes.

Long ton – A unit of mass weight, equal to 2240 lbs (Ton).

Luffing – A term which denotes the movement of a crane jib or derrick boom to move up or down – i.e. 'luff up' or 'luff down'.

Luffing derrick – A conventional single swinging derrick rigged in such a manner that permits the derrick head to be raised and lowered to establish any line of plumb. As opposed to static rigged derricks, as with a 'Union purchase rig'.

Mate's receipt – A receipt note issued by the Chief Officer for goods received on board. May act as a prelude to a Bill of Lading.

Measurement cargo – Cargo on which freight is usually charged on the volume occupied by the cargo. Such cargo is usually light and bulky stowing at more than $1.2m^2$/tonne (40 cu. ft./ton), but may also be heavy castings of an awkward shape where a lot of space is occupied.

Metacentric height (GM) – Expressed as GM and is the measured distance between the ships Centre of Gravity and its Metacentre.

Overhauling (i) – An expression used to describe the correct movement of a Block and Tackle arrangement, as with a lifting purchase of a derrick. The term indicates that all sheaves and wire parts are moving correctly and without restriction.

Overhauling (ii) – This term can also be used to describe a maintenance activity as when stripping down a cargo block for inspection and re-greasing. The block is being 'overhauled'.

Par-buckling – A method of moving a load in a controlled manner up or down by means of rope or wire hawsers in a roll or turning motion. Also employed to assist in salvage of the 'Costa Concordia' to cause the damaged ship to be tipped upright.

Passenger ship – A ship designed to carry more than twelve passengers.

Permeability – In relation to a compartment space, means the percentage of that space which lies below the margin line which can be occupied by water.

Note: Various formulae within the Ship Construction Regulations are used to determine the permeability of a particular compartment. Example values are:

Spaces occupied by cargo or stores	60%
Spaces employed for machinery	85%
Passengers and crew spaces	95%

Permissible length – of a compartment having its centre at any point in the ship's length, is determined by the product of the floodable length at that point and the factor of sub-division of the vessel.

Permissible length = Floodable length x Factor of sub-division

Plumb line – This is specifically a length of cord with a plumb-bob attached to it. However, it is often used around heavy lift operations as a term to describe the 'line of plumb', where the line of action is the same as the line of weight, namely the plumb line.

Preventer – A general term to describe a strength/weight bearing wire, found in a Union purchase rig on the outboard side, of each of the two derricks. They are also used as additional support for a mast structure when heavy lifting is ongoing by an attached Jumbo Derrick. Preventer stays are rigged in a position as per the rigging plan of the ship.

Proof load – That tonnage value that a lifting appliance is tested to. The value is equal to the SWL of the appliance + an additional percentage weight allowance.

Derricks/Cranes less than 20 tonnes SWL, proof load is 25% in excess
Derricks/Cranes 20–50 tonnes SWL proof Load equals + 5 ts in excess of SWL
Derrick/Cranes over 50 tonnes SWL, proof Load equals + 10% (of load)

Project cargo – A very large and/or heavy cargo that requires a detailed plan of operation to transport from place of manufacture to its operational position. Project cargoes are dominant within the offshore industries when establishing offshore installations and employing designated heavy-lift vessels.

Purchase – The term given to blocks and wire or fibre rope, when rove together. Sometimes referred to as a 'block n tackle'. Two multi-sheave blocks rove with flexible steel wire rope (FSWR) are found in common use as the lifting purchase, suspended from the spider band of a swinging derrick.

Rams horn hook – A heavy duty, double lifting hook capable of accepting slings or strops on either side. They are extensively engaged in the Heavy Lift sector of the industry.

Reduced eye – An eye splice made in a steel wire rope with only 3 of the six strands. It is sometimes referred to as a Flemish eye or Spanish Eye and is employed at the end of a wire runner to effect a 'U' clamp bolt to the barrel of a winch.

Register of ships' lifting appliances and cargo handling gear – A file kept and maintained by the ship's Chief Officer which contains all the certificates of lifting equipment employed on board the vessel, inclusive of all shackles, wires, hooks, chains etc.

Riding turn – An expression that describes a cross turn of wire around a barrel of a winch, or stag horn. It is highly undesirable and could cause the load to jump or slip when in movement. The condition should be cleared as soon as possible.

Rigging plan – A ship's plan which provides a diagrammatic view of all a ship's lifting gear. Derrick and crane outreaches and arcs of operation are included with 'Safe working loads' as applicable.

Ring bolt – A deck ring or 'pad eye' often used in conjunction with a doubling plate or screw securing. It is employed to provide an anchor point for associated rigging around a derrick position.

Risk assessment – A legal requirement under the SMS that an assessment of risks are made by a person in authority, for any work-based task that may generate harm to the wellbeing of human operators.

TABLE 1.1 Risk Assessment Grades

Grade of Risk	Category
1.	Trivial
2.	Tolerable
3.	Moderate
4.	Substantial
5.	Intolerable

Running rigging – A descriptive term used to describe wire or cordage ropes which pass around the sheave of a block (see also 'Standing rigging'). Where steel wire ropes are employed for running rigging they are of a flexible construction; examples include: 6 x 24 w.p.s. and 6 x 36 w.p.s.

Safe working load (SWL) – An acceptable working tonnage used for a weight-bearing item of equipment. The marine industry uses a factor of one-sixth of the breaking strain to establish the safe working value.

Safety tongue – A spring clip sealing device to cover the jaw of a lifting hook. It should be noted that these devices are not fool proof and have been known to slip themselves unintentionally. The tongue is meant to replace the need for 'mousing' the hook, and is designed to serve the same purpose as a 'mousing'.

Schooner guy – A bracing guy which joins the spider bands at the derrick heads of a 'Union purchase' rig.

Sheer legs – A large lifting device employed extensively within the marine industry. It is constructed with a pair of inclined struts resembling a crane, although the action when working is similar to a craning activity. (Smaller versions of sheer legs were previously used within the marine industry on tankers to hoist pipelines on board or more commonly found in training establishments for training cadets in rigging applications.) The modern-day sheer legs are now found on floating heavy lift (crane) barges and employed for extreme lifting operations, usually with 'project cargoes'.

Shore – A term used to describe a support, given to decks, bulkheads or cargo. They are usually timber, but may be in the form of a metal stanchion, depending on the intended use. (See 'Tomming off'.)

Slings – A term which describes the lifting strops to secure the load to be hoisted to the lift hook of the derrick or crane. Slings may be manufactured in steel wire rope, chains, rope or canvas.

Snatch block – A single sheave block, often employed to change the direction of lead of a wire or rope. The block has a hinged clamp situated over the 'swallow' which allows the bight of a wire or rope to be set into the block without having to pull the end through.

Snotter – A length of steel wire with an eye in each end. It is employed around loads as a lifting sling, with one eye passed through the other to tighten the wire around the load.

Speed crane – Modern derrick design with multi-gear operation which operates on the principle of the single jib, point-loading crane.

Spider band – A steel lugged strap found around the head of a derrick which the rigging, such as the topping lift and guys, are shackled onto. The equivalent on a mast structure is known as a 'Hounds band'.

Spreader – A steel or wood batten which effectively spreads the wire sling arrangement wider apart when lifting a large area load. Use of such a spreader generally provides greater stability to the movement of the weight. Formerly referred to as a lifting beam.

Stabilizers – Steel outriders, often telescopic in design and fitted with spread feet, which are extended from the base unit of a shoreside mobile crane. Prior to taking the load, the stabilizers are set to ensure that the load on the crane jib will not cause the crane to topple. (Not to be confused with ship stabilizers fitted to ships to reduce rolling actions of the vessel when at sea.)

Standing rigging – A term used to describe fixed steel wire rope supports. Examples can be found in ship's stays and shrouds. Construction of standing rigging is usually 6 x 6 w.p.s.

Steadying Lines – Cordage of up to about 24mm size, secured at adequate lengths to the load or a collar of the hoist when lifting a load. They provide stability and a steadying influence to the load when in transit from ship to quay and vice versa.

Stowage factor – The volume occupied by unit weight of cargo, usually expressed as cubic metres per tonne (m³/tonne) or cubic feet per ton (cu.ft./ton). It does not take account of any space which may be lost due to 'broken stowage'. A representative list of stowage factors is provided at the end of this book.

Sub-division factor – The factor of sub-division varies inversely with the ship's length, the number of passengers and the proportion of the underwater space used for passengers/crew and machinery. In effect it is the factor of safety allowed in determining the maximum space of transverse watertight bulkheads, i.e. the permissible length.

Tomming off – An expression which describes the securing of cargo parcels by means of baulks of timber, these being secured against the cargo to prevent its movement if and when the vessel is in a seaway and experiencing heavy rolling or pitching motions. (Alternative term is 'Shore'.)

Ton (Tonne) – Originated from the word 'tun' (ton) which was a term used to describe a wine cask or wine container, the capacity of which was stated as being 252 gallons as required by an Act of 1423, made by the English Parliament. It is synonymous that 252 gallons of wine equated to approximately 2240 lbs, 'one ton' as we know it today.

Triple swivel hook – A drop forged cargo hook used extensively with Union purchase rig employing two derricks. Sometimes referred to as a Seattle Hook.

Trunnion – A similar arrangement to the 'gooseneck' of a small derrick. The trunnion is normally found on intermediate-size derricks of 40 tons or over. They are usually manufactured in cast steel and allow freedom of movement from the lower heel position of the derrick.

Tumbler – A securing swivel connection found attached to the 'Samson post' or 'Mast table' to support the topping lift blocks of the span tackle.

'U' bolt – A bolt application which secures the reduced eye of a cargo runner to the barrel of a winch.

Union plate – A triangular steel plate set with three eyelets used in 'Union rig' to join the cargo runners and hook arrangement when a 'triple swivel hook' is not employed. It can also be used with a single span, topping lift derrick to couple the downhaul with the chain preventor and bull wire. Sometimes referred to as 'Monkey face plate'.

Union rig (alt. Union purchase rig) – A derrick rig which joins two single swinging derricks to work in 'union' with cargo runners joined to a triple swivel hook arrangement known as a 'Seattle hook' or 'Union hook'. The rig was previously known as 'Yard and Stay' and is a fast method of loading/discharging lighter parcels of cargo. Union rig operates at approximately one-third of the SWL of the smallest derrick of the pair.

Velle derrick – A moderate heavy lift derrick which can be operated as a crane by a single operator. The derrick is constructed with a 'T' bridle piece at the head of the derrick which allows topping lift wires to be secured to act in way of slewing guys and/or topping lift.

Walk back – An expression which signifies reversing the direction of a winch in order to allow the load to descend or the weight to come off the hoist wires.

Weather deck – The uppermost complete deck exposed to the weather and the sea.

Wires per Strand (w.p.s.) – An expression which describes the type of construction of the strands of a steel wire rope.

Wirex – A trade name given to a steel wire rope manufactured with a multi-plat type lay. The interwoven wire has a non-rotational anti twist property which is favoured as a crane wire.

Yard and Stay – Alternative descriptive term for 'Union purchase rig'.

CONVENTIONAL GENERAL CARGO HANDLING

Cargo Gear – Derricks, cranes and winches, together with their associated fittings, should be regularly overhauled and inspected under a planned maintenance schedule, appropriate to the ship. Winch guards should always be in place throughout winching operations and operators should conform to the Code of Safe Working Practice. Only certificated tested wires, blocks and shackles should be used for cargo handling and lifting operations.

Note: Wire ropes which have broken wires should be replaced, whenever 10% of wires are broken in any eight (8) diameter length, the wire should be condemned. Guy pennants, blocks and tackles should be maintained in good condition.

CONVENTIONAL FIVE (5) HATCH GENERAL CARGO VESSEL

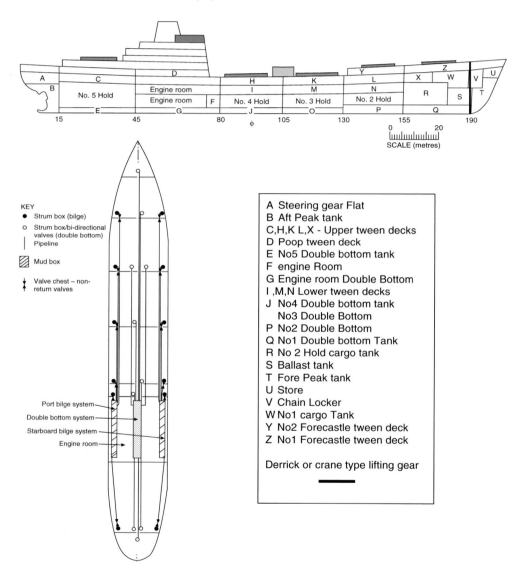

KEY
- ● Strum box (bilge)
- o Strum box/bi-directional valves (double bottom)
- | Pipeline
- ▨ Mud box
- ⇅ Valve chest – non-return valves

Port bilge system
Double bottom system
Starboard bilge system
Engine room

A Steering gear Flat
B Aft Peak tank
C,H,K L,X - Upper tween decks
D Poop tween deck
E No5 Double bottom tank
F engine Room
G Engine room Double Bottom
I ,M,N Lower tween decks
J No4 Double bottom tank
   No3 Double Bottom
P No2 Double Bottom
Q No1 Double bottom Tank
R No 2 Hold cargo tank
S Ballast tank
T Fore Peak tank
U Store
V Chain Locker
W No1 cargo Tank
Y No2 Forecastle tween deck
Z No1 Forecastle tween deck

Derrick or crane type lifting gear

FIGURE 1.1 Longitudinal profile plan of 'general cargo' vessel with hold bilge pumping and pipeline system.

Derrick Rigs – Union Purchase Method

The Union purchase method of rigging derricks is perhaps the most common with conventional derrick rigs. With this operation one of two derricks plumbs the hatch and the other derrick plumbs overside. The two runner falls of the two derricks are joined together at the cargo union hook. (This is a triple swivel hook arrangement sometimes referred to as a Seattle hook). The load is lifted by the fall which plumbs the load, when the load has been lifted above the height of the bulwark or ship's rail, or hatch coaming, the load is gradually transferred to the fall from the second derrick.

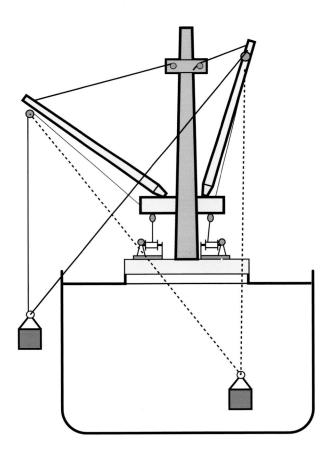

FIGURE 1.2 Union purchase rig.

Cargo movement is achieved by heaving on one derrick runner and slacking on the other. The safe working angle between the runners is 90° and should never be allowed to exceed 120°. There is a danger from overstressing the gear if unskilled winch drivers are employed or if winch drivers do not have an unobstructed view of the lifting/lowering operation. In the latter case, signallers and hatch foremen should always be employed within line of sight of winch operators.

The CSWP for Merchant Seaman provides a code of hand signals for use in such cargo operations and for the control of winch operations.

FIGURE 1.3 Triple swivel hook.

The Conventional Derrick Rig

FIGURE 1.4 General cargo vessel rigged with conventional 5 tonne SWL derricks and steel hatch covers. The derricks can be rigged to operate as single swinging derricks or rigged in 'Union purchase' with schooner guys secured between the derrick heads.

SWL (U) = 1.6 tonnes. Such vessels are in decline because of the growth in unit load container and Ro-Ro traffic.

Single Swinging Derricks

The conventional derrick initially evolved as a single hoist operation for the loading and discharging of weights. It was the basic concept as an aid, which became popular when combined within a 'Union rig'. However, improved materials and better designs have created sophisticated, single derricks in the form of the 'Hallen', and the 'Velle' and the more

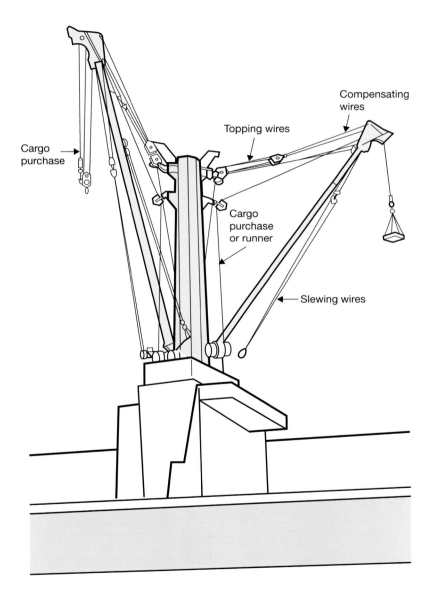

FIGURE 1.5 Speed crane. Employs the principle of the single swinging derrick, of which there are several variant designs.

popular speed cranes, all of which now work alongside shipboard cranes, where open stow cargoes remain within the diminishing number of General Cargo ships.

Where the single swinging derrick concept has been retained is in the arena of the heavier lifting operation. Examples are seen with the 'Hallen' and 'Velle' derrick systems, in that the topping lift arrangement and the slewing wires are incorporated together and secured aloft, clear of the lower deck. The outreach and slew are wide, achieved by the 'T' bar on the Velle derrick and by outriggers with the 'Hallen'.

Both systems are labour-saving and can be operated by a single controller, operating the luffing and slewing movement together with the cargo hoist movement.

The Hallen is distinctive by the 'Y' mast structure that provides the anchor points for the wide leads. The derrick also accommodates a centre lead sheave to direct the hoist wire to the relevant winch.

> It should be realised that the concept of the single swinging derrick has generally been superseded by the expansion of Container and Ro-Ro traffic. Operators found the maintenance of derricks time consuming and expensive. Cargo, tonnage for tonnage, could not compete with derricks/cranes, against containerisation.

Specialised Derrick Rigs

The many changes which have occurred in cargo handling methods have brought about extensive developments in specialised lifting gear. These developments have aimed at efficient and cost-effective cargo handling and modern vessels will be equipped with some type of specialist rig for operation within the medium to heavy lift range.

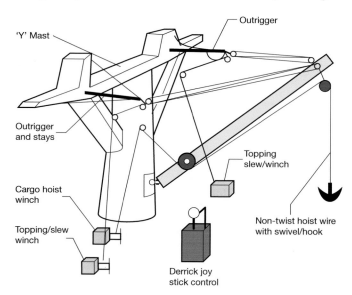

FIGURE 1.6 Hallen derrick with out-riggers.

The Hallen derrick is a single swinging derrick which is fast in operation and can work against an adverse list of up to 15°. They are usually manufactured in the 25–40 tonne Safe Working Load range and when engaged, operate under a single-man control.

Joystick control for luffing and slewing is achieved by the port and starboard slewing guys being incorporated into the topping lift arrangement. Use of the outriggers from a 'Y' mast structure provides clear leads even when the derrick is working at 90° to the ships fore and aft line. A second hoist control can be operated simultaneously with the derrick movement.

As a one-man operation, it is labour-saving over and above the use of conventional derricks, while at the same time keeps the deck area clear of guy ropes and preventers. Should heavy loads be involved, only the cargo hoist would need to be changed to satisfy different load requirements.

'Velle' Derrick

Similar in design to the 'Hallen' but without use of outriggers. The leads for the topping lift and slewing arrangement are spread by a cross 'T' piece 'yoke' at the head of the derrick. A wide spread structured mast is also a feature of this rig.

Again it is a single-man operation, with clear decks being achieved while in operation.

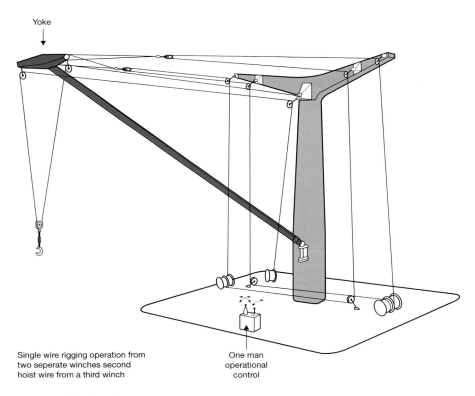

Yoke

Single wire rigging operation from two seperate winches second hoist wire from a third winch

One man operational control

FIGURE 1.7 Velle derrick.

Generally, the 'Velle' is manufactured as a heavier rig and variations of the design with a pivot cross piece at the derrick head are used with multi-sheave purchases to accept the heavy type load.

Luffing and slewing actions of the derrick are powered by two winches each equipped with divided barrels to which the bare ends of the fall wires are secured

WORKING WITH LIFTING PLANT

At no time should any attempt be made to lift weights in excess of the Safe Working Load (SWL) of the weakest part of the gear. The SWL is stamped on all derricks, blocks and shackles as well as noted on the 'Test Certificates'. Wire ropes are delivered with a test certificate on which will be found the SWL of the wire. All wires passing through a block or employed with a winch would be expected to be Flexible Steel Wire Ropes (FSWR). Topping Lifts would be Extra Flexible Steel Wire Rope (EFSWR).

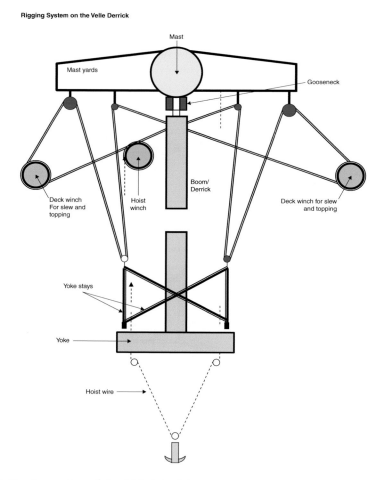

FIGURE 1.8 Rigging system of the 'Velle' derrick.

Assuming that the SWL is one-sixth of the breaking strain, the regulations require a minimum of one-fifth. The approximate SWLs of various materials can be obtained from the following formula.

Cordage

TABLE 1.2			
Material	Structure	Breaking Strain (BS) formula	SWL @ 1/6th BS
Manila	3 stranded Hawser laid	$2D^2/300$	$2D^2/1800$
Polypropylene	3 stranded Hawser laid	$3D^2/300$	$3D^2/1800$
Terylene	3 stranded Hawser laid	$4D^2/300$	$4D^2/1800$
Nylon	3 stranded Hawser laid	$5D^2/300$	$5D^2/1800$

Flexible Steel Wire Rope

TABLE 1.3				
Material	Construction	Breaking Strain (BS) formula	SWL @ 1/6th BS	Abbreviation
Flexible Steel Wire Rope	6 x 24 w.p.s	$20D^2/500$	$20D^2/3000$	FSWR
Grade 1, stud chain	12.5 mm to 120 mm	$20D^2/600$	$20D^2/3600$	–

When lifting loads in excess of about 1.5 tonnes, steam winches should generally be used in double gear. Electric winches are usually fused for a Safe Working Load of up to about 3 tonnes. For loads in excess of 2–3 tonnes it would be normal practice with conventional derricks to double up the rig, as opposed to operating on a single part runner wire.

Derricks may be encountered with two SWL marks on them. In such cases the lesser value is usually marked with a 'U' signifying the SWL for use in Union purchase rigs. In the event the derrick is not marked and intended for use in a Union rig, the SWL is recommended not to exceed one-third of the smaller of the two derricks.

Use of Lifting Purchases

The purchase diagrams shown are rigged to disadvantage. The velocity ratio is increased by '1' if the tackle is rigged to advantage.

The required purchase: the common ones are illustrated in Figure 1.9.

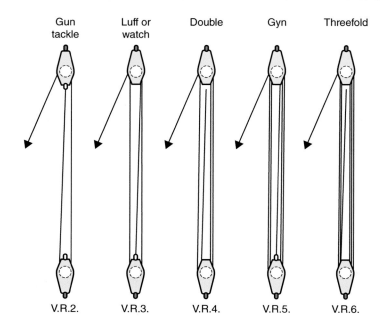

FIGURE 1.9 All tackles rove to disadvantage and velocity ratios stated for this rig. (When tackles are rove to advantage add +1 to the VR.)

The stress factors incurred with their use can be found by the following formula assuming 10% for friction:

$$S \times P = W + nW \ (10 \ / \ 100)$$

where: S is the stress in the hauling part.

P is the power gained by the purchase (this is the same as the number of rope parts at the moving block).

n is the number of sheaves in the purchase.

W is the weight being lifted.

10, represents the numerator of the fraction, is an arbitrary 10% allowance for friction.

CARGO HANDLING EQUIPMENT – CONDITION AND PERFORMANCE

Before any cargo operation takes place it is essential that the Chief Officer is confident that the ship's lifting equipment and associated loading/discharge facilities are 100% operational and free of any defects. Under the Lifting Plant Regulations (LOLER), the ISM Code and ship's planned maintenance schedule, all cargo handling equipment could expect to be inspected and maintained at regular intervals.

In the case of lifting plant, derricks, cranes, shackles, wires etc., the following test times would be required:

1   After installation when new.
2   Following any major repair.
3   At intervals of not more than 5 years.

Testing and Inspection of Lifting Plant

Cargo lifting appliances must be inspected to establish that they are correctly rigged on every occasion they are used. To this end the Chief Officers would normally delegate this duty to the Deck Cargo Officers to check the rig prior to commencing loading or discharge operations.

A thorough inspection would also take place annually by a 'Competent Person', namely the Chief Officer himself. This duty would not be delegated to a junior officer. This inspection would cause a detailed inspection to take place of all aspects, hydraulic, mechanical and electrical, of the lifting appliances. All wires would be visually inspected for defects and the mousing on shackles would be sighted to be satisfactory. The 'gooseneck' of derricks and all blocks would be stripped down and overhauled.

Thorough inspections would detect corrosion, damage, hairline cracks and excessive wear and tear. Once defects are found, corrective action would be taken to ensure that the plant is retained at 100% efficiency. These inspections would normally be carried out systematically under the ship's planned maintenance schedule. This allows a permanent record to be maintained and is evidence to present to an ISM Auditor.

Testing Plant

Lifting appliances are tested by a Cargo Surveyor at intervals of five years or following installation or repairs. The test could be conducted by either of two methods:

1   By lifting the Proof Load and swinging the load through the derrick or crane's operating arc, as per the ship's rigging plan. This test is known as the 'Proof Load Test' and concrete blocks of the correct weight are normally used to conduct this operation. (Dynamic (moving) Test: Test movement made in accordance with the ship's rigging plan.)
2   A static test is carried out by employing a 'Dynamometer' secured to the lifting point of the rig and an anchored position on the deck. The proof load weight is then placed on the rig and measured by the dynamometer, to the satisfaction of the cargo surveyor.

Load Test

Lifting gear when tested must provide the 'Proof Load' to meet the regulatory requirements. This is conveniently achieved by use of water bags filled to weight and measured to a digital load cell.

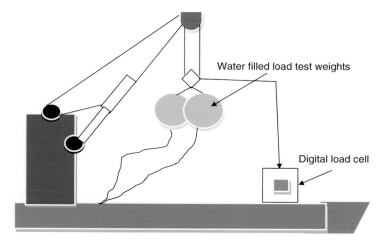

FIGURE 1.10 Testing lifting gear by water bags to match proof load.

FIGURE 1.11 Example of digital load cell as employed in modern shipyards.

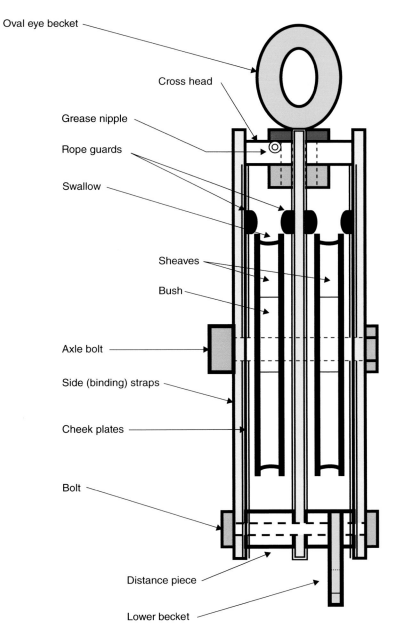

FIGURE 1.12 Parts of the double cargo block.

TABLE 1.4	
(Based on the Safe Working Load + an additional percentage)	
For derricks or cranes up to SWL 20 tonnes	Test load = 25% + SWL
For derricks or cranes between 20 and 50 tonnes	Test load = 5 tonnes + SWL
For derricks or cranes over 50 tonnes SWL	Test load = 10% + SWL

Certification

Once the testing has been completed satisfactorily, each lifting apparatus would be issued with a Test Certificate and the Chief Officer would retain all certificates in the Register of Ships' Lifting Appliances and Cargo Handling Gear.

In addition to these test certificates all shackles, wires, blocks, etc., would be purchased as proof tested and delivered to the vessel with its respective certificate. These would be retained in the Chief Officer's Register. In the event that a shackle or block is changed, the certificate in the register would also be changed, so keeping the ship's records up to date.

The Safe Working Load and the Certificate Number are found stamped into the binding straps of each block.

Grease recesses are found inside the bush and inside the inner bearing surface of the centre of each sheave.

The 'axle bolt' is of a square cross-section to hold the bearing 'bush', this allows the sheave to rotate about the bush.

Derrick Maintenance

As with many items of equipment, derrick rigs must similarly be checked and seen to be correctly rigged on every occasion prior to their engagement. It would be normal practice for the ship's Chief Officer to delegate this supervisory task to the duty deck/cargo officer before loading or discharge operations are allowed to commence.

In addition to the regular working checks, all lifting gear should undergo an annual inspection by a responsible person, namely the ship's Chief Officer. This annual inspection is never delegated but would be under the scrutiny of the ship's Mate.

The annual inspection would entail the overhaul and total inspection of all the derrick's moving parts inclusive of the head and heel blocks, the lifting purchase blocks, the topping lift and runner wires. The condition of the guys would also be inspected and the emphasis would be placed on the main weight-bearing element of the 'Gooseneck'.

The annual inspections do not usually require the derrick to be tested unless a degrading fault is found in the rig, necessitating a new part or a replacement part to be used. Testing normally takes place at five-yearly intervals or if repairs have been necessary, or in the event that the derricks are being brought back into use after a period of lay-up. If testing is required this would be carried out in the presence of a cargo surveyor and the lifting gear would have to show handling capability up to the proof load.

In order to conduct an annual inspection, the Chief Officer would order the complete overhaul of all the blocks associated with the derrick rig. Normal practice would dictate that the ship's boatswain would strip the blocks down and clean off any old grease and clear the grease recesses in the bush and the inside of the sheaves. The 'bolt' would be extracted and the bush bearing would be withdrawn. Inspection by the senior officer would take place and any signs of corrosion, hairline cracks or excessive wear and tear would be monitored. If the steelwork is found to be in good working order without any visible defects or signs of deterioration it would be re-greased and re-assembled for continued use.

It is a requirement of the ISM system that lifting gear is correctly maintained and inspected at regular intervals. Most shipping companies comply with this requirement by carrying out such inspections and maintenance under a 'planned maintenance schedule'. Such a procedure ensures that not only lifting gear, but mooring winches, gangways and any other mechanical or weight-bearing equipment is regularly maintained and continuously monitored. Inspections, tests and repairs are dated and certificates are retained in the Register of Ships' Lifting Appliances and Cargo Handling Gear.

PREPARATION FOR MAINTENANCE OF TOPPING LIFT BLOCKS

Prior to carrying out any overhaul of the topping lift blocks, the wire must be cleared from the sheaves. In order to strip the wire clear of the blocks, the derrick should be stowed in the crutch support at deck level. The bare end of the down haul should be crimped to a cable sock and joined to a heaving line. This will permit the wire itself to be pulled through the sheaves from the end of the wire which has the hard eye shackled to the block. This action will leave the heaving line (long length) rove through the sheaves of the two blocks.

FIGURE 1.13 On-going maintenance of topping lifts, of derricks aboard a tanker vessel. Tankers generally have either derricks or cranes in an amidships position to lift pipe line connections on board close to the manifold. The derrick systems are usually supported by 'Samson posts' to accommodate tank venting and upper block securings.

The blocks can then be lowered from the position aloft without bearing the excessive weight of the wire. At deck level the upper blocks can be overhauled in a safe environment. Once the topping lift wire has been lubricated at deck level, it can be re-rove by pulling the heaving line with the oiled wire back through the sheaves of the blocks. Alternative arrangements can employ a boatswain's chair operation to lubricate topping lift wires in situ.

DECK CRANES

FIGURE 1.14 A 25 ton, SWL Shipboard pedestal crane sited on the port side aboard the general cargo vessel 'Scandia Spirit' (IMO No. 8817837) while moored alongside in 'Dikili', Turkey.

Shipboard Heavy Duty Cranes

To state that cranes are more fashionable than derricks is not strictly correct. To say that they are probably more compact and versatile is more to the point. They tend to be more labour-saving than derricks, but if comparisons are made for heavier load capacity and greater lifting capability, then the modern H/L derrick must remain dominant.

Single-man drive and control is the key feature of the crane. They can achieve the plumb line quickly and accurately and for up to 40 tons SWL they tend to be well suited for shipboard operations. The main drawback for ship-mounted cranes is that the level of shipboard maintenance is increased, usually for the engineering department. They also need skilled labour to handle this increased maintenance workload.

CRANES ON CARGO SHIPS

FIGURE 1.15 The *Sir John* (IMO No. 9634646) general purpose/container cargo vessel lies starboard side to in the Port of Barcelona, fitted with two heavy duty deck cranes both situated on the starboard side of the vessel. Seen with the number one crane in the stowed position and the number two pedestal crane topped and working with shoreside cranage.

In this day and age, flexibility in shipping must be considered essential and such example cranes can be gear shifted into a faster mode of operation for handling containers up to 36 tons or other similar light general cargo parcels.

In the main shipboard cranes are in a fixed location, often located offset centre, to one side of the vessel. Offset centre cranes have the benefit of an extended outreach for the crane jib, the drawback being that the vessel is then conditioned to berth crane side to, at each docking, unless working into barges.

Cranes are generally operated with specialised wires having a non-rotational, non-twist property, sometimes referred to as 'wirex'. The lay of the wire is similar to a multi-plait design, woven around a central core in opposition to the directional lay of the core. Wires are tested in the normal manner as any other flexible steel wire construction. Despite these anti-twist properties, most incorporate a swivel arrangement over and above the hooking arrangement.

Where cranes are employed in tandem, they tend to be used in conjunction with a bridle or spreader arrangement to engage the total load volume. When lifting close to the crane capacity, such additional items need to be included in the total weight load for the purpose of calculation of the Safe Working Load. Note: Some bridle arrangements are often constructed out of steel section and in themselves can add considerable weight to the final load being lifted.

Most cranes operate within limits of slew, and with height luffing limitations. This is not to say that 360° rotational cranes are not available. Virtually all cranes are manufactured to operate through a complete circular arc, but limit switches are usually set with shipboard cranes to avoid the jib fouling with associated structures. Safe operational arcs are normally depicted on the ship's rigging plan and limit switches are set accordingly.

Operator cabs are usually positioned with aerial viewing and provide crane drivers with clear views of the lifting and hoist/ground areas. Topping lift arrangements generally passing overhead and behind the cabin space tend not to interfere with the driver's overall aspects. The hoist and topping lift wires are accommodated on winch barrels found in the base of the crane beneath the cab position.

Ships' cranes are versatile and have become increasingly popular since their conception. This is because of advanced designs having increased lift capacity and flexible features. They are manufactured in prefabricated steel and incorporate strength section members capable of accepting that heavier load, while at the same time retaining the ability to handle the more regular lighter load.

FIGURE 1.16 The general cargo bulk carrier 'Themis' seen with all deck cranes topped and working cargo from open steel hatches (IMO No. 9452543).

FIGURE 1.17 The cargo vessel 'Centaurus' seen in a seagoing capacity departing the River Mersey, all cranes lowered and secured and hatches battened down (IMO No. 8118592).

Arguments for Cranes or Derricks

Shipboard lifting gear has generally moved away from derrick use towards cranes, with the possible exception of the heavy lift sector. Here, the combined advantages of both the derrick and the crane, in the form of 'speed cranes', tend to have been amalgamated.

All cranes are provided with individual motors and as such lend to increased maintenance schedules. However they are probably more manoeuvrable with lighter cargoes and have the ability to directly plumb over the lifting point with ease.

The derrick boom, however, is generally notably stronger but was originally awkward to handle.

Innovations in the industry have generated fingertip control of derricks and the ability to provide an accurate line of plumb while at the same time increasing the overall lift capacity of derrick units. Previously it was only cranes that operated with a single controller and these were quickly seen as labour-saving, but the sophisticated derrick rig, especially in the heavy lift area, has now also achieved single-man control.

Shipboard cranes (single-man operation) are mounted to generally operate at about 35 tonne SWL up to an outreach of about 18–20 metres. These figures are flexible with specific designs but must be recognised as being suitable for the lighter, more general cargoes. They can sometimes be coupled to work in tandem, so increasing the load capacity and positioned to work two hatches as routine.

Cranes/derricks tend to be specific to the heavy lift/project cargo sector of the industry, having gained all the positive advantages of both the crane and the derrick rigs, taking few of the disadvantages. Wide prefabricated jibs and more powerful motors have removed the need for the old cluttered rigging of guys. Multi-sheave topping lifts have not only increased the load capacity but also outreach to permit increased loads at lower operating angles. The speed of operation, which has never been a contentious issue, has been improved with modern designs incorporating a smooth hoist, slew and luffing movements to current lifting apparatus.

Advantages/Disadvantages Cranes to Derricks

If the use of shipboard cranes is compared to the use of derricks, several factors lend to cranes being more popular and – depending on the intended loads – more preferable.

Crane advantage over derricks	**Crane disadvantage** over derricks
Simple operation	Comparatively high installation cost
Single-man operation, derricks are more labour-intensive	Increased deck space required, especially for 'gantry' type cranes
Clear deck operational views	Design is more complex, leaving more to go wrong
Clear deck space of rigging	
Versatility with heavy loads, and not re-quired to de-rig	Specialist maintenance required for hydraulics and electrics
360° slew and working arc when compared with limited operating areas for derricks	The SWL of cranes is generally less than that of specialist derrick rigs
Able to plumb any point quickly making a faster load/discharge operation	
Enclosed cabin for operator, whereas the majority of derrick operators are exposed, offering greater operator protection and comfort	
Cranes are acceptably safer to operate because of their simplicity, whereas derrick rigs can be overly complicated in rigging and operation	
Cranes can easily service two hatches, or twin hatches in the fore and aft direction because of their 360° slew ability	
Derrick rigs are usually designed to service a specific space	
Note: There are exceptions though. Some derrick designs with double-acting floating head rigs can work opposing hatches.	

FIGURE 1.18 The bulk carrier 'Yick Lee' (IMO No. 8025525) seen starboard side to the grain berth, Barcelona, Spain. The ship has turned her deck cranes outboard to port to allow access for the suctions of the grain elevators to discharge a bulk cargo of grain.

FIGURE 1.19 The 'Tasman Crusader' (IMO No. 9344667) renamed (*Cap Patton*) seen fitted with its own deck container cranes working containers by shore side cranes while starboard side to, in Napier, New Zealand. The ships own deck cranes are turned outboard to allow working space for cargo movement.

Gantry Cranes (Shipboard)

Gantry cranes are extensively found shoreside in the container terminals and these will be described in a later chapter. The use of gantries aboard ships has reduced dramatically on new tonnage because of the extensive facilities found at the terminal ports.

Where gantry rigs do operate, they tend to be 'tracked gantry rigs', which tend to travel the length of the cargo deck in order to service each cargo hold. They also use the rig for moving the hatch covers which are usually 'pontoon covers' that can be lifted and moved to suit the working plan of the vessel when in port.

FIGURE 1.20 Example of a mobile tracked gantry crane in operation on the ships foredeck. Suitable for a vessel with all aft accommodation.

FIGURE 1.21 Shipboard tracked travelling gantry crane fitted to the fore deck of vessel dedicated to the carriage and transport of nuclear waste material.

FIGURE 1.22 Stacked steel hatch covers aboard a modern coastal vessel. The ship is fitted with its own travelling mini-gantry crane, for the sole purpose of lifting on and off steel hatch covers.

FIGURE 1.23 The 'Pan Pac Spirit' (IMO No. 9186780) seen starboard side to the timber berth in Napier, New Zealand. The ship engages in loading finished package timber by means of her own ships gantry cranes.

Gantry Operations

Some shipboard gantry cranes are designed solely to remove and stow pontoon hatch covers, while others are suitably employed with outreach capability for working containers to the quayside as well as having the flexibility to remove pontoon covers.

The gantry structure tends to be a dominant feature and is subject to extensive maintenance attention, just as any other types of lifting appliances are.

GENERAL CARGOES – SLINGING ARRANGEMENTS

Although the majority of cargoes are carried in containers or unitised in one way or another, some cargoes and certainly ships' stores are required to be 'slung' with associated lifting gear. Many bagged cargoes employed 'canvas slings' but handling bagged cargoes proved costly in the modern commercial world and fewer bagged cargoes are used these days. Products are preferred to be shipped in bulk and bagged ashore if required at the distribution stage It should be realised that general cargo ships have declined considerably in number, with the main capacity going into the container or Ro-Ro trades. However, some

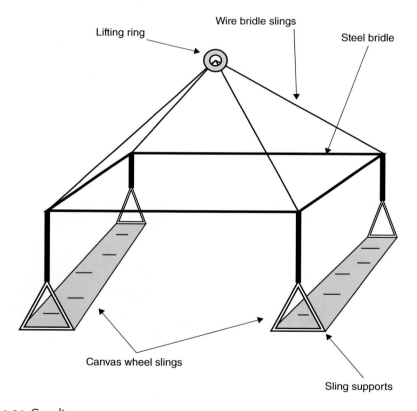

FIGURE 1.24 Car sling.

items like pre-slung packaged timber and palletisation have gone some way to bridge the ever-widening gap between general and containerised cargoes.

Single 'private' vehicles are still sometimes loaded and these are crated, containerised or require the customised 'car sling' for open stow. However, where cars (and trucks) are carried in quantity then 'Pure Car Carriers' (PCC) or 'Pure Car, Truck Carriers' (PCTC) are normally engaged.

Roped Cargo Sling Arrangements

Rope slings are probably the most versatile of slinging arrangement employed in the movement of general cargo parcels. They are made from 10–12 metres of 25–30 mm natural fibre rope. Employed for stropping boxes, crates, bales and case goods of varying sizes.

The board and canvas slings tend to be specialised for bagged cargo or sacks. With the lack of bagged cargoes being shipped these days, they have dropped away from general use, except in the smaller third world ports. Most bagged cargoes are now being containerised or shipped on pre-stow pallets.

FIGURE 1.25 Canvas and nylon strops are more frequently employed with soft package materials like cardboard cartons. The strops are user friendly, strong and pliable, and cause limited damage.

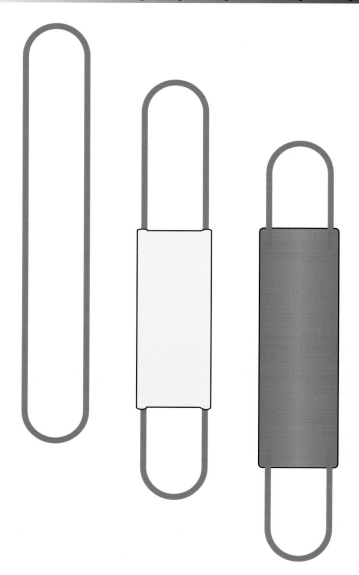

FIGURE 1.26 Roped cargo sling arrangements.

The permitted working load of a multi-leg sling, for any angle between the sling legs, up to a limit of 90°, is calculated by using the following factors:

2 leg sling 1.25 ⎤
3 leg sling 1.60 ⎬ times the SWL of the single leg
4 leg sling 2.00 ⎦

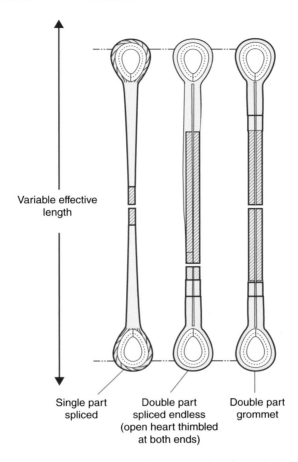

Variable effective length

Single part spliced

Double part spliced endless (open heart thimbled at both ends)

Double part grommet

FIGURE 1.27 Multi legged slings (Wire rope slings reproduced with kind permission of Bruntons (Musselburgh), Scotland.)

Where the angle between the sling legs has limitations and angles of 90° or less are too restrictive, a permissible working load for angles between 90° and 120° can be calculated as follows:

2 leg sling 1.00 ⎤
3 leg sling 1.25 ⎬ times the SWL of the single leg
4 leg sling 1.60 ⎦

Note: In the case of three-legged slings the included angle is that angle between any two adjacent legs. In the case of a four-legged sling the included angle is that angle between any two diagonally opposing legs.

Longer length slings can absorb shock load better than a shorter sling length.

NB. The greater the angle between the legs of the slings will effectively cause an increase in the stress that each leg of the sling will have to bear.

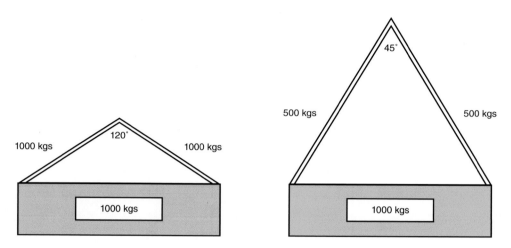

FIGURE 1.28 Alternative slinging methods: general cargoes.

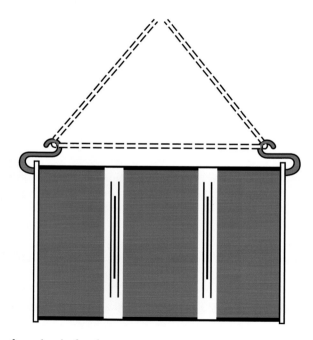

FIGURE 1.29 Use of can hooks for drums.

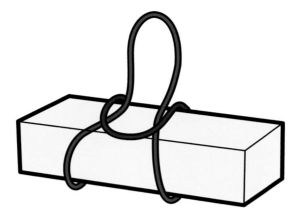

FIGURE 1.30 Bale sling strop use.

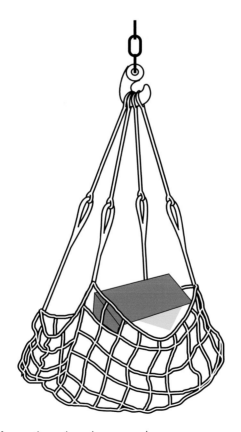

FIGURE 1.31 Cargo nets for mails and package goods.

FIGURE 1.32 Timber dogs for logs.

HANDLING AND SLINGING STEEL LOADS

Heavy Castings

Use of Chain Slings

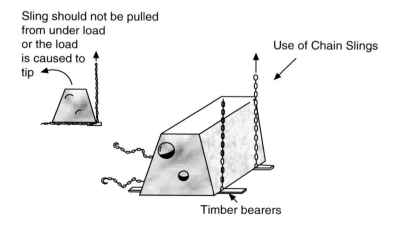

Sling should not be pulled from under load or the load is caused to tip

Use of Chain Slings

Timber bearers

FIGURE 1.33 Chain slings for use on steel castings.

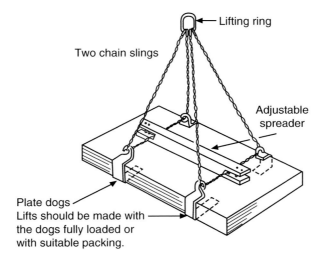

FIGURE 1.34 Use of plate dogs for packaged steel sheets.

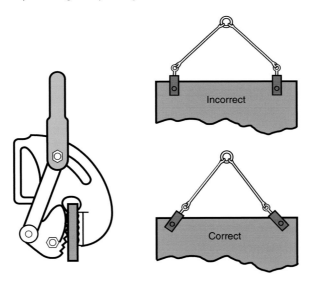

FIGURE 1.35 Use of steel plate clamps. Flexible steel wire snotters (eye each end) deployed each end of the bundle in opposite directions over an anti-slip dunnage piece.

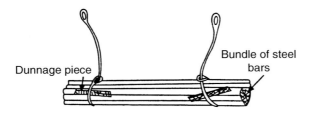

FIGURE 1.36 Bundles of steel rods/bars.

Example Slinging Arrangements for Heavier Cargoes

Weight and bulk often go together and many of the maritime heavy lifts are not only heavy in their own right but are often extremely bulky by way of having a large volume. Numerous methods have been employed over the years in order to conduct lifting operations in a safe manner. Many types of load-beams and bridle arrangements have been seen in practice as successful in spreading the overall weight and bringing added stability to the load movement during a load/discharge activity.

Heavy lifting operations – by use of heavy duty lifting beams.

See Heavy Lifting (Chapter 4) for further examples of the use of lifting beams for hoisting heavy loads.

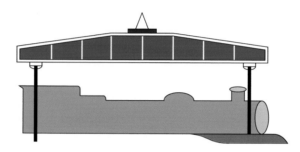

FIGURE 1.37 A steel construction in the form of a locomotive being lifted by means of a long strengthened lifting beam to accommodate the length and weight of the load. Similar vehicles like bulldozers, military armour/tanks, mechanical diggers, etc, can more often be lifted by use of a lifting beam.

Warning – case goods often arrive at the loading wharf with protruding, recommended lifting points. Cargo Officers should ensure that these lifting points are actually secured to the load inside the case and the hoist method is not just lifting the wooden case.

Opening the lid of the case will allow the prudent officer to establish that the load is secured as an integral unit of weight.

Use of the bridle with wrap around slings tends to eliminate the use of the designated lifting points

PALLETISATION

Prior to the massive expansion in the container trade, 'palletisation' became extremely popular as it speeded up the loading and discharging time for general cargo ships. This

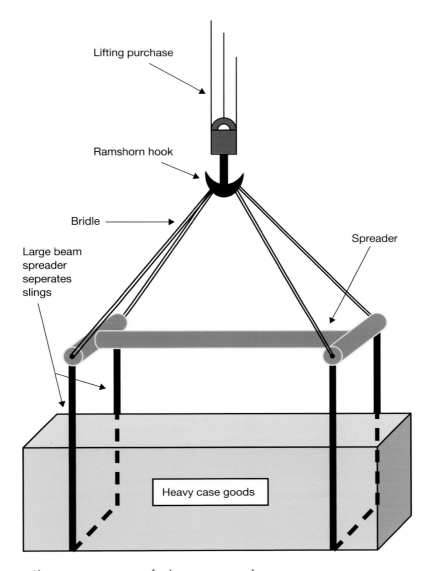

FIGURE 1.38 Slinging arrangements for large case goods.

meant that the time in port was reduced, together with associated port and harbour fees, a fact that was not wasted on shippers and vessel operators.

Cargo pallets are still extensively used with forklift trucks and employed to 'stuff' containers ashore prior to being loaded onto container vessels.

Pre-packed loaded pallets are still widely used around commercial ports and are packed in uniform blocks to minimise broken stowage. Typical cargoes suitable for loading to pallets are cartons, small boxes, crates, sacks and small drums.

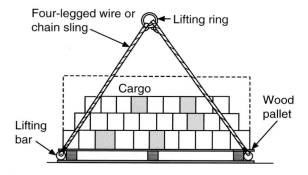

FIGURE 1.39 Alternative pallet sling (part load).

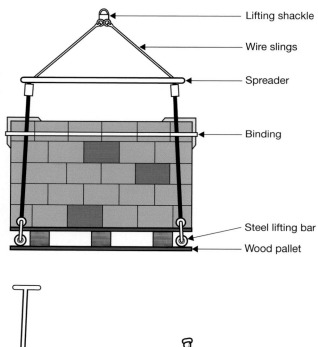

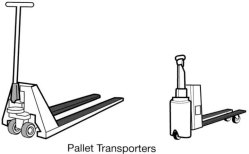

Pallet Transporters

FIGURE 1.40 Slinging of a fully loaded pallet and pallet transporters.

Palletisation has distinct advantages when compared with open stow, general cargo, break bulk handling methods:

1   Less handling of cargo
2   Less cargo damage (no hook use and limited pilferage)
3   Faster loading discharge times.

Vessels were designed specifically for the purpose of handling pallets and were usually fitted with large open hatchways which allowed spot landing by crane. The ships' design was often multi-deck and fitted with side elevators, shell doors, or roll systems to move cargoes into squared-off hatch corners.

Tween deck heights are such as to allow access and use of forklift trucks.

Pallet transporters, battery or manually operated, are useful for 'stuffing' containers where the modern container floor will generally have the capacity to support a forklift truck and its load.

USE OF FORKLIFT TRUCKS

Use of pallets and case goods often requires the use of forklift trucks, either on the quay-side or inside the ship's cargo hatches. They have the capability to move cargo parcels out from underdecks into the hatch square to facilitate easy lifting during discharge. Similarly they can stow heavy individual parcels into a tight stow into hatch corner spaces. It is appreciated that 'bull wires' could be employed for such movements, but rigging and operation of bull wires takes excessive time, while the forklift truck can be effective quickly. The main disadvantage of forklift truck use is that the vehicle requires manoeuvring space inside the hatch and as the hatch is loaded, available space becomes restricted.

Forklift trucks are manufactured in different sizes and are classed by weight. Ships' Officers are advised that the truck itself is a heavy load and will be fitted with a counter weight which provides stability to the working vehicle when transporting loads at its front end. It would be normal practice to separate the counter weight from the truck when lifting it into a ship's hold, especially so if the total combined weight was close to or exceeded the SWL of the lifting gear. Once on board the ship, the counter weight could then be reunited with the forklift truck for normal operation.

The use of forklift trucks is a skilled job and requires experienced drivers. Possible problems may be encountered if decks are greasy or wet, which could cause loss of traction and subsequent loss of control of the truck when in operation. Spreading sawdust on the deck as an absorbent can usually resolve this situation and keep operations ongoing.

Cargo Officers should exercise caution when working with these trucks aboard the vessel. Although the field of view for the driver is generally good, some cargoes could obscure the total vision and cause blind spots. The nature of the work is such that the number of men inside the hatch should be limited, thereby reducing the possibility of accidents.

FIGURE 1.41 Small forklift truck used on the quayside for the movement of small cargo parcels, pallets and stores, seen loading into a passenger vessel via its shell door. Lift capacity varies with the size of the forklift unit.

Forklift Truck – Alternative Uses

Probably the most versatile transporter for a variety of cargo parcels that the industry has ever used, the basic forklift can convert to a mini-crane, drum handler or clamp squeeze tool to suit package requirements. The main forks can be side shifted to work awkward spaces and working capacity can start from 2 tonnes upwards. Height of operations is dependent on model engaged.

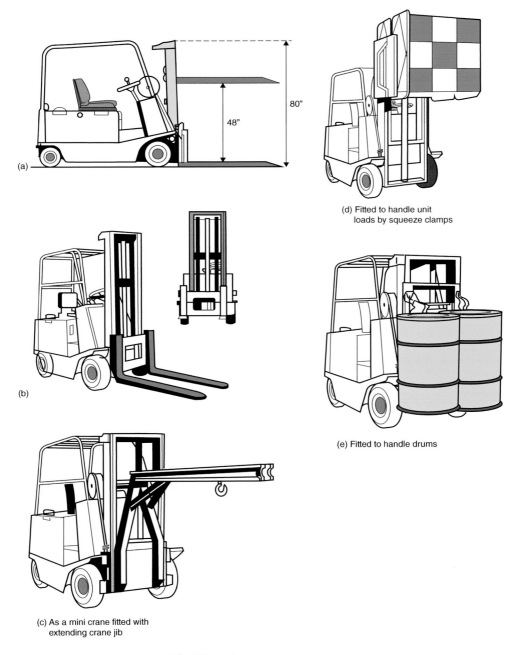

(a)

80"

48"

(b)

(c) As a mini crane fitted with
extending crane jib

(d) Fitted to handle unit
loads by squeeze clamps

(e) Fitted to handle drums

FIGURE 1.42 a–e The usage of forklift trucks.

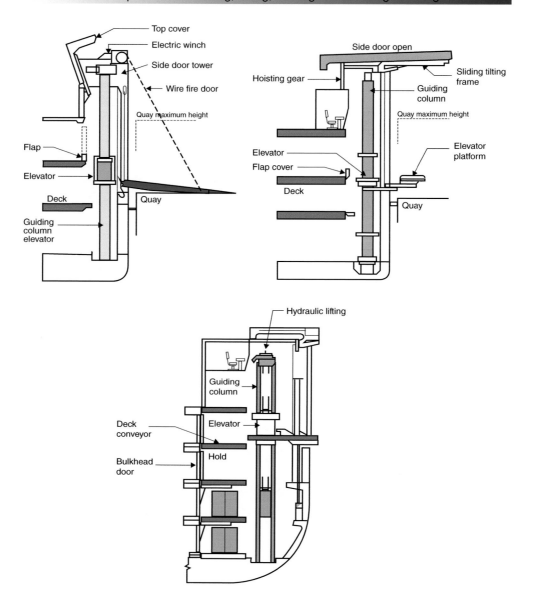

FIGURE 1.43 Side loading methods. Reproduced with kind permission of Transmarine, specialists in marine logistics, equipment systems and services, worldwide.

Side Loading Practice

Several ships have been constructed with side loading facilities for specific commodities, i.e. paper and forestry products on the Baltic trades. Watertight hull openings work in conjunction with internal elevators to move cargoes to differing deck levels. These openings, shell doors as such, may function as a loading ramp or platform depending on cargo and designation. Forklift trucks are engaged on board the vessel to position cargo parcels.

CHAPTER 2

Hatchwork and Carriage of Cargoes

INTRODUCTION

With the many changing trends of cargo transportation, it would be expected that the design and structure of cargo holds would change to meet the needs of modern shipping. This is clearly evident with container tonnage and the vehicle decks of the Ro-Ro vessels. Alongside the advances of Containers and Ro-Ro sectors, the carriage of open stow cargoes has been greatly reduced and is now considered comparatively small. This is possibly because most merchandise will suit the more popular container methods or similar unit load movement, as with palletisation.

Hold structures have tended to go towards square corners to reduce broken stowage, and suit pallets, pre-slung loads and forklift truck use inside the holds. Stowage by such vehicles has been aided by flush decks in way of the turn of the bilge, as opposed to the angle turn in the sides of the holds of older tonnage.

Hatch closure methods have continued to dominate by use of steel covers, usually hydraulically operated, both at weather and tween deck levels. Other trades especially where containers are a regular feature, still use pontoon hatch covers but these remain heavy and cumbersome.

Some specialist cargoes, such as steel coils, still suit conventional holds and clearly would not be compatible inside containers. Customised vessels have also been built to accommodate steel coils. Also large case goods or castings tend not to be compatible with containers, and transport better by means of conventional stowage in the more dated style of vessel. Such merchandise is clearly edging towards heavy lift type loads and these heavier loads are covered in more detail alongside the designated heavy lift ship and project cargoes in Chapter 4.

The objective of this chapter is to provide an overall picture of an industry sector which is an essential part of cargo handling and general shipping practice. It does not have such a high profile as the container or Ro-Ro movement, but it is, nevertheless, an indispensable arm to the practice of shipping.

DOI: 10.4324/9781003407706-2

HATCHWORK AND RIGGING (DEFINITIONS AND TERMINOLOGY)

(Employed with Cargo Operations)

Backstays – Additional strength stays applied to the opposing side of a mast structure when making a heavy lift. These stays are not usually kept permanently rigged and are only set as per the rigging plan when a heavy lift is about to be made.

Bearers – Substantial baulks of timber, used to accept the weight of a heavy load on a steel deck. The bearers are laid for two reasons: (i) to spread the load weight over a greater area of deck; and (ii) to prevent steel loads slipping on the steel deck plate.

Bilge suction – Pump suctions positioned aft in each cargo hold to allow the bilge pumps in the engine room to pump any accumulated water from the cargo holds. The suction itself is usually fitted with a 'Strum box', a sieve arrangement that allows the passage of fluid but not solids. The suctions are always tested prior to loading a cargo. In the event of a hold fire the hold can be deliberately flooded to extinguish the fire, and then the suctions can pump out the water. Bilge suction arrangements are fitted with non-return valves.

Booby hatch – Small access (inspection) hatch, found on the weather deck to allow entry into a tween deck/lower hold, without having to open the main hatch covers.

Breaking strength – Defined by the stress necessary to break a material in tension or compression. The stress factor is usually obtained by testing a sample to destruction.

Bridle – A lifting arrangement which is secured to a heavy load to provide a stable hoist operation when the load is lifted. Bridles may be fitted with a spreader to ensure that the legs of the bridle are kept wide spread so as not damage the lift and provide a balanced hoist operation.

Bulldog grip (Wire rope grips) – Screw clamps designed to join two parts of wire together to form a temporary eye or secure a wire end.

Bull wire – (i) A single wire, often used in conjunction with a 'lead block' rigged to move a load sideways off the line of plumb. Example use is found in dragging cargo loads from the sides of a hold into the hold centre.

(ii) A wire used on a single span topping lift, swinging derrick, to hoist or lower the derrick to the desired position. The bull wire is secured to a 'union plate' to work in conjunction with the chain preventer and the downhaul of the topping lift span.

Cargo battens – Usually referred to as 'Spar Ceiling'. They are timber lengths fastened to the sides of a cargo hold, fixed to prevent the cargo parcels touching the steelwork of the ship's sides. They allow ventilation around the stowed cargo and prevent cargo sweat, and ships sweat damaging cargo parcels with moisture.

Coaming – The upper surround of a hatchway opening position. Coamings are fixed from the upper weather deck and are usually a minimum of about one metre in height designed to prevent persons falling into an open cargo hatch.

Doubling up – A term used with a derrick which allows a load greater than the SWL of the runner wire but less than the SWL of the derrick to be lifted safely. It is achieved by means of a longer wire being used in conjunction with a floating block. This effectively provides a double wire support and turns a single whip runner wire into a 'gun tackle'.

Deep tanks – These are additional cargo spaces that are found at the bottom of cargo holds in some dry cargo ships. They provide the vessel with a capability to carry specialised liquid cargoes. Deep tanks are usually fitted with heating coils (Steam) for keeping cargoes like tallow at a warm temperature to prevent solidification during transport.

Hallen derrick – The commercial name given to a particular single-man operated derrick. It first came with a 'D' frame at the mast head. This was then superseded by a 'Y' mast structure fitted with outriggers. The function of the 'D' frame and outriggers was to prevent the upper guy and topping lift wires from fouling, while the derrick was slewing. A small 'Hallen' would have an SWL of 40 tonnes.

Hat box – A pump drain system found inside deep tank constructions and operates on the same principal as a bilge suction. Certain cargoes, like grain, would need the pipeline to the hat box to be 'blanked off' to prevent any back flow of liquid into the deep tank while loaded.

Kilindo rope – A multi-strand rope having non-rotating properties, of a type employed for crane wires (similar to a multi-ply wirex).

Lateral drag – The term describes the action of a load on a derrick or crane during the procedure of loading or discharging, where the suspended weight is caused to move in a horizontal direction, as opposed to the expected vertical direction. The action is often prominent when the ship is discharging a load. As the load is passed ashore the ship has been caused to heel over towards the quayside. As the load is landed, the weight comes off the derrick and the ship returns to the upright, causing the derrick head to move off the line of plumb. This change of plumb line causes the lifting purchase to 'drag' the weight sideways, e.g. lateral drag.

Lifting beam – A strength member, usually constructed in steel suspended from the lifting purchase of a heavy lift derrick when engaged in making a long or wide load lift. Lifting beams may accommodate 'yokes' at each end to facilitate the securing of the wire slings shackled to the load.

Lighting to cargo holds – Floodlights of one form or another are a fixture of cargo spaces. The lights themselves are usually cover protected by a wire mesh guard to prevent damage.

Line of plumb – A plumb line is specifically a cord with a 'plumb-bob' attached to it. However, it is often used around heavy lift operations as a term to express 'the line of plumb' where the line of action from a derrick/crane head is suspending a weight. The line of weight, namely being the 'line of plumb'.

Maximum angle of heel – A numerical figure usually calculated by a ship's Chief Officer in order to obtain the maximum angle that a ship would heel when making a heavy lift, to the maximum outreach of the derrick or crane, prior to the load being landed.

Preventor – A generic term to describe a strength, weight-bearing wire which prevents unwanted movement on a load. Preventors are common as fitted to a Union purchase derrick rig, but can be used to prevent movement of a cargo parcel.

Project cargo – Large heavy cargoes which require a detailed plan of operation for the safe handling, lifting and transport from the point of manufacture to the position of operation.

Purchase – A term given to blocks and rope (wire or fibre) when rove together. Sometimes referred to as a 'block and tackle'. Two (2) multi-sheave blocks are rove with flexible steel wire rope (FSWR) found in common use as the lifting purchase suspended from the spider band of a heavy lift derrick.

Saucer – Alternative name given to a collar arrangement set above the lifting hook.
The function of the saucer is to permit steadying lines to be shackled to it in order to provide stability to the load during hoisting and slewing operations. They can be fixed or swivel fitted.
Note: The term is also employed when carrying 'grain cargoes' where the upper level of the grain cargo is trimmed into a 'saucer' shape.

Snatch block – A lead block (metal or wood) which allows the bight of a rope or wire to run over the sheave without having to thread the end of the hawser all through the swallow of a block.
The snatch block is a convenient method when requiring the use of a 'bull wire'.

Steadying lines – Cordage of up to about 24 mm in size, secured in adequate lengths to the load being lifted in order to provide stability and a steadying influence to the load when in transit from quay to ship or ship to barge. Larger, heavier loads may use steadying tackles for the same purpose. However, these are more often secured to a collar/saucer arrangement, above the lifting hook, as opposed to being secured to the load itself. Tackles are rove with flexible steel wire rope, not fibre cordage.

Stülcken mast and derrick – Trade name for a heavy lift derrick and supporting mast structure. The patent for the design is held by Blohm & Voss A.G. of Hamburg, Germany.

This type of heavy lifting gear was extremely popular during the late 1960s and the 1970s, with numerous ships being fitted with one form or other of Stüelcken arrangement.

Tabernacle – A built bearing arrangement situated at deck level to accept the heel of a heavy lift derrick. The tabernacle allows freedom of movement in azimuth and slewing from port to starboard.

Ventilation – To Cargo compartments is essential for many different types of cargo. To this end most cargo holds will be fitted with either cowl or mushroom ventilators fitted above the main weather deck. Virtually all are provided with sealing flaps to be applied in the event of a hold fire, for shutting off the oxygen. Other types may be fitted with fans to provide forced ventilation.

The more modern vessel, probably operating with cranes, may be fitted with twin hatch tops to facilitate ease of operation from both ends of a hold. The construction of the hold tends to be spacious to accept a variety of long cargoes.

FIGURE 2.1 The *Mount France* general cargo vessel seen port side to on the timber berth in Napier, New Zealand. The vessel is loading 'logs' from quayside transports using its own deck cranes. The vessel has five hatches all fitted with hydraulic steel hatch covers, seen in the open position.

The ship's gunwales are fitted with upright stanchions ready to secure an expected timber deck cargo.

Conventional Ship Design – General Cargo Vessel

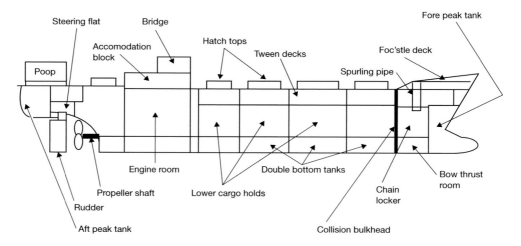

Athwartships – Half Profile

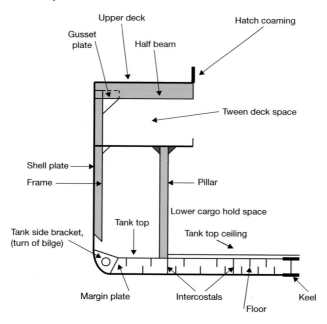

FIGURE 2.2 Conventional general cargo – half profile hold and tween cargo spaces

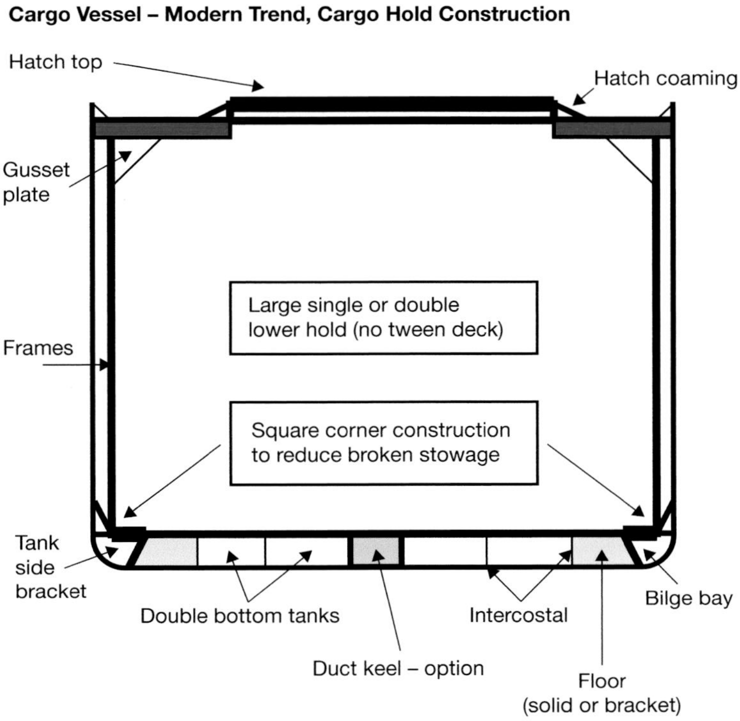

Cargo Vessel – Modern Trend, Cargo Hold Construction

FIGURE 2.3 Square cargo hatch construction used for new builds. Smaller tonnage may have additional/temporary athwartships bulkheads giving two holds as opposed to a through single space.

Double hold space, with or without temporary athwartships bulkheads which can section the hold depending on the nature of the cargo, provides flexibility to accommodate a variety of cargo types.

Square corner construction lends to reducing broken stowage especially with containers, pallets, vehicles or case goods. Flush 'bilge plate access' is generally a feature of this type of design, where steel bilge covers (previously limber boards) are countersunk into the deck so as not to obstruct cargo parcels being manoeuvred towards a tight side or corner stow.

Such clean cut construction lends easily to the use of forklift trucks inside the hold, when stowing box shaped cargoes/case goods, pallets etc.

HATCH COVERS (MECHANICAL)

Direct Pull (MacGregor) Weather Deck Hatch Covers

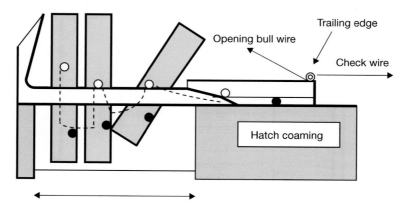

- ● Eccentric wheels lowered to track by manual levers or hydraulics.
- ○ Stowage bay wheels with interconnecting chain.

FIGURE 2.4 Direct pull (MacGregor) – weather deck hatch covers and stowage bay.

Operation

All hatch top wedges and side locking cleats are removed and the tracks are seen to be clear. The bull wire and check would be shackled to the securing lug of the trailing edge of the hatch top.

> Note: the bull wire and the check wire change function depending on whether opening or closing the hatch covers.

The eccentric wheels are turned down and the stowage bay is sighted to be clear. The locking pins at the end of the hatch would be removed as the weight is taken on the bull wire to open the hatch.

Once the hatch lids are open and stowed vertical into the stowage bay, the sections would be locked into the vertical position by lock bars or clamps to prevent accidental roll back.

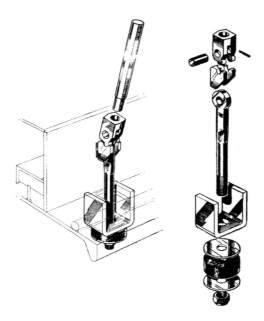

FIGURE 2.5 Side locking cleat.

Hatch Top Fittings

WARNING

Steel hatch tops are extremely heavy usually weighing several tonnes.

Caution should always be used when preparing to move hatch covers either opening or closing.

Turning down eccentric wheels

Must always be carried out as per the manufacturer's safety instructions.

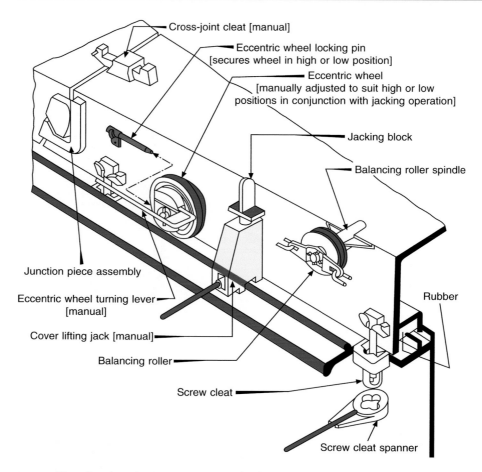

Cross-joint cleat [manual]

Eccentric wheel locking pin
[secures wheel in high or low position]

Eccentric wheel
[manually adjusted to suit high or low
positions in conjunction with jacking operation]

Jacking block

Balancing roller spindle

Junction piece assembly

Eccentric wheel turning lever
[manual]

Cover lifting jack [manual]

Balancing roller

Screw cleat

Rubber

Screw cleat spanner

FIGURE 2.6 Use of jack and turning eccentric wheels.

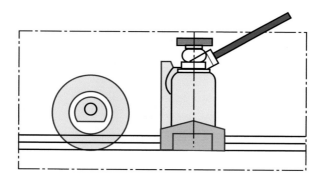

FIGURE 2.7 Raising and lowering of the eccentric wheels by use of hand-operated levers.

FOLDING (HYDRAULIC OPERATED) HATCH COVERS

The more modern method of operating steel hatch covers is by hydraulics, opening the sections in folding pairs, either a single pair, double or triple pair sections.

There are several manufacturers of steel hatch covers and they all generally achieve the same function of sealing the hatchways, quickly. Operationally, one man could close up five or six hatches very quickly by switching on the hydraulic pumps, releasing the locking bars to the stowed sections and operating the control levers designated to each set of covers.

A disadvantage of hydraulic operations is that the possibility of a burst pipe is always possible, with subsequent cargo damage due to hydraulic oil spillage.

FIGURE 2.8 Folding hydraulic operated double steel hatch covers, seen in the vertical open position. Securing cleating seen in position prevents accidental roll back when in open position.

FIGURE 2.9 Steel hatch pontoon type covers seen stacked on the part open hatch. The covers are lifted and stacked by the ship's own mini-gantry travelling crane.

FIGURE 2.10 The container ship 'Uranus' seen docking at Warren Point, Northern Ireland. The vessel is viewed with all pontoon hatch covers in place, prior to loading containers below deck and up to three high on top of the steel covers.

WEATHER DECK HATCH COVERS

Steel weather deck hatch covers now dominate virtually all sectors of general, bulk and container shipping. Conventional wooden hatch covers have been eclipsed by the steel designs which are much stronger as well as being easier and quicker to operate.

The advantages far outweigh the disadvantages in that continuity of strength of the ship is maintained throughout its overall length. Better watertight integrity is achieved and they are labour-saving, in that one man could open five hatches in the time it would take to strip a single conventional wooden hatch. The disadvantages are that they are more expensive, initially, to install and carry a requirement for more levels of skilled maintenance.

Once cleated down, a hard rubber seal is created around the hatch top perimeter providing a watertight seal on virtually all types of covers. Hydraulically operated covers cause a pressure to generate the seal, while mechanical cleating (dogs) provide an additional securing to the cargo space below. The engineering department of the ship usually caters to the maintenance of the hydraulic operations and the drawback is that a hydraulic leak may occur due to, say, a burst pipe, which could cause subsequent damage to cargo.

Extreme caution should be exercised when opening and closing steel covers and adequate training should be given to operators who are expected to engage in the opening and closing of what are very heavy steel sections. Check wires and respective safety pins should always be applied if appropriate, when operating direct pull types. Hydraulic folding 'M' types incorporate hydraulic actuators with a non-return capacity which prevents accidental collapse of the hatch tops during opening or closing. Whichever type is employed, they are invariably track mounted and such tracks must be seen to be clear of debris or obstruction prior to operation.

Strong flat steel covers lend to heavy lifts and general deck cargo parcels and have proved their capability with the strengthened pontoons which are found in the container vessels. The pontoons have specialised fittings to accept the deck stowage of containers over and above the cargo hold spaces. Similarly, specialised heavy lift vessels have adopted strengthened open steel decks in order to prosecute their own particular trade sector.

Alternative Weather Deck Hatch Covers

MacGregor – Direct Pull Type

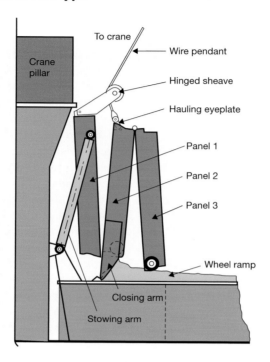

FIGURE 2.11 Direct pull weather deck hatch cover.

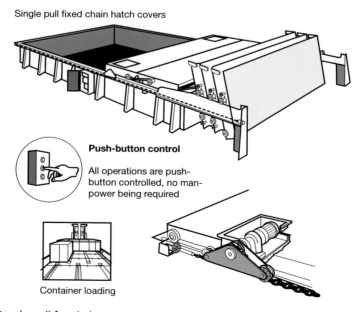

FIGURE 2.12 Single pull fixed chain MacGregor hatch cover.

Folding (Hydraulic Operated) Hatch Covers

FIGURE 2.13 Open hatchway seen exposing the two cargo holds of the 'Scandi Spirit' (IMO No. 8817837) with the folding 'M' type hydraulic hatch covers in the fixed open position. The hatches are serviced by two port side pedestal cranes.

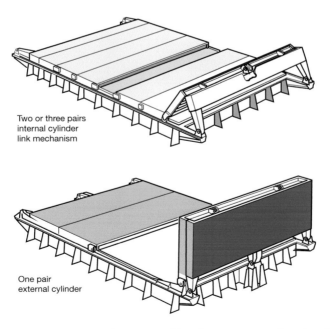

Two or three pairs
internal cylinder
link mechanism

One pair
external cylinder

FIGURE 2.14 Folding 'M' type hydraulic operated hatch covers reproduced with kind permission of MacGregor.

Operation of Steel Hatch Covers (Tween Decks)

Folding hatch covers are operated in pairs by hydraulic cylinders which actuate link mechanisms working from 0° to 180°. One, or two pairs of cover panels can be linked and stowed at the same end of the hatch if required.

To open the hatch, the leading pair of covers is first operated, immediately pulling the remaining pair(s) into a rolling position on the recessed side tracks and tow the hatch end. Once the leading pair is raised, the trailing pair(s) can follow into vertical stowage positions where they are secured to each other. The operational sequence is reversed when closing the hatch covers. Tween deck hatch covers are not required to be watertight and, unless specifically requested, they would have no additional cleating arrangement.

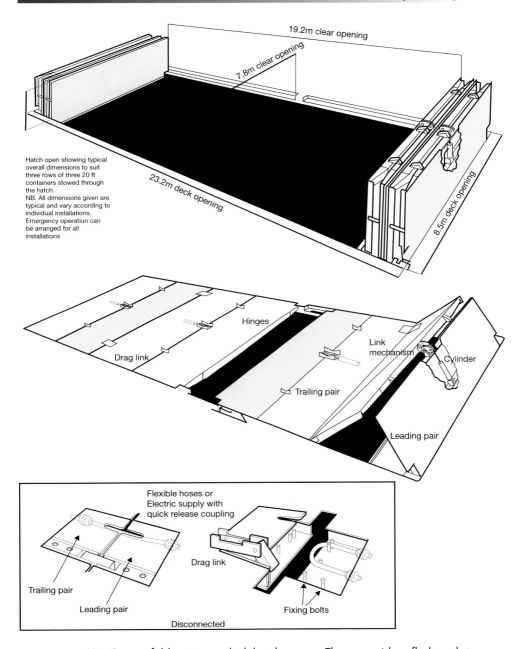

Hatch open showing typical overall dimensions to suit three rows of three 20 ft containers stowed through the hatch.
NB. All dimensions given are typical and vary according to individual installations. Emergency operation can be arranged for all installations

19.2m clear opening

7.8m clear opening

23.2m deck opening

8.5m deck opening

Hinges

Drag link

Link mechanism

Cylinder

Trailing pair

Leading pair

Flexible hoses or Electric supply with quick release coupling

Drag link

Trailing pair

Leading pair

Fixing bolts

Disconnected

FIGURE 2.15 MacGregor folding tween deck hatch covers. These provide a flush and strong deck surface which is ideal for the working of forklift trucks inside tween deck spaces. Hydraulically operated and user friendly.

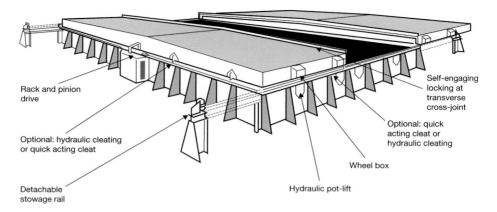

Rack and pinion drive

Optional: hydraulic cleating or quick acting cleat

Detachable stowage rail

Self-engaging locking at transverse cross-joint

Optional: quick acting cleat or hydraulic cleating

Wheel box

Hydraulic pot-lift

FIGURE 2.16 Side moving hatch covers. Usually employed for wide beam bulk carrier type vessels.

Partial Opening

As with many types of steel covers, partial opening is a feature and can be achieved comparatively quickly by the operation of limited hatch sections. Pontoon hatch tops can be hatch stacked on closed sections.

FIGURE 2.17 Two bulk carrying 'feeder' coasting vessels lie port side to, alongside the grain elevator in Barcelona. The lead ship is seen discharging with partially opened steel hatches, operated by its own mini-gantry crane. The one astern has folding weather deck hatches in the closed up position.

CHAPTER 3

Stowage Properties of General Cargoes

INTRODUCTION

General cargo is a term which covers a great variety of goods. Those goods may be in bags, cases, crates, drums or barrels or they may be kept together in bales. They could be individual parcels, castings or machinery parts, earthenware or confectionary. They all come under the collective term of 'general cargo'.

The Chief Officer is usually that person designated on board the vessel who is responsible for the handling and safe stowage of all cargoes loaded aboard the ship.

He/she is that person responsible for making sure that the holds are clean and ready to receive cargo parcels and ensure correct stowage and shipping of goods in a safe manner. Chief Officers are ultimately responsible for the carriage, ventilation and delivery, in good condition, of all the vessel's cargo.

In order to carry goods safely, the vessel must be seaworthy and the cargo spaces must be in such a condition as not to damage cargo parcels by ship's sweat, taint or any other harmful factor. To this end the Chief Officer would cause a cargo plan to be constructed to ensure that separation of cargoes is easily identifiable and that no contamination of products could take place during the course of the voyage. The Chief Officer's prime areas of duty lie with the well-being and stability of the vessel, together with the safe carriage of all cargo. Clearly, with the excessive weights involved with cargo parcels, the positive stability affecting the vessel's safe voyage could be impaired.

A correct order of loading, with the capability of an effective discharge, often to several ports, must be achieved to comply with the safe execution of the voyage and also to stay within regulatory conditions – i.e. loadline requirements.

This chapter is directly related to the details affecting stowage of particular cargoes and the associated idiosyncrasies affecting the correct stow and carriage requirements to permit a lawful and successful venture.

PREPARATION OF CARGO SPACES

The Chief Officer is generally that person aboard who is responsible for the preparation of the ship's holds to receive cargo. The preparations of the cargo compartments will usually

DOI: 10.4324/9781003407706-3

be the same for all non-containerised general cargo parcels with additional specific items being carried out for specialised cargoes.

1   Holds and tween deck spaces should be thoroughly swept down to remove all traces of the previous cargo. The amount of cleaning will depend on the type of the previous cargo and the nature of the next cargo to be carried. On occasions the hold will need to be washed (salt water wash) in order to remove heavy dust or glutinous residues, but the hold is only washed after the sweepings and wastes have been removed.

2   Bilge bays and suctions should be cleaned out and tested, while the hold is being swept down. Tween deck scuppers should also be tested and strum boxes should be sighted and clear. All non-return valves in the bilge lines should be seen to be free and operating normally.

Note: If the previous cargo was a bulk cargo, then any plugs at the bilge deck angle should be removed to allow correct drainage to bilge bays

3   Check that all limber boards or bilge bay covers are in good condition and provide a snug fit. If bilges are contaminated, say, from the previous cargo and have noticeable odours, these should be sweetened and disinfected.

4   The spar ceiling (sometimes referred to as cargo battens) should be examined and replaced where necessary. In specific cases, such as with an intended 'coal cargo', the spar ceiling would be totally removed from the compartment prior to loading.
   (With a coal cargo, spar ceiling is seen as a fuel and retaining oxygen in the event of a hold fire.)

5   All tween deck hatch coverings should be inspected for overall general condition and correct fitting. Tween deck guard rails, chains and stanchions should be fitted and seen to be in a good secure order.

6   Any soiled dunnage should be removed and, if appropriate, clean dunnage laid to suit the intended cargo to be loaded.

7   Checks should be made on the hold lighting, fan machinery, ventilation systems and the total flood fire detection/operation aspects.

8   Conduct a final inspection to ensure that cargo holds are ready to load. Some cargoes, such as foodstuffs, may require the compartments to be inspected by a surveyor prior to commencement of loading.

BILGE SUCTION ARRANGEMENTS

Bilge suctions are positioned at the aft end of cargo spaces and are pipeline connected via the valve chest to the pumps in the machinery space. The aft position of the bilge suctions assists the drainage of bilge water because most ships trim by the stern and this allows the accumulation of water in aft bilge bays, where the suctions are placed.

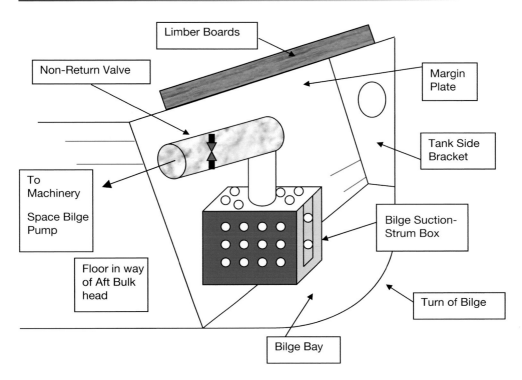

Limber Boards

Non-Return Valve

Margin Plate

Tank Side Bracket

To Machinery Space Bilge Pump

Bilge Suction-Strum Box

Floor in way of Aft Bulk head

Turn of Bilge

Bilge Bay

FIGURE 3.1 Bilge suction and strum box arrangement.

DUTIES OF THE JUNIOR CARGO OFFICER (DRY CARGO VESSELS)

Cargo Officers will have a variety of duties before, during and after cargo operations begin. He/she should be aware that monitoring the cargo movements and ensuring parcels remain in good condition is protecting the ship owners' interests. Extensive ship-keeping activities also go along with loading and discharging of the vessel's cargo.

Prior to Cargo Operation

1 Check that the designated compartments are clean and ready to receive cargo.
2 Inspect the drainage and bilge suctions are working effectively.
3 Ensure that cargo battens (spar ceiling) is in position and not damaged. (Some cargoes require cargo battens to be removed.)
4 Make sure the relevant hatch covers are open and properly secured in the stowed position.
5 Check the rigging of derricks and/or cranes are operating within respective SWL's.
6 Inspect the hatch lighting to ensure it is in good order.
7 Order engineers to bring power to deck winches. Inspect all lifting appliances to ensure that they are in good order.

8 Inspect and ensure all means of access to the compartments are safe.

9 Guard rails and safety barriers should be seen to be in place.

10 Ensure all necessary fire-fighting arrangements are in place.

11 Check that the ship's moorings remain taught.

12 Note the draughts fore and aft.

13 Check that the gangway is rigged in a safe aspect.

14 Monitor any ballast/bunker movements which may be expected or ongoing.

During Cargo Transfer

1 Note all starting and stopping times of cargo operations for reference into the log book.

2 Note the movement of cargo parcels into respective compartments for entry onto the stowage plan.

3 Refuse damaged cargo and inform the Chief Officer of the action.

4 Monitor the weather conditions throughout operations.

5 Note any damage to the ship or the cargo handling gear and inform the Chief Officer accordingly.

6 Maintain a security watch on all cargo parcels and prevent pilferage.

7 Tally in all special and valuable cargoes and provide lock-up stow if required.

8 Maintain an effective watch on the gangway and the vessel's moorings.

9 Ensure that appropriate dunnage, separation and securing of cargo takes place.

10 Monitor all fire prevention measures.

11 Check the movement of passengers' baggage (passenger carrying vessels).

12 Make sure all hazardous or dangerous cargoes have correct documentation and are given correct stowage relevant to their class (ref. IMDG code).

13 Inspect compartments and the transit warehouse at regular intervals to ensure cargo movement is regular.

14 Inform Chief Officer prior to loading heavy lifts.

15 After discharge operations, search the space to prevent parcels being over-carried.

16 Ensure that the local by-laws are adhered to throughout.

17 Note the ship's draughts fore and aft on the completion of loading/discharging.

18 Ensure cargo loads do not exceed the SWL of lifting gear engaged.

After Cargo Operations

1 Close up hatches and lock and secure access points.

2 Inform engineering department to shut down power to deck winches.

3 Secure all lifting appliances against potential damage or misuse.

4 Enter the day's working notes into the deck log book.

5   Adjust any ventilation/heating systems necessary which may be required to cargo compartments.

6   Inform the Chief Officer that the deck is secure and the current draughts.

7   Maintain fire and security watch.

Miscellaneous

The Chief Officer would ensure that the cargo stowage plan is kept updated when the vessel is in a loading situation. Officers' work books and tally sheets would be used at this stage. He would also at some time order the density of the dock water to be ascertained by means of the hydrometer.

Cargo loaded/discharged is then ascertained daily, by means of the deadweight scale, once the end of day draughts have been observed.

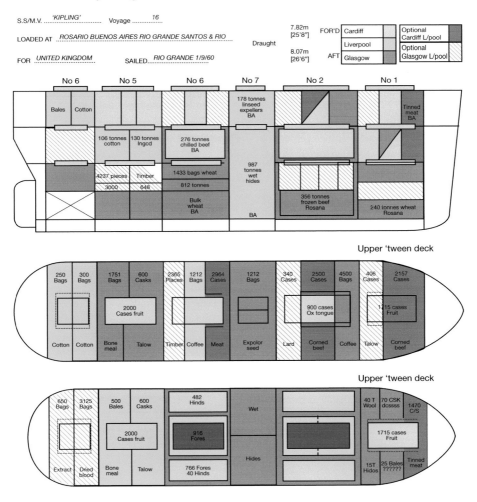

FIGURE 3.2 Example of open stow, general cargo plan.

THE CARGO STOWAGE PLAN

The function of the stowage plan is to identify the various cargo parcels by quantity, destination and nature of goods.

It permits the Chief Officer to assess the number of stevedore gangs for respective compartments and the times associated with cargo operations. Additionally it shows the location of special cargoes like 'heavy lifts' or 'hazardous goods', valuables and the lock-up stow goods.

Ventilation and fire-fighting procedures can be influenced by the disposition of respective cargoes. The owners/charterers are also provided with notification of available space between discharging ports, useful for diverting the vessel for further cargo.

Stowage plans provide the following relevant details in addition to the pictorial cargo distribution plan:

Deadweight Particulars

The above information with the ship's name and port(s) of loading together with date of sailing are all included on the plan; fuel, ballast and fresh water are usually depicted in alternative colours to colour codes as used for discharge ports.

The distribution table is usually forwarded to the ship's port agents ahead of the ship's arrival, in order to ensure dock labour requirements for discharge/loading are met.

Tanker Stowage Plans (Profile + Pipeline) – Ref. Chapter 6

These are particularly useful with product tankers where the disposition of grades of cargo can be clearly illustrated. The plan can ensure that adjacent tanks are not likely to generate

Table 3.1 Cargo Distribution Summary (Tonnes)

Port of Discharge	Colour Code	No.1 Hold	No.2 Hold	No.3 Hold	No.4 Hold	On Deck	Port Total
L'pool		tonnage	Tonnes	Tonnes	Tonnes	xxx	XXX
Cardiff		tonnage		Tonnes	Tonnes	xxx	XXX
Glasgow		tonnage		Tonnes	Tonnes	xxx	XXX
Cardiff/L'pool		tonnage	Tonnes	Tonnes	Tonnes	xxx	XXX
Glas/L'pool			Tonnes	Tonnes	Tonnes	xxx	XXX
Totals		Xxx	xxx	xxx	xxx		XXX
							XXX

Table 3.2 Ships Tonnages and Deadweight Distributions

Draughts:
Forward _____
Aft _____
Mean _____
Density correction _____ Scale D/W
S.W. Draught _____ Tons
Cargo _____
Fuel _____
Fresh water _____
Ballast _____
Stores _____

_____

Total D/Weight =
Scale D/Weight =

_____

Difference
_____ Tons _____

contamination. The pipeline system is often employed in conjunction with the plan to ensure that correct lines are operational with the correct grade of product. Quantities and type of each product can be easily identified, but clearly this plan is not as detailed as with, say, an open stow general cargo vessel carrying many different types of cargo.

Roll On, Roll Off Stowage Plans (Plan View) – Ref. Chapter 8

These are generally computer-generated and, like other stowage plans, help to identify individual units. This is specifically required for any units carrying dangerous/hazardous products. It also permits the order of discharge to be pre-arranged. Modern software is usually involved with the planning of cargo stowage with Ro-Ro vessels. They permit known weights aboard the vessel to be pre-programmed and the C of G of each unit with its respective stowage space can be entered to provide the ship's overall GM very quickly.

Container Stowage Plans (Elevation + Cross Section) – Ref. Chapter 9

Container stowage plans are a proven way of tracking specific units during the sea passage. The plan identifies each unit and allows shippers to estimate arrival times and the whereabouts of their goods during every stage of shipment. It is also an effective security aspect for knowing which unit is where and tracing what goods are in what particular unit.

Pre-Load Plans

These provide a provisional distribution plan for the intended cargo parcels. They may be accompanied by capacity space set against cargo capacity to reflect unused space. They can determine access points or detail pipeline arrangements prior to commencing cargo operations. They are generally used on all types of vessel.

STEEL CARGOES

Steelwork

Probably the most physically dangerous of all cargoes is steelwork. Steel cargoes tend to come in all shapes and sizes, from the biggest 'casting' to the long steel 'H' girders used extensively in the construction industry. Long and heavy loads are difficult to control and the slightest contact with surrounding structures could generate extensive damage or injury to personnel. The fact that in most cases they are rigid and heavy makes safe handling extremely difficult. One of the exceptions to this, among the steel cargoes, is the loading and carriage of bulk scrap metal – also a dangerous cargo, but for different reasons. It is loaded/discharged by heavy grabs which generally cause some fall-out between quay-side and shipside.

Steel Coils

Another type of steel cargo is in the form of heavy steel coils. The round shape makes this cargo a high risk to shifting in a seaway, especially if it is not properly secured. In the event of the cargo shifting, the ship could expect to take on a list which, if considered dangerous, could necessitate the vessel altering course to a port of refuge. The prime purpose of this is to discharge and then re-secure cargo, a costly business.

Steel coils are normally stowed in a double tier with the bottom coils on athwartships dunnage and wedged against athwartships movement, each coil being hard-up against the next. The turn of the bilge is protected by vertical dunnage and the second tier of coils is then placed on top. As the second tier is filled, it should be recognised that the stow will have 'key' locking coils and these should be lashed into position by steel wire lashings, while remaining accessible. It is also worth noting that the size and weight of individual coils is not always uniform and, as such, differences create small gaps between the cargo stow. These gaps, where substantial, should be 'chocked' with baulks of timber, while wedges can be used to prevent movement between smaller gaps.

Cargo Officers should be wary when working this cargo, as the method of lifting during the loading process will usually be by a standard crane with adequate Safe Working Load. However, when discharging, the odd coil may be of a heavier variety and cause the lifting gear to be overloaded. (Some coils go up to 20 tonnes.)

The main concern for any Ship's Master with coils within his cargo is that they are correctly secured and to this end it is not unusual to hire a rigging gang during the period of loading. The ship's stability must also be taken into consideration with such a heavy cargo and, consequently, coils tend to be always loaded in lower holds, as opposed to tween deck spaces. Such a loading pattern would tend to generate a favourable GM.

The quantity of cargo is usually restrictive, because its overall weight will soon bring the vessel to her loadline marks. Geographically, the loading port will dictate which load-line zones the vessel must route through to reach her port of discharge and, as such, the passage plan should reflect a route that would minimise the ship's rolling pattern wherever possible.

The objective is to form a large immovable stow with any void spaces between coils chocked off with dunnage. End of stows would be fenced with timber battens and 'locking coils' together with the top tier of coils would most certainly be lashed with steel wire rope lashings. With such a heavy cargo, the ship could be expected to reach her loadline marks quickly, leaving some considerable broken stowage with this type of cargo.

Steel Coils – Secure Stowage

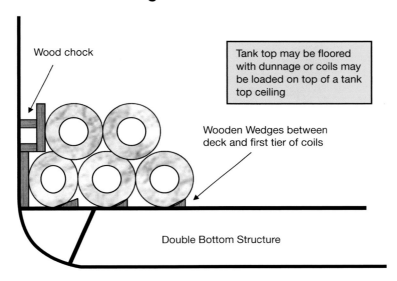

FIGURE 3.3 Turn of the bilge stowage of steel coils. Wedges and wood chocks used.

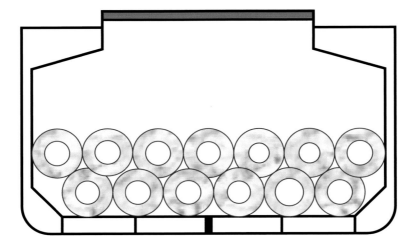

FIGURE 3.4 Steel coil loading – diamond bulker design with complete double hull is able to stow two tiers of heavy steel coils (each up to 20 tonnes) across the hatch.

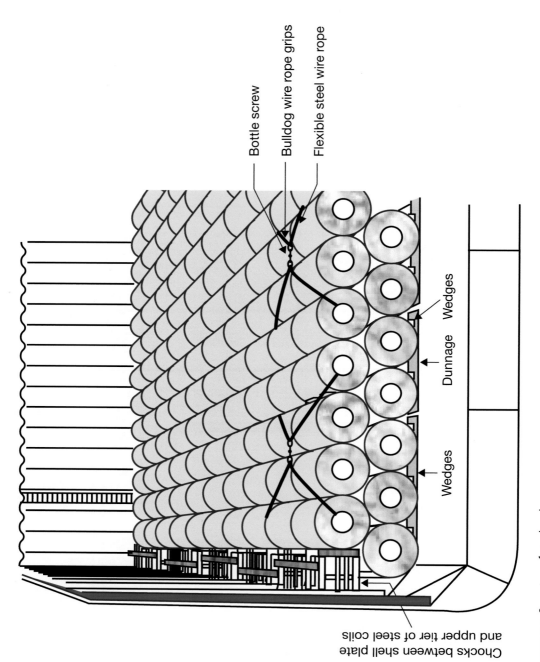

Bottle screw

Bulldog wire rope grips

Flexible steel wire rope

Wedges

Dunnage

Wedges

Chocks between shell plate
and upper tier of steel coils

FIGURE 3.5 Securing of steel coils.

Steel Plate

Steel plate comes in a variety of sizes, from very long to heavy bundles of about 2 metres in length. Various methods are employed to handle this commodity from plate clamps on chain slings to electromagnetic expandable beams. The weight and overall size tend to make this an awkward and dangerous cargo to load or discharge and once stowed, requires chain securings. Being heavy, it is usually given bottom stow and floored with dunnage to accommodate any over stowing. Large plates may incur damage to the vessel and/or other cargoes during movement and careful handling should be the order of the day.

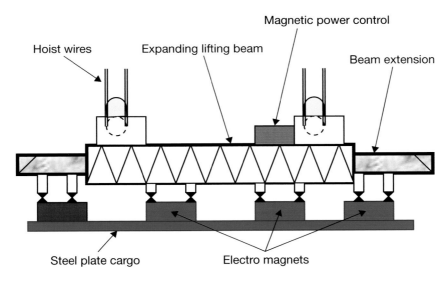

FIGURE 3.6 Operating principal of telescopic magnetic lifting beam.

Modern handling techniques where steel plate is a regular cargo tend to employ gantry cranes working with electromagnetic expandable beams. These are similar to steel stockyard cranes and are now seen in port terminals (e.g. Immingham, UK) working in a similar manner to container gantry cranes.

BAGGED CARGOES

There are many examples of bagged cargoes: fishmeal, grain, beans, cocoa, etc., to name but a few. They may be packed in paper bags, as for cement, or hessian sacks, as for grain or bean products. Loading takes place either in containers or on pallet slings. Sizes of bags tend to vary depending on the product and were seen as a regular type of package for general cargo vessels.

Handling bagged cargo is expensive by today's standards and many of the products lend more easily and more economically to bulk carriage or container stow. Where bags are stowed in a ship's hold they should be on double dunnage, stacked either bag on bag or stowed half bag as shown:

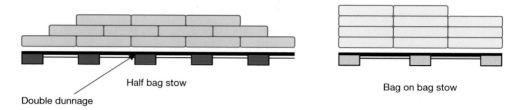

Half bag stow

Double dunnage

Bag on bag stow

FIGURE 3.7 Examples of bagged stowage.

When receiving bagged cargo, the bags should be seen to be clean and not torn. Neither should they be bled in order to get a few extra bags into the compartment. Such an action would only increase the sweepings after discharge and lead to increased cargo claims.

Slings should be made up in, or close to, the square of the hatch. If they are made up in the wings then bags are liable to tear, as the load is dragged to the centre. Stevedores should not use hooks with paper bags and bags should not be hoisted directly by hooked lifting appliances.

Shippers frequently provide additional unused bags to the ship to allow for residual sweepings. All bags are discharged ashore (even torn bags) to ensure that a complete tally is achieved.

Bags containing oil seeds of any type must be stowed in a cool place as these are liable to spontaneous combustion.

Example Products for Bag Stowage

Bone meal – Other than keeping dry, no special stowage precautions are required.

Cattle food – Should be kept dry and away from strong-smelling goods.

Cement – Paper bags require care in handling. Stow in a dry place and not more than fifteen (15) bags high. Alternative carriage can be as a bulk cargo in specifically designed ships. Bilges should be rendered sift proof and compartments must be thoroughly clean to avoid contamination which could render the cement useless as a binding agent.

Chemicals – Prior to loading, check the IMDG code and provide suitable stow.

Cocoa – Stow away from heat and from other cargoes which are liable to taint.

Coffee – Requires plenty of ventilation and susceptible to damage from strong-smelling goods.

Copra – Dried coconut flesh, liable to spontaneous combustion. It could taint other cargoes and cause oxygen deficiency in the compartment. Requires good surface ventilation.

Dried blood – Used as a fertiliser and must be stowed away from any cargoes liable to taint. (Similar stow for bones.)

Expeller seed – Must be shipped dry. It has an extremely high risk of spontaneous combustion and must not be stowed close to bulkheads, especially hot bulkheads.

Fishmeal – Gives off an offensive odour and requires good ventilation. This cargo is liable to spontaneous combustion and requires continuous monitoring of bags and surrounding air temperatures. Bags should not be loaded in a wet or damp condition or if they are over 35° or +5° above ambient temperature, whichever is higher.

Flour – Easily tainted. Any stowage must be kept dry and clear of strong-smelling goods.

Potatoes – Loaded in paper sacks. Require a cool, well ventilated stow.

Quebracho extract – This is a resin extract used in the tanning industry. Bags are known to stick together and should be separated on loading by wood shavings.

Rice – See below.

Salt – Requires a dry stowage area.

Soda ash – Should be stowed away from iron work and foodstuffs and must be kept dry.

Sugar – Also carried as bulk cargo. Bagged green sugar exudes a lot of syrup. Stowage should be kept clear of the ship's side as the bags are susceptible to tearing as the cargo settles. Dry refined sugar and wet or green sugar must not be stowed together. Cover steelwork with brown paper for bulk sugar and keep dry.

Rice

Rice cargoes are now usually carried in bulk. This eliminates the costs of handling bags for the shipping phase. It is more economical, and common, to bag rice products at the distribution stage.

Rice contains a considerable amount of water and is liable to sweat. It must be well ventilated and not allowed to become moist or it will start to rot and give off a pungent smell which could affect other rice cargoes in the vicinity. It is also known to give off carbonic acid gas (a weak acid formed when CO_2 is dissolved in water).

Ventilators should generally be trimmed back to wind, although matured grain rice will require less ventilation than new grain rice. In any event, a void space should be left between the deck head of the compartment and the surface of the stow, bearing in mind the possibility of cargo movement and the necessity to employ shifting boards. Surface ventilation should be ongoing to remove warm air currents rising from the bulk stow.

Prior to loading rice, the compartments should be thoroughly cleaned, bilges sweetened and made sift-free. A lime coating is recommended together with a cement wash. Their condition must be sufficient to pass survey inspection. The hold ceiling should be stain-free and covered by a tarpaulin or separation cloth. To this end an adequate supply of matting and separation cloths is to be recommended.

Rice is considered as a 'grain' cargo and would need to meet the requirements of the Grain Regulations affecting stowage. A ship's condition format would be required to show the cargo distribution and a curve of Statical Stability for the condition would need to be constructed.

If compartments are only partly filled, then bagged rice with suitable separation cloths may be used to secure the stow. Bags for rice are usually of a breathable man-made, interwoven fabric. A ship loading rice would need a Certificate of Authorisation or alternatively the Master would need to show that the vessel can comply with the carriage regulations to the satisfaction of an MCA Surveyor.

Modern loading methods usually employ chutes, while pneumatic suction systems are often engaged for the discharge process. Working capacity of distribution and suction units is up to about 15,000 tons per hour. (Stowage factor for rice in bags = 1.39 m³/tonne, or bulk stow = 1.20 m³/tonne)

Note: Also see Chapter 5, Bulk Cargoes – Grain.

BALE GOODS

Various types of goods are carried in bales, either in open stow or containerised.

Bales in open stow are normally laid on thick single dunnage of at least 50 mm in depth. Bales are expected to be clean with all bands intact. Any stained or oil-marked bales should be rejected at the time of loading. All bales should be protected against ship's sweat and the upper level of cargo should be covered with matting or waterproof paper to prevent moisture from the deck head dripping onto the cargo surface.

Examples of Bale Cargoes

Carpets – A valuable cargo which must be kept dry. Hooks should not be used. More commonly carried in containers these days.

Cotton/Cotton waste – Bales of cotton are highly inflammable and stringent fire precautions should be adopted when loading this cargo. A strict no smoking policy should be observed. If the bales have been in contact with oil or are damp, they are liable to the effects of spontaneous combustion. Generally a dry stowage area is recommended.

Esparto grass – Along with products like hay and straw, these bales have a high risk of spontaneous combustion, especially if wet and loosely packed. Poorly compressed bales should be rejected. If carried on deck, these bales should be covered with tarpaulins or other protective coverage.

Fibres – Fibres such as jute, hemp, sisal, coir, flax or kapok are all easily combustible. A strict no smoking policy should be observed at all stages of contact. Bales must be kept away from oil and should not be stowed in the same compartment as coal or other inflammable substances or other cargoes liable to spontaneous combustion.

Oakum – This is hemp fibres impregnated with pine tar or pitch. It is highly inflammable and strict no smoking procedures should be adopted. It is also liable to spontaneous combustion.

Rubber – If packed in bales these give an unstable platform on which to overstow other cargoes, other than more bales of rubber. Crêpe rubber tends to become compressed and sticks to adjacent bales and talcum powder is dusted over the bales to prevent this stickiness between bales. Polythene sheeting with ventilation holes is also used and is now in more common use for the same purpose. Up-to-date methods tend to wrap the whole bales separately in polythene to eliminate the sticking element.

Tobacco – Usually stowed in bales in open stow. It is liable to taint other cargoes and is also susceptible to taint from other cargoes in close proximity. The stowage compartment should be dry and kept well ventilated or there is a risk of mildew forming.

Wood pulp – Must be kept dry; if it is allowed to get wet it will swell and could cause serious damage to the steel boundaries of the compartment. Notice M 1051 recommends that care should be taken to ensure that no water is allowed to enter the compartment. To this end all air pipes and ventilators should be sealed against the accidental ingress of water.

Wool – Can be shipped in either scoured or unscoured condition. The two types should not be stowed together. Bales should be laid on ample clean dunnage and provided with good ventilation. Slipe and pie wools are liable to spontaneous combustion and should if possible be stowed in accessible parts of the hold.

LOADING, STOWAGE AND IDENTIFICATION OF CARGO PARCELS

In order to ensure correct handling and stowage of goods, cargoes tend to be labelled and marked with instructions on the side of respective packages. Cargo is shipped all over the globe and not all countries of discharge are English-speaking; to this end labels are pictorially displayed. Cargo Officers can monitor from the labelling that instructions are complied with and that stowage practice is as per shipper's instructions.

Stowage of Wine

Wine was often carried in barrels and in some cases still is. However, bulk road tankers and even designated wine carriers are engaged in the shipment of large quantities of wine in bulk. Where barrels are transported they should be stowed on the side (bilge) with the 'bung' uppermost. The stow should not be greater than eight high and the first height level should be laid on a bed to keep the bilge free. 'Quoins', a type of wedge arrangement, are used to support the barrels and prevent them from moving.

Fragile handle with care

This way up

Use no hooks

Heavy weights this end

Sling here

Keep cool stow away from boilers

FIGURE 3.8 Example labels provide instructions to stevedores for correct stowage practice.

Barrels are heavy, with a capacity of 36 imperial gallons (164 litres), and normally require two men to handle and stow in a fore and aft direction. Modern aluminium casks have to some extent replaced the old wooden barrels but some companies still use the old-fashioned wood barrels for their product.

Barrels are given underdeck stowage and would not generally be taken as deck cargo.

Where wine is not shipped in bulk holding tanks or barrels, the more popular method in this day and age is to pre-bottle the commodity and export in cartons, usually in a container. Distinct advantages are associated with this method, in that pilferage is reduced with the bottled wine under lock and key. Containers are easily packed and sealed under customs controls. Mixed commodities like spirits or beer can also be packed into the same container. Once sealed, transport and shipping via a container terminal is usually trouble-free.

Barrels are used more these days to allow wines to mature, rather than as transport vessels. They are awkward to handle and have difficulties in stowage. The art of the 'cooper' is also becoming scarce and if barrels are damaged in transit it becomes expensive to effect repairs.

Occasionally barrels are still employed, but with specialist commodities or shipped from one wine cellar to another

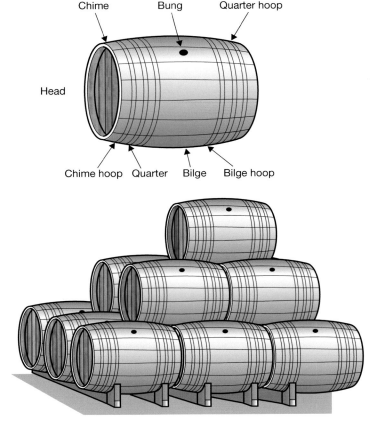

FIGURE 3.9 Wine barrel stow, bungs uppermost.

BULK FLUID PRODUCTS

Some products like wine and fruit juices have generated the construction of specialised transports, specifically for the carriage. An example of this is seen with the *Carlos Fischer*, fitted with free-standing, stainless steel tanks for the purpose of shipping bulk orange juice.

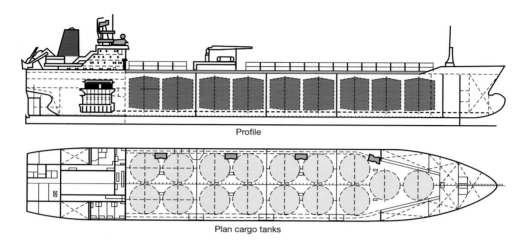

Profile

Plan cargo tanks

FIGURE 3.10 Profile and tank disposition of *Carlos Fischer* (IMO No. 923095).

The ship is 42,500 dwt, and is engaged in shipping concentrated orange juice from estates in Brazil. It is double hulled but not classed as a tanker, having four holds each with four vertical cylindrical fruit juice tanks. Cargo piping running through the holds is led to manifolds in lockers in the deck house.

CASE GOODS

Case goods lend particularly to a general cargo open stow but can be containerised depending on size. Heavy cases should always be given bottom stow with the lighter cases on top. If the contents of the case are pilferable then they should be loaded into a lock-up stow and tallied in and tallied out.

Slinging of case goods will be directly related to their weight and may be fitted with identified lifting points. Care should be taken that such lifting points are attached to the load and not just to the package.

Specific case goods may have special stowage requirements, such as glass. This would probably be marked as 'Fragile', 'This way up' and require side, end-on stowage. Crated cars or boats, would expect to be loaded on level ground and generally other crated goods are treated as case goods depending on the nature of the contents.

Forklift truck operations are often employed with the movement and stowage of heavy case goods, both in the warehouse, on the quayside and aboard the vessel.

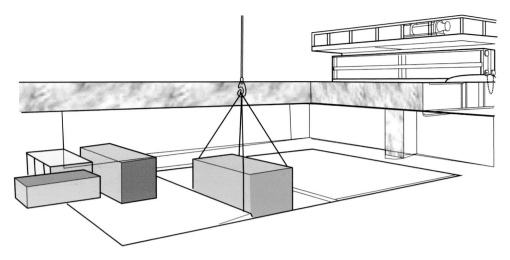

FIGURE 3.11 Example of case goods/general cargo being loaded/discharged.

However, the use of forklift trucks inside the hold tends to be restrictive with case goods because they are so bulky. They need open deck space to allow manoeuvring and large cases quickly start to fill the manoeuvring space and make land only as the only method of continuation of loading.

THE LOADING AND CARRIAGE OF DRUMS

Cargo in drums is not unusual and can be varied by way of chemicals, oils, paints, dyes, even sheep dip. Drums may differ in size, but a 50-gallon drum is probably the most common size for oils and is often used for ship's own stores of lubricating or diesel oil.

Drums are often taken as deck stow. In such an event, they would be protected by nets or a timber-built compound to keep the stow tight, depending on the number of drums carried. Where upper decks are covered, this may necessitate a catwalk being built over the drums in order to provide accessibility to all parts of the vessel when at sea. In any event, drum cargoes are placed on single dunnage and are invariably secured by wire lashings, with or without nets, to prevent movement of the cargo when at sea.

Concern with such cargo arises with the obvious problem of having a leaking drum. If such an occurrence did take place, the action would depend on the contents of the drum, the associated effects on other cargoes, the potential fire risk, and the ability to get at the affected drum(s). To this end, where corrosives are carried in large numbers, it may be better to stow the drums in smaller batches to allow accessibility to damaged units, as opposed to a total block stow of many drums together. Such a block stow may prove difficult, if not impossible to get at the affected drums when in transit.

When substances with a flash point below 23°C (73°F) are carried below decks adequate ventilation should be given to prevent the build-up of any dangerous concentrations of flammable vapours. Low flashpoint cargoes having a wide flammable range are extremely hazardous. Any such cargoes that are likely to present a health hazard or increased fire risk to the vessel should initially be checked against the advice offered by the IMDG code and any precautions followed accordingly.

Underdeck stowage of drummed commodities often tend to run a high fire risk with or without explosion risk. The compartment should be well ventilated and any gases or fumes should not be allowed to build up into dangerous concentrations. Prudent use of cargo hold fans should be exercised while on the voyage to ensure a continued safe atmosphere within the compartment and a no smoking policy observed.

Casks are manufactured in aluminium and are used extensively for beer. They are comparatively light and may be full or empty. They require a compact stow and are often netted to prevent movement when in open stow.

A cargo of ingots – Copper, lead or tin ingots are all very heavy concentrated cargo parcels and require bottom stow, similar to the iron cargoes of castings, iron billets and long steel work.

Lighter goods may be stowed on top of ingots but a secure separation between cargoes is desired. Ingots cannot be stowed high and are difficult to work on top of the cargo without a dunnage floor. Ingots are often baled and banded, but are sometimes shipped as single bar elements being floor stacked. Ingots can be considered a valuable cargo and are usually tallied in and tallied out at discharge.

Cable reels – Large wooden reels with power cable rove around a central core are carried as general cargo. They are stowed in the upright position, on a firm deck, and should be secured against any pitching or rolling of the vessel when in a seaway. They can be quite large, 3–4 metres in diameter, and consequently may be considered as a heavy load, especially if the cable contains a steel construction element.

Designated 'cable ships' with telegraph cable tend to load the cable directly into specially constructed cylindrical tanks in specialised cable holds. Such cables should not be confused with the cable reels discussed as general cargo.

Paper cargoes – Paper may be carried in many forms, from waste paper to newsprint. The compartment in whatever form the paper is to be carried must be in a dry condition and well ventilated. Newsprint is carried in rolls which are normally stowed on their ends to avoid distortion, preferably on double dunnage.

Ships' steelwork would normally be protected with waterproof paper to prevent ships' sweat from damaging the rolls. Hooks should not be used during the loading or discharge periods. On occasions, like in tween decks, the rolls may be stowed on their sides. If this does occur, they should be chocked off to prevent friction burns and movement when the vessel is at sea.

Rolls of paper should be sighted as being unmarked by oil or other similar stains on loading. Once on board, the cargo should be kept clean and not allowed to become contaminated by any form of oil or water.

Bales of paper pulp must be seen to be dry on loading and stowed clear of any source of moisture.

Dried fruits – These include apricots, currants, dates, figs, prunes, raisins and sultanas. They may be shipped in cases, cartons, small boxes or even baskets. However carried, they must be stowed away from cargoes which are liable to taint. Dried fruits tend to give off a strong smell and generally may contain drugs and insects which could contaminate other cargoes, especially foodstuffs. The fruit itself is liable to taint from other strong odorous cargoes and stowage should be kept separate in cool, well-ventilated compartments. Tween deck stowage is preferred but if stowed in lower holds, adequate ventilation must be available through the course of the voyage. If in open stow, good layers of dunnage are recommended to assist air flow and the cargo should not be overstowed.

Garlic and onions – These are shipped in bags, cases or crates and give off a pungent odour and must be stowed clear of other cargoes liable to taint. It is essential that onions and garlic are provided with good ventilation, similar to fresh fruit. Considerable moisture will be given off onions and adequate drainage facilities would be expected.

Fresh fruit – Apples, apricots, pears, peaches, grapefruit, grapes, lemons and oranges can be carried quite successfully in non-refrigerated compartments, the proviso being that adequate dunnage is used and with good ventilation. In the event that mechanical ventilation is not used, then hatches should be opened (weather permitting). Fruit, especially green fruit, gives off a lot of gas and extreme care should be exercised before entering any compartment stowed with fresh fruit. Following the discharge of fruit, the holds should be well aired and deodorised.

CARGO MONITORING AND TALLYING

Tallying – All cargoes are tallied on board the vessel and for monitoring the cargo parcels in this manner, specialised 'Tally Clerks' are employed. These clerks tend to reflect the shipper's interest, while others so engaged by the ship may represent the owner's or ship's operator's interests. Cargo parcels are not only tallied into the ship but also tallied out at the port of discharge.

Tally counts are important especially in the case of valuable effects, or short quantities being delivered to the ship. Cargo claims draw on tally information to substantiate quality and quantity as and when disputes evolve between the ship and the shipper, bearing in mind that the ship's personnel are there to protect the ship owner's or charterer's interest.

Mate's receipts tend to be the supporting document which denotes the quantity, marks, description and the apparent condition of goods received on board. It is usually signed by the ship's Chief Officer, hence the name 'Mate's receipt'.

It is important that the details of the cargo are correctly stated on the Mate's receipts as it is from these that the Bills of Lading (B/L) could be prepared. The Bills of Lading are sent to the various consignees, who will in turn present them to the Master before the cargo is handed over. The Bills of Lading are the consignee's title to the goods stated and he therefore can expect to receive those goods as described. In the event of the goods not being in the same condition as stated on the B/L by way of quantity or quality, then he (the shipper) could make a claim against the ship for any discrepancy.

Ship's officers should bear in mind that they are temporary custodians of goods which belong to a third party. As such they must endeavour to keep them in the same condition as that in which they were received aboard the vessel. As far as possible damaged cargo or damaged package should be rejected for shipping.

CARGO shipped on Board " _____ "

In good condition except where otherwise stated

Port of Shipment _____ Date _____

Destination _____ Hatch No _____ Ex _____

NB Ship's Tally Clerk to record all visible damage

MARKS	PACKAGES	SEPARATE NUMBERS	TOTAL

Ship's Tally Clerk _____

FIGURE 3.12 Example of tally clerk's account.

CARGO SWEAT AND VENTILATION

A great number of cargo claims are made for merchandise which has been damaged in transit. Much of this damage is caused by either 'ship's sweat' or 'cargo sweat' and could be effectively reduced by prudent ventilation of cargo spaces.

Sweat is formed when water vapour in the air condenses out into water droplets once the air is cooled below its dew point. The water droplets may be deposited onto the ship's structure or onto the cargo. If the former, it is known as 'ship's sweat' and this may run or subsequently drip onto the cargo. Where water droplets form on cargo this is known as 'cargo sweat' and will occur when the temperature of the cargo is cold and the incoming air is warm.

To avoid sweat and its damaging effects it is imperative that 'wet and dry' bulb temperatures of the air entering and the air contained within the cargo compartment are taken at frequent intervals. If the temperature of the external air is less than the dew point of the air already inside the space, sweating could well occur. Such condition gives rise to 'ship's sweat' and is commonly found on voyages from warm climates towards colder destinations. Similarly if the temperature of the air in the cargo compartment (or the cargo) is lower than the dew point of incoming air, sweating could again occur, giving rise to 'cargo sweat'. This would be expected on voyages from cold places towards destinations in warmer climates.

If cargo sweat is being experienced or likely to occur, ventilation from the outside air should be stopped until more favourable conditions are obtained. However, it should be noted that indiscriminate ventilation often does more harm than no ventilation whatsoever. It is also of concern that variation in the angles of ventilators away from the wind can cause very different rates of air flow within the compartment. The angle which the ship's course makes with the wind also affects the general flow of air to cargo compartments. In general the greatest air flow occurs when the lee ventilators are trimmed on the wind and the weather ventilators are trimmed away from the wind.

This is known as through ventilation (see Figure 3.13).

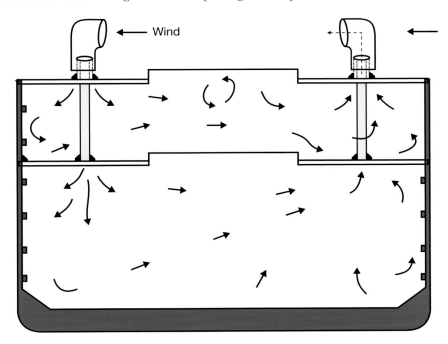

FIGURE 3.13 Showing air circulation with lee ventilators on the wind and weather vents off. This is known as through ventilation.

Forced ventilation – If the dew point temperature in the cargo compartment can be retained below the temperature of the ship's structure, i.e. decks, sides, bulkheads and the cargo, there would be no risk of sweat forming. Such a condition cannot always be achieved without some form of mechanical (forced) ventilation from fans or blowers.

There are several excellent systems on the commercial market which have the ability to circulate and dry the air inside the cargo holds. Systems vary but often employ 'baffle' plates fitted in the hold and tween decks so that air can be prevented from entering from the outside when conditions are unfavourable. Systems recirculating the compartment's air can also operate in conjunction with dehumidifying equipment to achieve satisfactory conditions pertinent to relevant cargo.

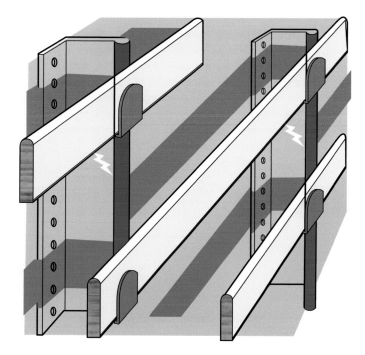

FIGURE 3.14 Cargo battens fitted horizontally to allow separation of the cargo from the ship's inner steel hull.

The purpose of the wooden cargo battens, which can be fitted horizontally or vertically, is to keep the cargo off the ship's inner steel hull. This arrangement produces an air gap of about 230 mm between the steel work and the cargo surface and subsequently reduces the risk of cargo sweat damaging cargo parcels.

It is normal practice with some bulk cargoes (e.g. coal), when carried in holds fitted with spar ceiling, to remove the wood battens to reduce the damage incurred to the wood, prior to loading.

Dunnage

This comprises timber slats of a thickness of about 35/40 mm which are ordered in bundles by the ship's Chief Officer. The purpose of dunnage, which can be laid either singularly or in a criss-cross double dunnage pattern, is to provide an air gap to the underside of the cargo. This allows ventilation around all sides of the cargo stow. This is again to effectively reduce the risk of sweat damage to cargo. Dunnage should be in a clean condition and not oily or greasy as this could cause contamination to sensitive cargoes.

Tank Top Ceiling

This is a wood sheathing which covers the steelwork of the tank top, in way of the hatchway in the lower hold. This timber flooring not only protects the tank tops but lends to a non-skid surface in the hold. It generally assists drainage of any moisture in the space and can be used in conjunction with a single dunnage layer.

ADDITIONAL GENERAL CARGO TERMS

Contamination

Cargoes which taint easily – e.g. tea, flour, tobacco, etc. – should be kept well away from strong-smelling cargoes. If a pungent cargo has been carried previously – e.g. cloves or cinnamon – the compartment should be deodorised before loading the next cargo.

Dirty cargoes should never be carried in the same compartment as clean cargoes. A general comparison of dirty cargoes would include such commodities as oils, paints or animal products, whereas clean cargoes would cover the likes of foodstuffs or fabrics. Obviously some notable exceptions in each of the two classes are to be found.

Separation of Cargoes

It is often a requirement when separate parcels of the same cargo are carried together that a degree of separation between the units is essential. The type of separation method employed will depend on the type of goods being shipped. Examples of separation materials include colour wash, tarpaulins, burlap, paper sheeting, dunnage, chalk marks, rope yarns or polythene sheets. The idea of separation is to ensure that the cargo parcels, although they may look the same, are not allowed to become inadvertently mixed.

Optional Cargo

Optional cargo is cargo which is destined for discharge at either one, two or even more ports. Consequently it should be stowed in such a position as to be readily available for discharge, once the designated port is declared.

Over Carried Cargo

If cargo which is meant for discharge is not discharged, it is said to be over carried to the next port. Such an event causes inconvenience, extra cost and additional paperwork. To this end, hatches are searched on the completion of discharge to ensure that all the designated cargo for the port of discharge has indeed left the ship. A method of checking can be made against the cargo plan and the cargo manifest and comparing figures with the tally clerks – although it must be said this is not fool proof, especially if pressures are being applied to finish cargo operations and sail, possibly departing before the holds have been properly examined for over carried cargo pieces.

Pilferage

Certain cargoes always attract thieves. Notable items include spirits, beer, tobacco or high-value small items. To reduce losses such cargoes should be tallied in and tallied out. Lock-up stow should be provided throughout the voyage from the onset of loading to the time of discharge. Shore watchmen and security personnel should be used whenever it is practical and good watch-keeping practice should be the order of the day.

DEEP TANK USE

Many vessels are fitted with 'deep tanks' employed as ballast tanks or for the carriage of specialised liquid cargoes such as vegetable oils – i.e. coconut oil, bean oil, cottonseed oil, linseed oil, palm oil or mineral oils. Other cargoes include tallow or bulk commodities like grain, molasses or latex. The specialisation of such cargoes often requires rigid temperature control of the cargo and to this end most cargo deep tanks are fitted with heating coils which may or may not be blanked off as the circumstances dictate.

> Note: Some vessels with a shaft tunnel may be fitted with additional deep tanks aft in a position either side of the shaft tunnel, but these are not common.

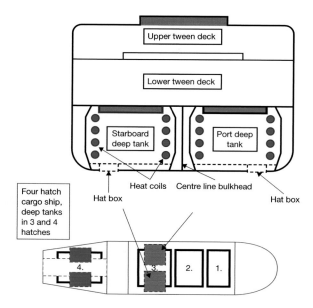

FIGURE 3.15 Typical cargo hatch arrangement showing upper and lower tween decks with twin deep tanks in place of a lower hold space.

Preparation of Deep Tanks

The need for absolute cleanliness with deep tanks is paramount and Cargo Officers are advised that they are virtually always subject to supervision and survey prior to loading example cargoes. Claims for contamination of these cargoes are high and meticulous cleaning of the tank itself and the pipelines employed for loading and discharging must be a matter of course. To enable the Classification Surveyor to certify that the tank has watertight integrity and is clean, Chief Officers should, depending on the previous cargo carried, carry out the following procedures:

> Note: All safety precautions for the entry into an enclosed space must be taken prior to carrying out any and all maintenance inside deep tanks.
> All work inside the tanks must be covered by a permit to work.

After the carriage of a general cargo, sweep the tank down completely and remove any waste. In the event of a liquid cargo (assuming of a non-hazardous nature), puddle any residual fluids to the suction and allow the tank to dry.

If the tank is uncoated (they are often coated in epoxy covering), the bulkheads, decks and deckhead should be inspected for rust spots. These should be scraped and wire brushed and all traces of corrosion removed.

Heating coils should be rigged and tested. These coils may be 'side coils' or 'bottom coils' or a combination of both.

Hat boxes should be cleaned out and the suctions tested.

Fill the tank with clean ballast and pressure test the tank lid. (Tanks are to be tested to a head of water equal to the maximum to which the tank will be subjected but not less than 2.44 m above the crown of the tank.)

Empty the tank to just above the heating coils. Add a cleansing agent and heat the residual water by means of the coils. Carry out a wash down using a hose and submersible pump.

After cleaning, turn off the heating element and sluice down with fresh water. Pump dry and allow the tank to dry, mopping up any residual puddles.

Finally, clean and blank off bilge suctions.

> Note: Personnel so involved inside deep tanks should be provided with protective clothing and footwear, together with goggle eye protection.
> Breathing apparatus may also be a requirement. A risk assessment would need to be carried out prior to commencing any work inside the tank(s).

DEEP TANK CARGOES

Vegetable oils – when shipped in bulk, the tank must be thoroughly cleaned and all traces of previous cargoes must be removed. Tank suctions will be blanked off, and the overall condition will be inspected by a cargo surveyor. The tank itself would be tested for oil tightness prior to loading. Heating coils will probably be in operation depending on the required shipping temperature. Some oils solidify at 0°C; others, such as palm oil or palm nut oil, solidify at between 32°C and 39°C; cottonseed oil and kapok seed oil solidify at about 10°C to 13°C. Chief Officers could expect to be supplied with relevant shipping criteria for the oil.

Care must be taken that the heating is not too fierce or applied too quickly as the cargo could scorch. Such an occurrence would be noticeable by some discolouration of the oil, which could result in a cargo claim being filed.

Contamination is avoided by use of shoreside cargo pumps when discharging, while monitoring on passage is conducted by taking ullages and temperatures at least twice per day for oils kept in the liquid state.

Following discharge of the cargo, the tank would probably be steam cleaned and washed with a caustic soda-type solution to ensure cleanliness.

> Warning: During the loading and heating of vegetable oils there is a risk of Carbon Monoxide being generated and suitable precautions should be taken by involved personnel close to the cargo and tank stow area.

Latex – The 'sap' from rubber trees which rapidly solidifies when exposed to air. It is retained in liquid form by added chemicals, usually ammonia, and shipped in bulk. Note: Ammonia attacks brass and copper metalwork and latex tanks should not have such metals as part of their construction.

Prior to loading latex, the tank would be tested and inspected to be thoroughly clean. All steelwork would be coated with hot paraffin wax. The heating coils would be removed as they are not needed for the carriage. Ventilators, air pipes and sounding pipes are all sealed to prevent ammonia loss due to evaporation. Fire extinguishing pipes, if fitted, should also be plugged. Gas relief valves are fitted to ease any pressure build-up inside the tank.

Discharge of the cargo is carried out by shoreside pumps and the tank would then be washed down with water to remove all traces of ammonia. The wax coating is often left in place unless the tank is to be used immediately for another cargo.

Molasses – A syrup obtained from the manufacturing process of sugar. It may be carried in deep tanks, similar to vegetable oils, with heating coils operational to retain the cargo in a liquid state. It is discharged by shoreside pumps and the tanks would be scrubbed and washed down with plenty of water as soon after discharge as is practical. Most contamination claims develop from dirty pipelines.

Note: Specially designated vessels are employed for the carriage of molasses so the use of deep tanks has diminished with this type of cargo.

RANCIDITY

The possibility of products turning rancid is always present, especially with fatty oils and fats which contain strong flavours and odours. These elements become developed by being exposed to light, moisture and air and move towards a condition we know as rancidity. A by-product following excessive exposure and subsequent chemical reaction is that fatty acids are produced. These then decompose and form other compounds which are dramatically increased by temperature rise. Such action means that less refined, pure oil is recoverable.

> Note: Fats are considered as products which are solid at ordinary temperatures, e.g. 15°C. Fatty oils are those which are liquid at that temperature. The difference between fats and fatty oils is that fatty oils are more chemically reactive than fats.

Hides – These may be shipped in either a wet or dry condition, either in bundles or in casks or even loose. They are often carried in deep tanks, usually because there is not enough of them to fill a tween deck or lower hold space. Another factor that is against stowage in a tween deck is that wet hides require adequate drainage which would be difficult to achieve in exposed stow. Pickling and/or brine fluid can expect to find its way to the bilges which will necessitate pumping probably twice daily at the beginning of a voyage with hides in the cargo.

> *Handling precautions* – Hides must only be handled with gloves as there is a high risk of contracting anthrax, which could prove fatal. Neither should stevedores use hooks in the handling, because of damage to the product. In the case of dry hides, these are often brittle and any person being scratched or cut should receive immediate hospital treatment.

The stowage of hides must be away from dry goods and iron work. They have a pungent odour and should be stowed well away from goods that are liable to spoil. They should not be overstowed.

BALLASTING AND BALLAST MANAGEMENT

As cargo is loaded it is general practice for most types of vessels to de-ballast. Some tanks are retained for the purpose of trimming the ship and adjusting the stability conditions, but overall if the ballast was kept on board, the ship could well be seen to be overloaded.

It is now expected that participating governments to the IMO Convention will have to restrict discharge of ballast water because of the impurities and invasive species it may contain. Local marine environments have introduced legislation to protect their waters from destructive and polluting elements.

To this end, a Ballast Management Plan and Record Book is required to be kept, (under SMS), indicating which tanks are filled/emptied, the ship's position of the ballast movement and details of quantities and any treatment, e.g. ultraviolet light, that the water may be submitted to. Coastguard and similar authorities inspect Ballast Management records of inbound vessel's to ensure that ships abide by the regulations and protect the marine environment overall.

Stability

When loading/discharging cargo, due regard must be taken of the ship's condition of stability at every stage and position of the voyage. A reasonably positive GM must be appropriate throughout the passage and loadline zones must not be infringed. Most modern vessels would engage the flexibility of a customised computer software program for working relevant stability criteria. Associated software of this nature would also provide bending and shear force stresses incurred and take account of total weights of stores, bunkers, fresh water and ballast content to provide example conditions. (Older vessels may still use a shipboard loadicator, but improved software has virtually made these obsolete.)

(See also Appendix II for: Stability Examples.)

> Note: Free surface moments have a negative effect on the ship's GM, especially when loading or discharging heavy lift cargoes which may cause the vessel to heel. To this end 'slack tanks', should be avoided if at all possible. (Press up or pump out as best suits.)

Loadlines

Ships' Cargo Officers must take care that the vessel is not overloaded beyond the appropriate loadline. Overloading endangers the safety of the vessel and would incur the risk of a heavy fine against the Ship's Master. When loading certain cargoes, especially bulk cargoes like bulk ore and oil, the vessel is liable to become hogged or more probably adopt a sagged position. If the vessel is sagging the apparent mean draught will be less than the actual mean draught. This situation does not permit overloading.

The various loadlines are shown and they are assigned to the vessel following a Loadline Survey by an assigning authority, e.g. Lloyds Register.

S	The Summer loadline mark is calculated from the Load Line Rules and is dependent on many factors including the ship's length, type of vessel and the number of superstructures, the amount of sheer, minimum bow height and so on.
W	The Winter mark is 1/48th of the summer load draught below S.
T	The Tropical mark is 1/48th of the summer load draught above S.
F	The Fresh mark is an equal amount of $\Delta/4T$ millimetres above S. where Δ represents the displacement in metric tons at the summer load draught and T represents the metric tons per centimetre immersion at the above. In any case where the displacement cannot be ascertained, F is the same level as T.
TF	The Tropical Fresh mark relative to T is found in the same manner as that of F relative to S.
WNA	The Winter North Atlantic mark is employed by vessels not exceeding 100 metres in length when in certain areas of the North Atlantic Ocean, during the winter period. When it is assigned it is positioned 50 mm below the Winter W mark.

Principle of Loadlines

The purpose of loadlines and loadline zones is to ensure vessels have adequate reserve buoyancy to survive the prevalent weather in the respective zones.

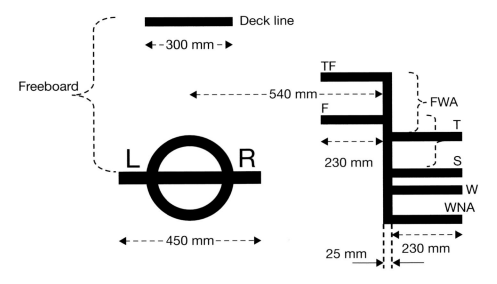

FIGURE 3.16 Loadline marks (starboard side shown).

Should the ship carry a Lumber Loadline this would be positioned aft of the Plimsoll Mark and identity marks would be prefixed with an 'L' (see Figure 3.17).

e.g. LTF = Lumber Tropical Fresh.

Timber Loadlines

Certain vessels are assigned Timber Freeboards when they meet certain additional conditions. One of these conditions must be that the vessel must have a forecastle of at least 0.07 extent of the ship's length and of not less than a standard height (1.8 m for a vessel 75 m long or less in length and 2.3 m for a vessel 125 m or more in length, with intermediate heights for intermediate lengths). A poop deck or raised quarter deck is also required if the length of the vessel is less than 100 m.

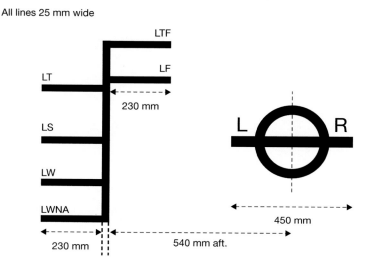

FIGURE 3.17 Timber loadlines (starboard side shown).

The letters denoting the assigning authority LR (Lloyds Registry), should be approximately 115 mm in height and 75 mm in width.

LS is derived from the appropriate tables contained in the Load Line Rules.

LW is $1/36^{th}$ of the summer timber load draught below LS.

LT is $1/48^{th}$ of the summer timber load draught above LS.

LF and LTF are both calculated in a similar way to F and TF except that the displacement used in the formula is that of the vessel at her summer timber load draught. If this cannot be ascertained these marks will be one forty-eighth ($1/48^{th}$) of LS draught above LS and LT, respectively. LWNA is at the same level as the WNA mark.

Ships with timber loadlines and carrying timber deck cargo in accord with the M.S. (Loadlines) (Deck Cargo) regulations 1989 must observe the applicable loadline that she would use if she were not marked with timber loadlines, i.e. Lumber Summer (LS) in the Summer Zone. However, if the timber is not carried in accord with the regulations, the ordinary loadline should be applied.

Deadweight Scale

Once cargo has been loaded the ship's draughts would normally be ascertained and it would be the Chief Officer's practice to employ the Deadweight Scale (part of the ship's stability documentation) to ascertain the ship's final displacement. The known figures of fuel, stores and fresh water can then be applied to provide a check against total cargo loaded from the Scaled deadweight figure.

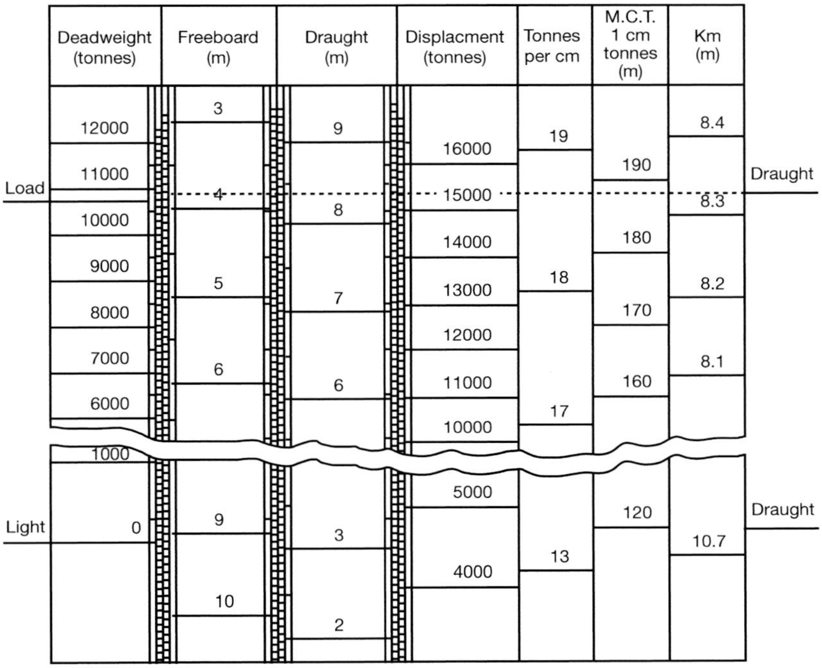

FIGURE 3.18 Deadweight scale.

Note: The Dock Water Allowance would be applied for vessels which are loading in waters other than sea water of 1.025 kgs per cubic meter.

Dock Water Allowance formula:

$$DWA = \frac{1025 - \text{Hydrometer Value } (P_P)}{25} \times FWA$$

The Dock Water Allowance value in millimetres is the amount a ship is allowed to submerge her summer loadline disc, in a dock water density other than that of salt water of 1025.

Offence to Overload

Cargo Officers should be aware that it is an offence to overload a vessel beyond her legal marks and attempt to proceed to sea. The owner or Master will be liable on summary conviction to a fine not exceeding the statutory maximum (£50,000) or on conviction on indictment, 2 years prison. The ship may also be detained. The contravention will also carry in addition to the stated fine a further £1,000 per centimetre of the amount of overload.

Example

In November 2010, a ship arrived in Liverpool from Canada carrying 'Rock Salt' and was observed to be overloaded by 39.5 cm. Following a Port State Control inspection the ship owners were taken to court for the overloading offence. Following a guilty plea and the fact that the overload was not for gain, the Magistrates reduced the £42,000 fine to £28,015 + £5000 costs to the MCA.

Alternative Tonnage Marks

Vessels which regularly trade in fresh or tropical waters may be assigned an 'Alternative Tonnage' mark to each side of the vessel.

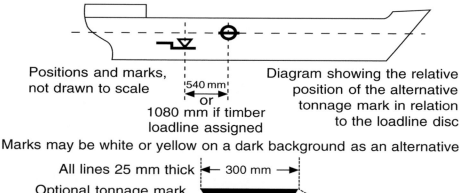

Positions and marks, not drawn to scale

540 mm
or
1080 mm if timber loadline assigned

Diagram showing the relative position of the alternative tonnage mark in relation to the loadline disc

Marks may be white or yellow on a dark background as an alternative

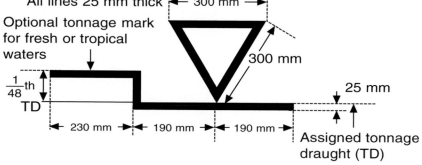

All lines 25 mm thick ← 300 mm →

Optional tonnage mark for fresh or tropical waters

$\frac{1}{48}$th

TD

300 mm

25 mm

← 230 mm → ← 190 mm → ← 190 mm →

Assigned tonnage draught (TD)

FIGURE 3.19 The Alternative Tonnage mark must be carved at a position below the load line disc and at a surveyed vertical distance from the upper deck.

Should the vessel be loaded as to submerge the Alternative Tonnage Mark, then the normal gross and registered tonnage would apply.

When a vessel loads a full cargo of light commodities, low density cargo, the ship's owners can request a modified tonnage from the Certifying Authority, in addition to its normal gross and registered tonnage

Restrictions to Loading

The loadline regulations provide various zones around the world's ocean/sea areas. These zones reflect permanent and seasonal areas which are depicted on a chart which accompanies the regulations. There are three permanent zones, namely a summer and winter zone in each hemisphere of the globe and a tropical zone across the equatorial belt; while the ship is passing through these zones the appropriate loadline would be used.

A ship cannot load deeper than her summer loadline in the summer zone, neither can a vessel load deeper than her tropical mark when in the tropical zone.

There are five 'Winter Seasonal Areas', usually found confined by land masses: the Black Sea, the Baltic Sea, the Mediterranean, the Sea of Japan and the special 'Winter' area in the North Atlantic, applicable for ships 100 metres or less in length.

Cargo Officers will frequently find themselves loading in dock water of less density than sea water and such a situation would warrant use of the Dock Water Allowance formula which would permit a vessel to load beyond her marks, knowing that the vessel will rise to the permitted loadline once entering the sea water of the respective zone or seasonal area.

CHAPTER 4

Heavy Lifting and Project Cargo

INTRODUCTION

With the increase in overall ship size and the need for greater loads it was to be expected that a comparable increased role, in marine heavy lifts, would become necessary to meet larger and greater demands. Such fundamentals caused massive expansion to take place in the 'Project Cargo Sector' of the industry, especially with the increased association with the offshore sector. It also generated the need for safer and more detailed planning for what used to be termed the conventional 'Heavy Lift'.

Previous operational practice relied heavily on single swinging derrick use, leading to specialist rigs like Hallen, Velle and onto the Stüelcken rigs. Although adequate for the day, they were not compatible for the flourishing unit load systems. Floating cranes and sheer leg platforms became entwined within the heavy lifting sector for the occasional heavy load. Computer assisted lifting grew with tandem lifts and direct control achieved via a personal joystick operator for heavy lift movement.

Such advances brought in the more intimate use of the designated heavy lift ship, using hydrostatic forces to allow Float on Float off techniques to dominate. Progress from what was known as a conventional 'Heavy Lift' which needed to overcome the forces of gravity, expanded into that of major project cargo movements.

This comparatively new sector has become a welcome associate of the 'Offshore Oil/Gas Industry', inclusive of offshore Wind Farms and the world of Marine Salvage. These developments are still ongoing, and in many areas counter the threats of Global Warming and the need for greener energies.

The early initial searches for hydrocarbons caused the generation of many offshore installations. However, many of the rigs and platforms are now reaching their viable life span as oil/gas fields become exhausted. So much so, that major companies are charged with clearing offshore sites, in order to return the sea-bed to its original natural environment. Such clearances, probably into the next century, will see the increased needs for heavy lifting operations, by not only floating shear legs, floating cranes and the semi-submersible heavy lifting vessels.

DOI: 10.4324/9781003407706-4

LOADING AND DISCHARGING HEAVY LIFTS

It is normal seagoing practice for the Chief Officer of the vessel to supervise the movement of heavy lifts both in and out of the vessel. This is not to say that the Mate of the ship will not delegate specific functions to the more junior cargo officer or to the stevedore supervisor.

Prior to commencing any lift, the derrick and associated lifting gear needs to be prepared. Several vessels were fitted with the large 'Stüelcken' type derricks, or specialised Hallen or Velle derricks as opposed to the more conventional 'Jumbo' derrick. Rigging of these lifting appliances usually necessitates reference to the manufacturer's instructions and to the ship's rigging plan.

Preparations of setting up the lifting gear, especially when officers are unfamiliar with the style of rig, can be a detailed process and is usually co-ordinated by a check list and would have a corresponding risk assessment as a standard procedure.

Preparation time for a heavy derrick can vary, depending on the type, but a period of up to two hours would not be unusual. Man management of the rigging crew and advance planning with regard to the number of lifts and in what order they are to be made, in relation to the Port of Discharge and order of reception of cargo parcels, would be the expected norm.

Note: Where a load is outside the working capacity of the ships lifting gear, it would be usual practice to engage a floating crane or a floating sheer leg platform to take on the load. Chief Officers would then need to assure themselves that the Port of Discharge has a suitable heavy lifting capability in order to carry out discharge to land the load.

With loading or discharging any heavy cargo parcels the ship retains total responsibility for the vessels stability. It is usual for the Chief Officer (Chief Mate) to assess the positive stability of the vessel throughout any operation to load or discharge heavy lifts. Particular attention being paid to any angle of heel the vessel may incur when engaging ships' own lifting gear. Any added weight to the vessel, above the ships C of G, will effectively reduce the vessel's GM, unless pre-compensated for.

THE SHIPBOARD HEAVY LIFT OPERATION (CONVENTIONAL 'JUMBO' DERRICK)

It must be assumed that the ship's Chief Officer will order all preparations for the derrick to be rigged in accord with the ship's Rigging Plan and to any instructions of the derrick/crane manufacturer. The boom would normally be broken from its stowage position and this is usually from a clamped position against the ships mast or 'Samson (Goal) Post' support structure.

Unless the vessel is regularly engaged in Heavy Lift operations the use of the 'Jumbo' derrick may be considered an exceptional occasion. If this is the case it would be essential that a 'Risk Assessment' is carried out prior to breaking out the derrick from stowage and engaging in such an operation. Consideration of any risk assessment should include the weather conditions for the time of the lift, the stability criteria, the experience of personnel engaged and the equipment involved with the lift, together with any pertinent specifics regarding the type of load being loaded/discharged.

Early preparations to ensure correct rigging of the boom would need to ensure that all other small derricks are clear of the operational area. Additional 'Preventer Backstays', if required, are rigged to supporting mast structures as per the rigging plan.

Additionally, all personnel involved should be informed and made aware of the activities of the task on hand, before releasing the securing clamp holding the boom in position.

If the derrick is not in regular use, the 'topping lift' would generally be protected by a canvas cover sheeting while in the stowed position and this would require to be removed. Men ordered to remove this cover would also connect steadying guys (rope gantlines) to each side of the head of the derrick (prior to releasing the clamp), to permit the derrick to be controlled while lowing the boom on the topping lift. Although different ships have different practices, when the derrick is stowed for sea, it is normally stripped of its operational guys and associated rigging, other than the topping lift. As such, the bare boom would need to be lowered to deck level in order to secure the slewing guys and the lifting purchase to the derrick head.

Note: Long sea passages sometimes necessitate the removal of the topping lift as well. Such action would require the topping lift to be re-rove in situation, to support the initial lowering of the derrick.

Derrick Rigging

It should also be realised from the onset, that the rigging of the boom, usually constructed of tubular steel, is not an overly heavy task in itself. It is only when the boom is fully rigged and accepting the cargo load that the operation enters the heavy lift arena. This often becomes noticeable when breaking out the boom from its clamped position. The comparatively light tubular steel boom may require a wire messenger to be shackled at the head and lead to a winch in order to heave the derrick away from the vertical stow and allow the topping lift blocks to subsequently overhaul, causing the derrick to be lowered. The combined weight on the gantlines and the wire messenger should be enough to lower the derrick in a controlled manner and allow the topping lift to overhaul. Once at the deck position, all additional rigging can be secured to the derrick and should be seen to be well greased and in an effective operational condition.

THE CONVENTIONAL HEAVY LIFT – 'JUMBO' DERRICK

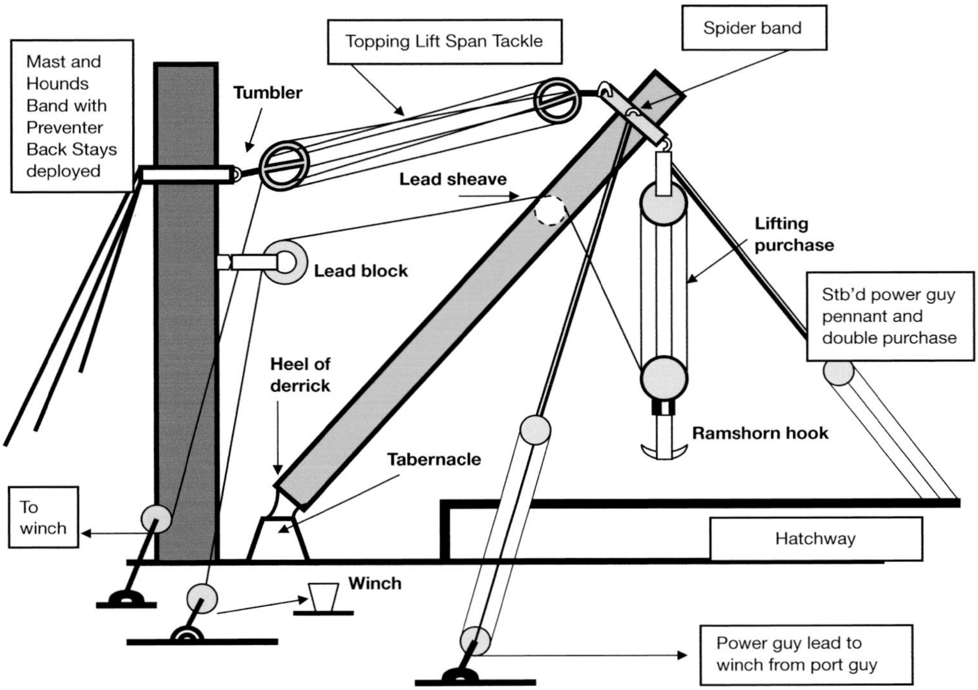

FIGURE 4.1 Basic working design of a conventional heavy lift shipboard derrick, found up to a size of about 150 tonnes SWL, usually stowed in the vertical clamped against a mast.

Many older dry cargo ships were equipped with a heavy duty crane or a heavy derrick, for that occasional lift. The lifting appliance was not always kept in the rigged condition. When needed they had to be rigged as per the ships rigging plan.

Operations of such appliances were expensive in time and maintenance and their use has been in decline with the expansion of designated heavy lift ships.

PRECAUTIONARY CHECK LIST FOR HEAVY LIFT DERRICK OPERATION

1 Carry out a 'Risk Assessment' prior to commencing the operation to ensure that all possible areas of hazard are taken account of and that all risks are at an acceptable, tolerable risk level.

2 Ensure that the stability of the vessel is adequate to compensate for the anticipated angle of heel that will be experienced when the load is at the maximum angle of outreach. All free surface elements should be reduced or eliminated if possible, to ensure a positive value of 'GM' throughout the operation.

3 Any additional rigging, such as 'Preventer backstays', should be secured as per the ship's rigging plan.

4 A full inspection of all guys, lifting tackles, blocks, shackles and wires should be inspected prior to commencing the lift, and should be made by the Officer in Charge. All associated equipment should be found to be in correct order with correct Safe Working Load shackles in position and all tackles must be seen to be overhauling.

5 Men should be ordered to lift the gangway from the quayside, and then ordered to positions of stand-by, to tend the vessels moorings at the fore and aft stations.

6 Ships 'fenders' should be rigged overside to prevent ship contact with the quayside at the moment of heeling.

7 Ensure that the deck area, where the weight is to be landed (when loading) is clear of obstructions and the deck plate is laid with timber bearers (heavy dunnage) to spread the weight of the load. The ship's plans should be consulted to ensure that the limitations of the load density plan and deck load capacity are not exceeded.

8 Check that the winch drivers are experienced and competent and that all winches are placed into double gear to ensure a slow moving operation.

9 Remove any obstructive ship's side rails if appropriate and check that the passage of the load from shore to ship is clear of obstructions.

10 Release any barges or small boats moored to the ships sides before commencing any heavy lift operation.

11 Secure steadying lines to the load itself and to any saucer/collar connection fitment attached to the lifting hook.

12 Inspect and confirm the lifting points of the load are attached to the load itself and not just secured to any protective casing.

13 Ensure that the area is clear of all unnecessary personnel and that winch drivers are in sight of a single controller.

14 Set tight all power guys, and secure the lifting strops to the hook and load, respectively.

15 When all rigging is considered ready, the weight of the load should be taken to 'float' the weight clear of the quayside (loading). This action will cause the vessel to heel over as the full weight of the load becomes effective at the head of the derrick boom.

NB. Some lateral drag movement must be anticipated on the load and it is important that the line of plumb is not lost with the ship heeling over.

16 Once the load is suspended from the derrick and the Chief Officer can check that the rigging of the equipment is satisfactory, then the control of the hoist operation can be passed to the hatch controlling foreman.

Assuming that all checks are in order, the Chief Officer would not normally intervene with the lifting operation being controlled by the hatch foreman, only if something untoward happened which would warrant intervention by the ship's officer. This is strictly a case of too many cooks could spoil a safe loading operation.

NB. The main duties of the Chief Officer are to ensure that the vessel has adequate positive stability and this can be improved by filling double bottom water ballast tanks.

Additionally he should ensure that the derrick is rigged correctly and that all moving parts are operating in a smooth manner and all tackles are overhauling.

Operation

It would be normal practice for the Chief Officer to assess the existing ballast within the vessel and, if appropriate, add additional water ballast to eliminate free surface and provide bottom weight to increase the range of 'GM'. Reference to the ballast management plan and record book would be anticipated before and after making the heavy lift.

The actual stowage position for the Heavy Lift should have been assessed for total cubic capacity and the deck load density plan should reflect acceptable values for the intended load.

Heavy loads tend to generate unwanted ship stresses and these should be eliminated by the structural fabric of and around the stowage area of the load.

Measurement of the load extremities must allow load access to the intended stowage area/compartment. Where any height restrictions exist inside a compartment, allowance should also be made for any additional thickness of timber bearers. (Point loading can be avoided by timber bearers spreading the load over a larger area.)

Adequate manpower should be available in the form of competent winch drivers and the supervising controller. Winches should be set into double gear for a slow operation and steadying lines of appropriate size should be secured to points on the load to allow position adjustments to be made.

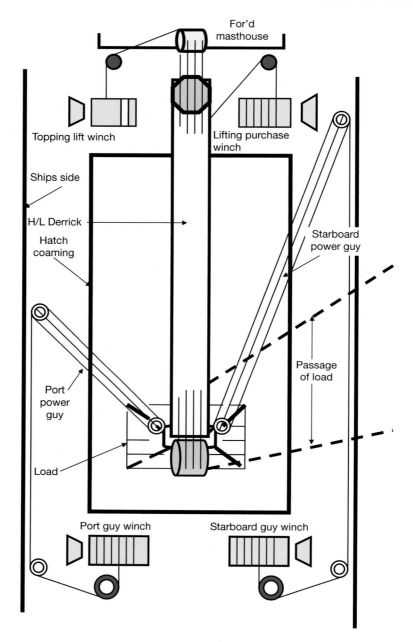

For'd
masthouse

Topping lift winch

Lifting purchase
winch

Ships side

H/L Derrick

Hatch
coaming

Starboard
power guy

Passage
of load

Port
power
guy

Load

Port guy winch

Starboard guy winch

FIGURE 4.2 Heavy lift – Jumbo derrick.

FIGURE 4.3 Conventional Jumbo heavy lifting derrick, seen in the vertically stowed position clamped to a double Samson Post (goal post) structure. The Samson posts also support smaller 10 tonne S.W.L. derricks for working general cargoes other than heavy lifts.

THE STÜELCKEN MAST CARGO HANDLING SYSTEM

The heavy lift Stüelcken systems are noticeable by the prominent angled support mast structure positioned either side of the ship's centre line. The main boom is usually socket mounted and fitted into a tabernacle on the centre line. This positioning allows the derrick to work two hatches forward and aft and does not restrict heavy loads to a single space, as with a conventional derrick.

The Stüelcken posts, set athwartships, provide not only leads for the topping lifts and guy arrangement but also support smaller five and ten ton derricks with their associated rigging. The posts are of such a wide diameter that they accommodate an internal staircase to provide access to the operator's cab, usually set high up on the post to allow overall vision of the operation.

The rigging and winch arrangement is such that four winches control the topping lift and guy arrangement while two additional winches control the main lifting purchase. Endless wires pay out/wind on, to the winch barrels, by operation of a one-man, six-notch controller.

Various designs have been developed over the years and modifications have been added. The 'Double Pendulum' model which serves two hatches operates with a floating

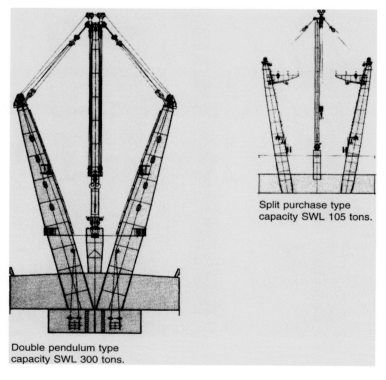

Split purchase type
capacity SWL 105 tons.

Double pendulum type
capacity SWL 300 tons.

FIGURE 4.4 Stüelcken heavy lifting mast structure.

head which is allowed to tilt in the fore and aft line when serving respective cargo spaces. A 'Rams Horn Hook' with a changeable double collar fitting being secured across the two pendulum lifting tackles. The system operates with an emergency cut off which stops winches and applies electro-magnetic locking brakes.

Stüelcken derrick rigs are configured with numerous anti-friction bearings which produce only about 2% friction throughout a lifting operation. These bearings are extremely durable and do not require maintenance for up to four years making them an attractive option to operators, in their day.

All blocks are of a multi-sheave construction often running to ten (10) or twelve (12) sheaves. They tend to have an extremely smooth, slow moving operation, virtually frictionless movement when engaged in loading or discharging cargo parcels.

The standard wires for the rig are 40 mm and the barrels of winches are usually spiral grooved to safeguard their condition and endurance. The length of the span tackles are variable and will be dependent on the length of the boom.

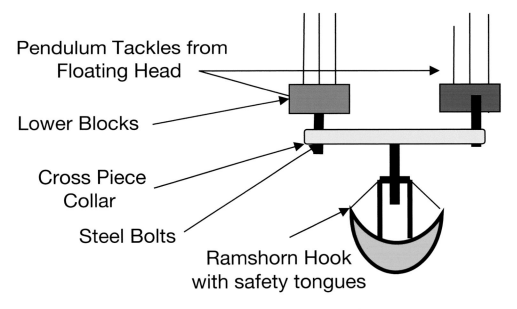

FIGURE 4.5 Lifting hoist arrangement for a double pendulum Stüelcken cargo gear system.

STÜELCKEN MAST: RIG TYPES FOR HEAVY LIFTING

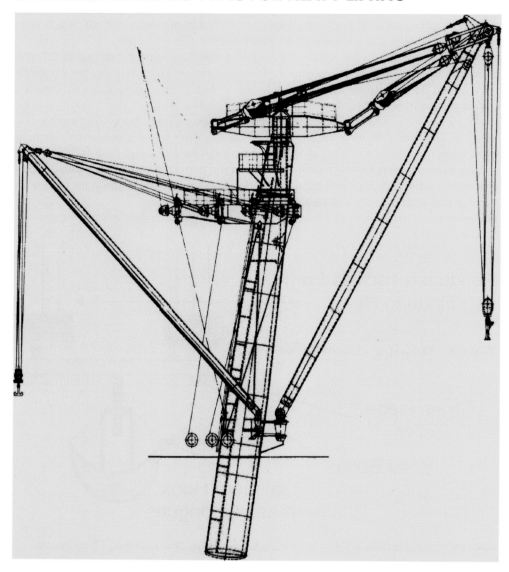

FIGURE 4.6 Stüelcken mast with heavy lift swinging derricks.

Note: Although Stüelcken rigs still remain operational, their use has diminished with the improved designs of designated Heavy Lift Vessels, which tend to have dominated the 'Heavy Lift Sector' of the industry over the last decade.

FIGURE 4.7 A heavy lift operation of an inshore and harbour work boat being lifted by a Stüelcken mast arrangement using the ramshorn hook with a double pendulum lifting purchase arrangement.

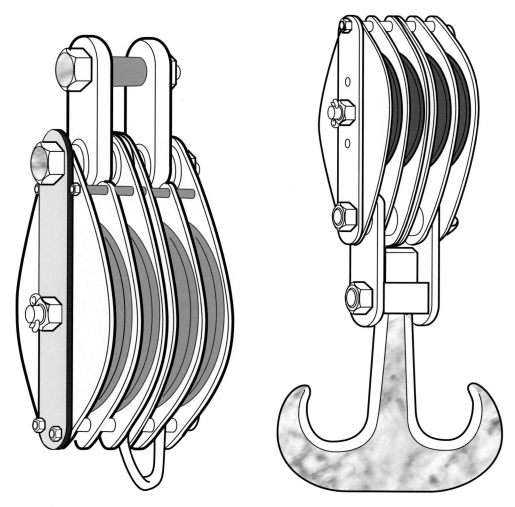

FIGURE 4.8 Example of multi-sheave heavy duty blocks used extensively with heavy lifting plant. The Stüelken lifting rigs usually engaged with 12 sheave heavy duty blocks. Where large rigging fitments are employed full account must be made for the weight of slings blocks and shackles when assessing the Safe Working Load capability of the rig.

THE ORDERING OF CARGO LIFTING GEAR

It is normal practice for the Chief Officers of ships to ensure that the ship's lifting gear is maintained in an efficient manner. To this end voyage requirements usually require the ordering of new shackles, blocks, wires etc. Manufacturers and suppliers need to be appraised of all relevant facts to which rigging equipment will be applied.

To this end the following details are required to accompany orders of:

Blocks – Description and Position of Block

Quantity required.
SWL of the derrick/crane.
Number of sheaves.
Diameter of sheave.
Size of wire or rope.
SWL of block required.
Proof load.
With or without becket fitment.
With or without shackle.
Specified fitment, i.e. duck bill or oval eye, swivel yes/no
    (with respective dimensions).
Catalogue number if available.

Shackles – Description and Position of Shackle

Quantity required.
Type of shackle (bow or 'D').
SWL of shackle.
Proof load.
Catalogue number if available.

Wires – Description and Position of Wire

Quantity required.
Construction of wire (i.e. 6 x 24 wps F.S.W.R.).
SWL of wire.
Size of wire diameter (i.e. 24 mm).
Proof load.
Length of coil.
Catalogue number if available.

When ordering new rigging equipment they are normally supplied with the required certificate. Chief Officers would normally file these for insertion into the Register of Lifting Appliances and Cargo Handling Gear as required for ISM audits.

STABILITY DETAIL (LIFT ON/LIFT OFF OPERATIONS)

It must be anticipated that the vessel will move to an angle of heel when making a heavy lift with the derrick extended. This angle of heel should be calculated and the loss of 'GM' ascertained prior to commencing the lift. Clearly any loss of positive stability should be kept to a minimum and to this end, any free surface effects in the ship's tanks should be eliminated before any movement of the load, wherever possible.

Stability with Heavy Lifts

Whenever a vessel is engaged in loading heavy lifts, mean sinkage will occur and the vessels draughts can expect to change. Depending on the weight of the load itself and the method of loading/discharging the vessel can expect to be heeled over (port or starboard) to a greater or lesser degree. Conventional lifts by shipboard derrick or crane would generally be expected to generate an angle of heel.

If it is realised from the onset that once a heavy lift is taken up by a crane or derrick the centre of gravity of the load is deemed to act from the head of that derrick or crane jib. When calculating the ships stability criteria, this assumption is, for all intents and purposes, like loading a weight above the ships centre of gravity.

Mariners who find themselves involved in ship stability calculations will appreciate that when a weight is loaded on board the vessel, a movement of the ships 'G' will be expected. This movement (GG^1), will be in a direction towards the weight being loaded. It therefore follows that once a weight is lifted and that weight effectively acts from the head of the derrick, the ships position of 'G' will move upwards towards this point of action.

The outcome of lifting the load and causing an upward movement of 'G' is to cause G to move towards M (the Metacentre). This action would be to effect a reduction in the ships GM Value (GM = Metacentric Height).

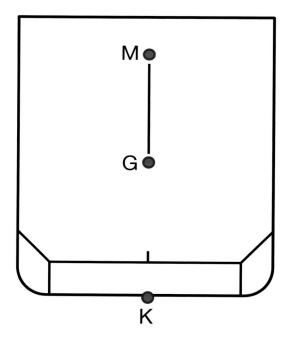

FIGURE 4.9 Condition 1 shows the reflective positions of M, the Metacentre. G, the ship's centre of gravity, and K, the position of the keel with the vessel in an upright aspect (double bottom tanks empty).

Once the weight of the load is taken by the ships derrick, chief officers should appreciate that the ships 'G' will rise towards ' M', possibly even rising above 'M' causing an unstable condition. It would therefore make sense to lower the position of 'G', in anticipation of the rising 'G' prior to a heavy lift being made.

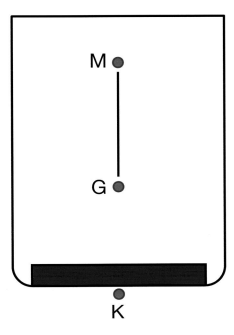

FIGURE 4.10 Condition 2, the vessel is still in the upright, but the double bottom tanks have been filled, adding weight below 'G'. This action causes 'G' to move down and generates an increase in the ship's GM value.

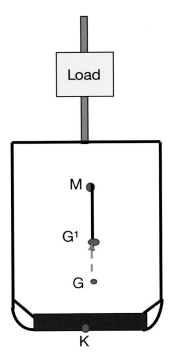

FIGURE 4.11 Condition 3, where the derrick lifts the load on the centre line of the vessel causing 'G' to move upwards, towards the new G^1 position (vessel stays in the upright).

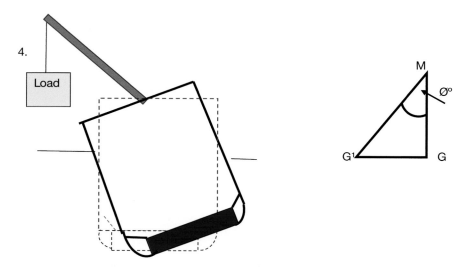

FIGURE 4.12 Condition 4, where the derrick and the load is swung over side causing the vessel to heel over to Ø°.

The Chief Officer would normally be charged with the task of ascertaining the maximum angle of heel that would affect the vessel during the period of lifting.

If the GM can be increased before the lift takes place, i.e. by filling double bottom tanks, the angle of heel can be seen to be less.

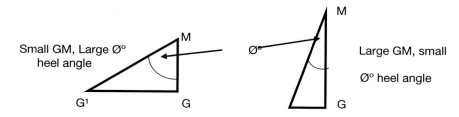

FIGURE 4.13 Angle of heel and the relation to GM.

MOVEMENT OF THE CENTRE OF BUOYANCY 'B'

When the weight is being lifted overside, either in a loading or discharging situation, the vessel will experience an angle of heel. This angle of heel will effectively increase the 'Waterplane Area' of the ship causing a shift in the position of 'B'.

A 'Righting Couple' is generated, which will increase if the vessel continues to list further towards the quay.

To this end, some designated heavy lift ships have deployed a tracked water tank to the off-shore side of the ship. Once lowered on the tracks, this tank is allowed to fill with water as the tank is lowered past the waterline. This action effectively increases the ships water plane area, thereby increasing the righting couple established from the shift of the Centre of Buoyancy 'B'.

As the position of 'B' is defined as the geometric centre of the underwater volume it should be realised that if the ship is inclined, for any reason, the underwater volume will change and 'B' will move off the ships centre line, towards the low side, to a new position of the new Centre of Buoyancy 'B¹'. The underwater volume being the same size but of a different shape.

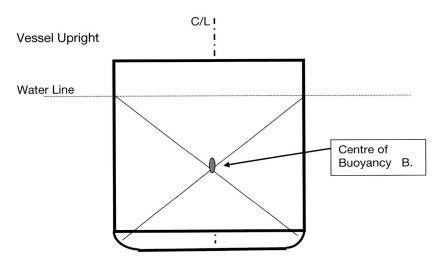

FIGURE 4.14 Position of centre of buoyancy (geometric Centre of the underwater volume).

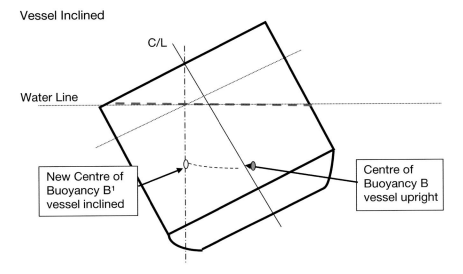

FIGURE 4.15 Centre of buoyancy movement with an angle of heel increasing waterplane area.

Shipboard (Occasional Heavy Lift) Operations

Not all ships are suitably equipped with heavy lifting rigs but can still take on heavy loads by engaging floating cranes or sheer leg barge platforms. Although generally expensive to employ a suitable freight rate can be set to the shipper, to ensure such costs are covered.

FIGURE 4.16 The heavy lift, floating crane, 'Mersey Mammoth' (IMO No. 8521622) seen underway and making way on the River Mersey, Liverpool, England.

Many ports and harbours around the world have their own heavy lifting plant in the form of a floating crane. Their use in loading and discharging to merchant ships is only one of their activities. A harbour will often engage a floating crane in civil engineering projects like extending break waters or bridge construction. They are regularly engaged in all forms of the port's operations.

HEAVY LIFT FLOATING CRANE

Conventional heavy cargo loads, which are scheduled for carriage by sea, are often required to be loaded by means of a floating crane. When the load is too great to be handled by the ship's own lifting gear, the second option is usually the next immediate choice. Most major ports around the world have this facility as an alternative option for heavy specialist work. The type of activity is twofold, because if loaded by heavy lift plant like this at the port of departure, the same load must be discharged at its destination by similar or equivalent methods. (Ships' Chief Officers need to ascertain that if the weight of the load is above the ship's lifting gear capability, the discharge port has adequate means of lifting the load out.)

Ships' Cargo Officers need to ensure that the heavy load is accessible and that the floating crane facility is booked in advance in order to make the scheduled lift. Booking of a special crane would normally be carried out via the ship's agents, and ships' personnel are very much in the hands of external parties – the Port Authority often controlling the movement of all commercial and specialist traffic in and around the harbour.

The 'floating crane' should not be confused with the specialist 'crane barge'. Floating cranes differ in that they may not be self-propelled and may require the assistance of tugs to manoeuvre alongside the ocean transport, prior to engaging in the lift(s). The construction of these conventional cranes is such that the crane is mounted on a pontoon barge with open deck space to accommodate the cargo parcel. The pontoon barge is a tank system that can be trimmed to suit the necessity of the operation if the case requires.

The main disadvantage against the more modern, floating sheer legs is that, generally speaking, the outreach of the crane's jib is limited in its arc of operation. Also the lift capacity can be restrictive on weight when compared with the heavier and larger units which tend to operate extensively in the offshore/shipyard arenas.

Agents when booking the facility need to be made aware of the weight of the load and its overall size, also its respective position on board the vessel, together with its accessibility. Hire costs of the unit are usually quite high and with this in mind, any delays incurred by the ship not being ready to discharge or accept a scheduled load on arrival of the crane could become a costly exercise.

The Crane/Sheer Leg Barge (Self–Propelled)

FIGURE 4.17 The Smit 'Cyclone' (IMO No. 7603688) floating sheer leg barge, engages in a general cargo heavy lift operation on a container vessels offshore side.

Derrick/crane barges tend to work extensively in the offshore sector of the marine indus-try but their mobility under own propulsion, together with thruster operations, provide flexibility to many heavy lift options. Some builds incorporate dynamic positioning and depending on overall size have a lifting capacity up to and including 6,000 tonnes with main crane jib operations.

Alternative to the floating cranes are these seagoing, floating sheer leg platforms, with their extended outreach and high capacity weight lifting capability. These usually self-propelled platforms are extensively employed inside the offshore energy sector, but are also used worldwide in civil engineering projects, shipyards and salvage operations.

FIGURE 4.18 Heavy lift 'Sheer Leg' platform the 'Musashi' floating crane barge, seen off Kobe, Japan. This barge platform is towed to its working position. It has excessive ballast capacity for making extreme lifts up to 3,700 tonnes.

FIGURE 4.19 The 'Seal Teal' (IMO No. 8113566), showing the exposed steel cargo deck. The semi-submersible is seen without cargo and would submerge the deck to allow a float on cargo lift to take place.

DESIGNATED HEAVY LIFT SHIPS/BARGES

Float On Float Off – Methods for the Transport of Heavy Lifts

The methodology of floating a load onto the upper deck of a transport vessel is clearly the basic principle of the floating dock. Initially submerged to allow a vessel to drive into a dock, then de-ballast the dock to bring the dock and its ship load to the surface. The use of hydrostatics has been gainfully employed in canal locks, floating docks and now extensively with semi-submersible heavy lift vessels. Many of which are currently involved in the movement of project cargoes.

FIGURE 4.20 The heavy lift vessel 'Super Servant 3' (IMO No. 8025331) loaded by float on float off methods with the crane barge *Al-Baraka 7*, seen lying at anchor in the Arabian Gulf region.

Designated heavy lift ships generally have the capability to carry out lift on and lift off, for heavy or awkward cargo parcels, inside its own lifting capacity. While the semi-submersible type vessels engage in float on, float off methods to load their cargoes.

This is not to say that this latter type of transport cannot also accept direct load on to its deck via alternative (shoreside) lifting gear. However, the semi-submersibles do not usually carry any heavy lifting cranes, derricks or sheer legs of their own.

FIGURE 4.21 The semi-submersible 'Super Servant 3' as seen from astern carrying the crane barge *Al-Baraka 7*, on its main weather deck.

These ships are initially constructed with high capacity ballast tanks as well as trimming tanks to submerge and level the upper cargo deck, when required.

Combined use of wing tanks and double bottom tanks, depending on the size of load, can adjust the GM value during cargo operations and during the period of the voyage. However, it should be remembered that slack tanks can generate free surface effects which would ultimately be detrimental to the positive stability of the ship.

Once the load is on board the vessel, the wing tanks can be emptied to suit, retaining the double bottoms (full), such action would eliminate any free surface effect from both side wing tanks.

Heavy Lifting

Several shipping companies operate purpose built heavy lift ships but probably the most high profile is the 'Big Lift' shipping company operating numerous dedicated heavy lift vessels.

These ships are frequently engaged in tandem lifting using two heavy duty cranes.

Such lifts are usually controlled by a single operator working with a remote controller. The control unit is operated by the individuals chest mounted portable box, with each crane being represented often by a joystick control to reposition crane heads to establish the plumb line, causing luff and slew motions to the load.

FIGURE 4.22 The *Happy Sky* (IMO No. 9457220) engages in a lift on lift off tandem lift of a long steel pipe section.

During a loading operation any vessel could be caused to heel, once the weight is taken up by an off centre line lifting crane. Heavy lift ships anticipate and expect this with lift on and lift off methods. Counter measures to correct any heeling are usually generated by use of incorporated trimming tanks.

Tandem Lift

FIGURE 4.23 Big Lift vessel 'Happy Delta' (IMO No. 9551935) engages in a tandem lift, employing its two heavy duty Huisman cranes to discharge a long steel barge (cargo) into a calm water landing. Heavy duty lift beams work with four bridle spreaders to accept the load along its full length.

FIGURE 4.24 The 'Happy Star' (IMO No. 9661259) one of the Big Lift shipping company's vessels engages in lifting a topside module by means of heavy duty lifting beams and a tandem lift by its two on board Huisman heavy lift cranes.

AN INTRODUCTION TO PROJECT CARGOES

Heavy Lift Ships – Semi-Submersibles and Project Cargoes

The need for heavy lift ships, semi-submersibles and floating sheerleg platforms has expanded considerably alongside the immense size of the loads required, especially within the development of the offshore industries.

These heavy duty transports all tend to have operational tank systems which allow them to work by employing the same hydrostatic power of Archimedes principles and the laws of flotation. The semi-submersibles submerging themselves to allow a load to float over its deck prior to de-ballasting and lifting the load clear of the water line (float on float off). While the heavy lift ship, loading into holds or direct to its weather deck (load on load off).

Both methods can expect to change the draught levels of ships so involved by a considerable amount after accepting distinctive cargoes. So much so that under-keel clearance will always be a concern for any contemplated passage plan, as in some cases with deck cargoes, will be the air draught.

When semi-submersibles are engaged their cargo may often be mounted to a raft to achieve a float on, float off procedure. The size and weight of the raft itself, may also influence not only the stability of the vessel, but also the securement of the load, aboard the carrier.

EXAMPLE OF HEAVY LIFT SEMI-SUBMERSIBLE VESSEL IN OPERATION

(Example of loading/discharging by float on float off method)

FIGURE 4.25 The *Dockwise Vanguard* (IMO No. 9618783) seen in a semi-submersible position lifting the Goliat (FPSO), a turret loading system, for delivery to an offshore site. Such project loads are frequently of several thousand tonnes per transport. The vessel has since been renamed as 'Boka Vanguard'. It has a lift capacity of 110,000 tonnes.

Designated Heavy Lift Ships

The tanking system of a designated heavy lift vessel is so arranged that the stability of the ship at any period of the loading or discharging procedure is retained within the safety parameters. This is achieved by manipulation of the ship's centre of gravity.

Heavy lift vessels are not all of the same design and the tank systems differ to suit their main mode of operation.

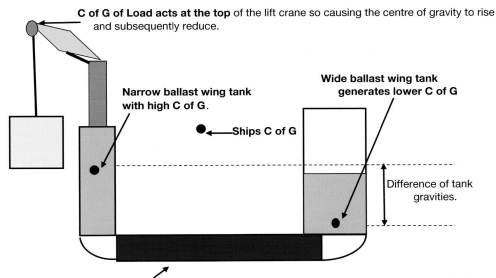

FIGURE 4.26 Tank structure of designated heavy lift ship.

The centre of gravity of the ship is seemingly artificially kept low by the use of such tanks, effectively keeping a satisfactory value of metacentric height throughout the loading period.

Once any angle of heel is established water ballast from the narrow wing tank can be pumped/transferred to the wider ballast tank on the opposite side.

Elements for Consideration for Heavy Lift Transports

1 Overall size – dimensions of the load.
2 Weight of the load.
3 Weight of lifting accessories.
4 SWL of lifting elements.
5 Weather conditions.
6 Positive stability of transporting vessel (at acceptable draughts).
7 Density of water in load and discharge ports.
8 Ballast arrangements for trim and list of vessel.
9 Passage plan of transport route.

10 Fuel burn on route.

11 Speed and ETA of passage.

12 Loadline zone requirements not infringed.

13 Method of discharge.

14 Facilities at the port of discharge.

15 Manpower requirements for loading/shipping/discharging.

16 Documentation for the load.

17 Specialist handling personnel.

18 Communication facilities (Walkie Talkies) to accommodate loading/discharge.

19 Securing arrangements for load on route.

20 Load management on voyage.

Large heavy loads tend to accrue logistical problems from the time of construction to that moment in time when the load arrives at its final destination. The shipping element of the load's journey is just one stage during the transportation. Cargo surveyors, safety experts, company officials and troubleshooters of various kinds tend to move alongside the passage of the load up to that time of final delivery.

The Project Cargo

The question is often asked 'what is a project cargo?' To say it is a specialised load is certain, but in any event it is not an item that can be easily defined.

To the majority, a 'project cargo' is by their very nature, usually large and very heavy and will need considerable time and man hours to achieve transportation. Although this is not strictly true, in every case. Could a project cargo not equally be a hazardous chemical in bulk that requires extraordinary precautions or an item that is extremely valuable that extra security becomes necessary? Are not these extenuating circumstances taking the transport of such goods into the defined region of falling into the project cargo sector?

That large heavy load, that is testing to load successfully, is easily recognised as being different to the run of the mill cargoes. But the associated questions of whether the ship and its load will be capable of a safe transition of the Panama Canal locks, or not, turns the transport operation into a project worthy of investigation. Will the heavy load aboard the vessel cause a change in ships draught, so that its under-keel clearance is compromised? Can such a load be delivered and landed safely? All questionable aspects of conducting a successful outcome to the satisfaction of the shipper.

In other words, that extraordinary load which requires excessive precautionary actions, to allow the ship to deliver a specific cargo parcel, safely. To achieve such an outcome a detailed plan must be considered in the light of the idiosyncrasies that accompany the project cargo parcel. Inclusive of transport to and from the loading/discharge ports from place of manufacture or assembly. Namely the transfer from shore to ship, with vice versa systems in place, to achieve ship to shore at the delivery port.

Once loaded aboard a ship, will the ship and its cargo be able to pass under any bridge obstructions on its route? Will the vessel and its load be able to navigate through the geographic hazards that might be presented on the forthcoming voyage?

Has the so-called cargo been cleared of any inherent dangers? As say with a nuclear product, is it clear of radiation hazards? Does the cargo have inherent dangers that could affect personnel at the loading position, on route or at discharge? Will the port of delivery accept the cargo without the imposition of limiting conditions? Will the cargo affect the political climate, as with armaments? The unknown questions are many that can occur with that most unusual of loads.

The transport of that extraordinary cargo must be incident free and covered by a capability risk assessment. It must be acceptable to all parties inclusive of the ship owner, the ship's Master, charterers, the shipper, and the receiving customer. Additional parties with vested interest will also include marine insurance, regional agents and customs officials.

FIGURE 4.27 Scheuerle multi-wheel land driven heavy lifting transport platforms. These combination platforms can be adapted in size to suit smaller and larger loads at shippers needs. They are extensively used to transport heavy lift parcels and project cargoes.

SHORESIDE ADMINISTRATION FOR HEAVY LIFT/PROJECT CARGO

Every heavy lift/project cargo operation will pass through various degrees of administration prior to the practical lift taking place. The manufacturers/shippers will be required to provide clear information as to dimensions, weight, lifting and securing points and the position of the centre of gravity, before the load can be accepted by the ship, while the ship may be required to give details of its crane capability inclusive of outreach and load capacity.

The loading operation itself, as to whether it will be from the quayside or from a barge, must also be discussed together with the detail of use of ship's gear or floating crane. Weather conditions and mooring arrangements may also be featured at this time. Once loading is proposed, the stability data and the maximum angle of heel which will be attained would need to be calculated. Ballast arrangements pertinent to the operation may well need to be adjusted prior to contemplating the actual lift or move.

The ship would be consulted on voyage and carriage details, as to the securing of the load, the deck capacity to accommodate the load and the stability criteria. The ship would also require assurances regarding the port of discharge and the capabilities of said port. If the load is beyond the capacity of ship's lifting gear, then the discharge port must have access to a floating crane facility or alternative lift equipment and ensure that this facility will be available at the required time.

Where road transport is involved to deliver the load to the quayside, road width and load capability would need to be assessed. A 500-ton load on the back of a low-loader may well cause land slip or subsidence of a roadside, which must be clear of obstructions like bridges and rail crossings. Wide loads or special bulky loads may require police escort for movement on public highways to and from loading/discharge ports.

Once loaded, the weight will need to be secured and to this end a rigging gang is often employed. However, prudent overseeing by ship's officers is expected on this particular exercise, bearing in mind that the rigging gang are not sailing with the ship and once the ship lets her moorings go, any movement of the load will be down to the ship's crew to effect re-securing.

Customs clearance would also be required as per any other cargo parcel and this would be obtained through the usual channels when the manifest is presented to clear the vessel inwards. Export licences are the responsibility of the shipper, together with any special details where the cargo is of a hazardous nature, covered by special clearances, e.g. armaments.

MOVEMENT LOGISTICS OF THE LARGE LOAD

The task of transporting 'project cargoes' does not lend itself easily to the use of public roads. Fortunately, the building sites for such items are often located by coastlines and generally do not encroach on public highways. For example, shipyards build and transport modules or installations within their own perimeters and transport within those same perimeters. However, occasionally, a one-off project requires a specialised route. Timing is critical at all stages of the journey to ensure minimal disruption to the general public, and police escort must be anticipated when going door to door.

Transport of Heavy Loads (Ground Handling Equipment)

Further reference should be made to the IMO publication on 'The Safe Transport of Dangerous Cargoes and Related Activities in Port Areas'.

Planning for Project Cargo Transport

It would be natural for the layman to assume that the extreme heavy load just moves on its own with the help of a police escort. This is clearly not the case for extreme loads, needing that larger-than-life transport. Extensive planning, usually from the building stage must be considered for the delivery of such a special and unusual cargo parcel.

A company may be able to build for the customer, but if the load cannot be transported safely, because of weight or size then the actual building becomes an expensive 'white elephant', in more ways than one.

Also costs for the transportation could be considerable and these must be accounted for when contracts and financial agreements are drawn up. Such costs should take into account any external franchise required, insurance premiums and external charters as for tugs, escorts or similar additional overheads.

FIGURE 4.28 A Scheuerle multi-wheel transport platform, seen engaged in ground handling of a steel installation structure. The movement of such a load must be logistically planned, prior to the building/construction phase of the project.

Transportation and Planning Considerations for Project Cargoes

Measurement of the load – Not only weight measurement of the cargo, but its overall length, breadth and depth will be required. If the load is structured to pass through a float on float off procedure during the loading or discharging period, then additional information would be required, such as:

a) Draught of the load at floatation

b) Freeboard of the load and/or the load platform

c) Centre of Gravity of the load

d) Centre of buoyance of the load

e) Density of the dock water at the loading port

f) Density of the dock water at the port of discharge

g) Draughts of the load or the load platform in fresh and salt waters

h) Air Draught of vessel when loaded

i) Centre of gravity of ship when loaded (ship's G)

j) Seasons and long range weather forecasts.

Tidal considerations at the load and discharge positions should also be calculated for the designated operational periods.

Security and safety throughout the movement must be detailed to ensure that personnel involved with each phase are briefed and made aware of respective duties and responsibilities. Additional consideration may need to take account of the environment and that it is left or returned to its original state, following the movement of passage.

A risk assessment must be completed to ensure each phase is completed within general safe procedures, local by-laws and international regulations. Example: load lines must not be infringed.

Transport vehicle – Capability of the carrier to carry out the task. In the case of a ship: is the vessel capable of accepting the load? What is the displacement and physical size of the vessel and its capability to accept the deck load? (Reference made to the vessel's deck load density plan and its overall deadweight.)

A general assessment of the ship's stability, throughout all stages of the passage, would be required – the value of the ship's GM prior to loading and with the load added being a prime consideration.

The endurance of the vessel with regard to the capacity of the ship's tanks for fuel and water and the effects of the same from burning bunker oil and consuming water on route affects the overall stability of the transport.

Ballast movement and the ability to trim or list the vessel for the purpose of loading/discharging would need to be considered along with the draughts in the loaded and light/ballast conditions.

The ship's machinery and its manoeuvring capability, together with operational speed through the various voyage phases, would need to be considered.

Fuel consumption and the bunker availability on passage must be established.

The ships complement, experience of the Master, charter rates and not least the availability of the vessel must all be taken into full account.

FIGURE 4.29 The offshore installation *Polar Pioneer* is seen being manoeuvred with tug assistance over the submersed platform deck of the semi-submersible *Dockwise Vanguard*.

Voyage Planning

The movement of project cargoes is, by the very nature of the task, generally carried out at a slow speed. This is especially so as in the next example shown as an extreme lift being made by the *Pioneering Spirit*. Often tug assistance is employed and the operation must be conducted at a safe speed for the circumstances. Operational speed(s) would be covered by the charter party and at the Master's discretion.

The movement from the loading port, towards the discharge position, being carried out under correct navigation signals, appropriate to each phase of the passage.

As with any passage/voyage plan, the principles of 'passage planning' would need to be observed but clearly specialist conditions apply over and above those imposed on a conventional ship at sea.

Passage planning involves the following phases:

Appraisal – The gathering of relevant charts, publications, information and relevant data to enable the construction of a charted voyage plan.

Planning – The actual construction of the voyage plan to highlight the proposed route.
To provide details of way points, bunkering stations, navigation hazards, margins of safety, currents and tidal information, monitoring points, contingency plans, traffic focal-points, pilotage arrangements, under-keel clearances, etc.

Full use of nautical charts and navigation references would be applied at this stage.

Execution – The movement of the transport to follow the plan through to its completion. The positive execution of the plan by the vessel.
Hazards on route would normally be highlighted for any voyage but where project cargo is being carried, particular hazards from the aspect of the load may cause exception to the norm. Bearing in mind that the load may restrict passage through canals, under bridges or through narrow waterways and areas of reduced under-keel clearance (UKC & air draught).

Monitoring – The confirmation that the vessel is proceeding as per the designated plan. Monitoring of the ships position, it's under-keel clearance, communications, weather conditions and operational speed must be under continual observation.
The movement of the vessel proceeding through the various stages of the voyage, would need full support from the shipping company and associated agencies, with a designated person ashore (DPA), as a first contact.
In this day and age electronic GPS monitoring of all and any movement would be expected to be continuous, through to the port of discharge.
NB. A passage plan is equally meant to highlight the areas where the vessel should not go, a particularly important aspect to vessels engaged with 'project cargoes'.

EXTREME LIFTS IN THE OFFSHORE SECTOR

Historical Note: In 1996, the Smit Transport Cargo Barges worked in conjunction to carry out the combined transportation of 9,600 tonne module for the 'Sleipner Vest' platform. This operation was completed as a Float-Over, where the deck module was positioned over the supporting unit, (previously established on site), in the 'Cobo Field', West Africa.

FIGURE 4.30 The world's currently largest ship *Pioneering Spirit* (IMO No. 9593505) has the equivalent deck space of six football pitches and has a lift capacity of 48,000 tons. In April 2017, the ship was engaged to lift the *Brent Delta* topside accommodation module, off the oil rig platform in the North Sea. The lift was 24,000 tons and was achieved in 12 hours by positioning the twin ballast hulls of the *Pioneering Spirit* under the rig and de-ballasting the ship to rise. (This lift holds the world's lifting record.)

The *Pioneering Spirit* has been designed and built to carry out platform installation and decommissioning of existing installations. Entire platform topside sections, up to 48,000 tonnes can be lifted and installed in a single step movement. Such activity, inclusive of pipe laying ability, can directly limit operational time offshore leading to significant cost savings in field development.

The vessel works in conjunction with a cargo barge 'Iron Lady' which accepts the load when in close to the shore where shallow water might restrict movement of the deeper draughted *Pioneering Spirit*.

Voyage Plan Acceptance

Once the plan is constructed, it would warrant close inspection by the project manager, any super cargo and the ship's Master. Such a plan would need to incorporate a considerable number of special features prior to being considered acceptable to relevant parties.

Passage plans are made up to ensure 'berth to berth' movement. Such a plan must ensure safe execution for movement of any project cargo and would expect to include the following special features:

Risk assessment – *completed on the basis of the initial plan.*
   (Passage plans are meant to be flexible and circumstances may make a deviation from the proposal a necessary action when en route.)

Communications (primary sat/coms) – Methods: VHF channels (ship to shore), secondary methods, advisory contacts, coastguard, VTS, hydrographic office, meteorological office, agents, medical contingency.
   Most towing operations and project movements would normally be accompanied by a navigation warning to advise shipping likely to be affected. Such warnings could be effected by coast radio stations, port and harbour controls, and/or the hydrographic offices of the countries involved. Preliminary notices to mariners may be appropriate.

Loading procedures (following acceptable stability assessment) – Method (various examples): lift on lift off, float on float off, etc., tug assistance, marine pilots, rigging and lifting personnel as required. Tidal conditions and continuous weather conditions monitored.

Securing procedures – Personnel and associated equipment. Warranty surveyor/project manager inspection. Contractors: riggers, lashings, welders may be required for steel cargo parcels.

Risk assessment – tolerable.

Safety assessment – LSA/manpower, navigation equipment test. Engine test.
   Weather forecast 48 hours, long range forecast.

Route planning – Weather, ports of call, mooring and bunker facilities, UKC, width of channel, position monitoring methods, communications to shore to include progress reports, navigation hazards, command authority, natural or physical obstructions by way of canal transit or bridge obstructions. Traffic focal points. Seasonal weather and loadline considerations. Under-keel clearance with sea and air draughts monitored throughout.

Contingencies – Endurance, bunkers, manpower, emergency communication contacts. Weather, mechanical failure, steering failure, tug assistance, on-route ports of refuge. Use of anchors, safe anchorages. special signals. Support services (shore based). Emergency docking facilities

Schedule – Timing to effect move, speed of move relevant to each movement phase. Charter party, delivery date, 'penalty clauses'. Sailing plan, monitoring and tracking operations, daylight/night passage times with posreps and progress reports.

Risk assessment – per phase of voyage.

Discharge procedure – Method and stability assessment. Ground handling equipment, secondary transport. Specialist personnel and equipment. Quayside facilities and tidal considerations. Risk assessment.

Personnel requirements – (To be within the safe manning requirements and capable of meeting the required and approved watchkeeping system.) Master and crew (company employment). Surveyors, specialist handlers various contractors. Super cargo (shipper's representative) .

Log in with blockchain technology – Computing and IT contractors. Security personnel. Port Agents addresses and contact numbers.

Documentation – Manifest.

Insurance – Shoreside administration/shipping company.

Continuous Synopsis – ongoing from outset of the voyage and access/gangway log.

Customs/Quarantine Clearances – Reception, delivery communications, export licenses.

Ancillary Units – Tugs, lifting units, support vessels, bunker barges, equipment, consumables.

Specialist Equipment – Ice regions, cold weather stores and personnel gear (if required).

Accommodations – Airports, hotels, local transport facilities, labour force and expenses.

Security – ISPS declaration between ship and shore. ISPS watch and stowaway searches. Piracy, road transport, in port, at sea, communications, police, customs, military, security codes effecting contingencies.

Costs – Market assessment, freight, political considerations.

Administration via agency contacts – Communications, allocation of funds, mails, pilotage/tugs bookings, store replenishments, employments.

PROJECT CARGO – SUMMARY

These extreme cargoes can often take a lengthy period of time to reach a safe delivery to the consignee. Every activity associated with the movement of the load needs to go through a fail-safe process of risk assessments, and may be years in meticulous planning. The activity of load movement must be legal within a home or foreign environment. Whether the movement involves a lift on lift off operation or a float on float off activity each must contend with the elements of weather and the laws of the land.

The execution of the loading, in itself, may only be over a few hours or days, but all stages must adhere to a safety schedule to ensure a successful outcome. The so-called project cargo is not the run of the mill cargo parcel, either from the point of view of weight or size or just awkwardness to load, carry and safely deliver.

FIGURE 4.31 The 'Happy Sky' (IMO No. 9457220) of the Big Lift shipping company engaged in transporting two container gantry cranes secured to its upper deck. With such a high load, the centre of gravity of the cargo above the ship's centre of gravity, effective lower ballast was used to maintain the ships stability. Obvious concerns for the passage plan for such a cargo would also be for the air draught through narrow waterways and if passing under low bridges.

CHAPTER 5

Bulk Cargoes

INTRODUCTION

The demand for raw materials continues to sustain a major sector of the shipping industry. Bulk products are shipped all over the world from their point of origin to that position of demand. The 'bulkers' transport everything from grain and coal to chemicals and iron ore. The bulk trades involve vast tonnage movement of any one commodity and such movement can present its own hazards and problems associated with the cargo.

Designs of ships' holds have evolved to maximise capacity while at the same time generating a safer method of carriage. The Maritime Safety Committee of the IMO has adopted amendments to Chapter XII SOLAS (Additional Safety Measures for Bulk Carriers) which entered into force on 1 July 2004, affecting all bulk carriers regardless of their date of construction. These amendments include the fitting of dry-space, water ingress monitors. These level detectors are alarmed to monitor any fluid levels that could lead to an adverse flooded condition. Additionally, means of draining and pumping dry space bilges located forward of the 'collision bulkhead' are also a requirement.

Further recommendations have been agreed for bulk carriers (built after 1999) over 150 metres in length to be constructed with 'double hulls', where vessels are carrying solid bulk cargoes having a density of more than 1000 kg/m^3. Effectively, the double hull bulk carrier is the future for bulk cargoes. How these cargoes are loaded, managed and discharged in the types of vessels involved is as follows.

References for Bulk Cargoes

International Code for the Safe Carriage of Grain in Bulk (IMO Grain Regulations)
International Code for the Construction and Equipment of Ships Carrying Dangerous Chemicals in Bulk (IBC Code)
The International Maritime Solid Bulk Cargoes Code (IMSBC Code)
Code of Practice for the Safe Loading and Unloading of Bulk Cargoes (BLU Code)
Resolutions of the 1977 SOLAS Conference, regarding the Inspection and Surveys of Bulk Carrier Vessels
Ballast Management Plans and Procedures. (shipboard)
MSC/Circ 908 (June 1999) Appendix C, Uniform Method of Measurement of the Density of Bulk Cargoes

DOI: 10.4324/9781003407706-5

MSC/Circ 646 (June 1994) Recommendations for the fitting of Hull Stress Monitoring Systems (also MGN 108 M)

Code of Safe Working Practice for Merchant Seafarers, consolidated 2015 inclusive of amendments 1–6

MGN 60

MGN 144 (M) Merchant Shipping (Additional safety measures for Bulk Carriers)

MGN 335(M) Bulk Carriers: Guidelines on early assessment of hull damage and need for abandonment. Includes IMO MSC / 1143.

MGN 512 (M)

MGN 513 (M)

DEFINITIONS AND TERMINOLOGY EMPLOYED WITH BULK CARGOES

Angle of repose – The natural angle between the cone slope and the horizontal plane, when bulk cargo is emptied onto this plane, in ideal conditions. A value is quoted for specific types of cargo, results being obtained from use of a 'tilting box'. The angle of repose value is used as a means of registering the likelihood of a cargo to shift during the voyage.

An angle of repose of 35° is taken as being the dividing line for bulk cargoes of lesser or greater shifting hazard and cargoes having angles of repose of more or less than the figure are considered separately.

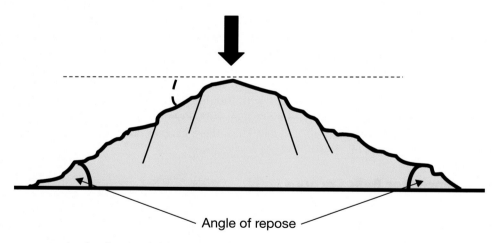

Angle of repose

FIGURE 5.1 Angle of repose of the cone of a bulk cargo.

Bulk density – The weight of solids, air and water per unit volume. It includes the moisture of the cargo and the voids, whether filled with air or water.

Cargoes which may liquefy – These are cargoes which are subject to moisture migration and subsequent liquefaction if shipped with a moisture content in excess of the transportable moisture limit.

Combination carriers (OBO or O/O) – A ship whose design is similar to a conventional bulk carrier but is equipped with pipelines, pumps and inert gas plant so as to enable the carriage of oil cargoes in designated spaces.

Concentrates – These are materials that have been derived from a natural ore by physical or chemical refinement or purification processes. They are usually in small granular or powder form.

Conveyor system – This means the entire system for delivering cargo from the shore stockpile or receiving point to the ship.

Flow moisture point – The percentage of moisture content when a flow state develops.

Flow state – This occurs when a mass of granular material is saturated with liquid to such an extent that it loses its internal shear strength and behaves as if the whole mass was in liquid form.

Incompatible materials – Those materials which may react dangerously when mixed and are subject to recommendations for segregation.

Moisture content – The percentage proportion of the total mass which is water, ice, or other liquid.

Moisture migration – The movement of moisture contained in the bulk stow when water is progressively displaced as a result of settling and consolidation in conjunction with vibration and the ship's movement. Part or all of the bulk cargo may develop a flow state.

Pour – The quantity of cargo poured through one hatch opening as one step in the loading plan, i.e. from the time the spout is positioned over a hatch opening until it is moved to another hatch opening.

FIGURE 5.2 An overhead view of a general cargo vessel engaged in the discharge of concentrates by means of a free-standing crane using a mechanical grab. The ship's own deck cranes are turned outboard to allow easy access for the shoreside crane operation.

Transportable moisture limit – The maximum moisture content of a cargo that may liquefy which is considered safe for carriage in ships other than those ships which, because of design features of specialised fittings, may carry cargo with a moisture content over and above this limit.

Trimming – A manual or mechanically achieved adjustment to the surface level of the form/shape of a bulk stow in a cargo space. It may consist of altering the distribution or changing the surface angle to the point, perhaps, of levelling some or all of the cargo, following loading.

The International Maritime Solid Bulk Cargoes (IMSBC) Code

The IMSBC Code is an amendment to Chapter VI of SOLAS and made the code mandatory from 1 January 2011. It replaced the Code of Safe Working Practice for solid bulk cargoes, previously known as the BC Code.

Its aim is to facilitate the safe stowage and shipment of bulk cargoes, by providing information on the dangers that are associated with the shipment and carriage of types of cargo in bulk form. It highlights associated hazards and provides guidance on various procedures which should be adopted when example products are shipped in bulk.

Recommendations are made about the stowage of the cargoes and include suggested maximum weights to be allocated to lower holds as found from the formula:

0.9 LBD

where   D represents the ship's summer load draught
         L represents the length of the lower hold
         B represents the average breadth of the lower hold

The height of the cargo pile peak should not exceed:

1 x D x SF (m³/tonnes) metres

Legislative, Unified Requirements (UR) for Bulk Carriers

Water ingress alarms – are now required under SOLAS XII Regulation 12, by first survey after 1 July 2004. Such alarms must be fitted to all cargo holds and be audible and visual to the navigation bridge.

Existing bulk carriers are also required to have, in addition to the water level alarms stated above, permanent access for close-up inspection and the use of green sea loads on deck for the design of hatches and deck fittings. Such measures are expected to ensure that a well-maintained single hull bulk carrier will remain satisfactory for the remainder of its lifetime.

New bulk carriers – will be of 'double hull' construction (May 2004).
They will also require harmonised class notation and standard design loading conditions together with 'double side shell', water ingress alarms to cargo holds and forward spaces.

Note: The distance between the inner and outer hulls should be 1000 mm

Additional increased strength and integrity for foredeck fittings and a free-fall lifeboat with immersion suits for all crew members are now statutory requirements..

New structural changes will incorporate the permanent means of access for close-up inspection, an amendment to the loadline which will allow the building of stronger and more robust vessels but reduce deadweight capacity by approximate estimates of 0.5–1.5%, depending on size.

Structural standards – as per SOLAS Chapter XII, applying to single hull, side skin bulk carriers will also apply to new double hull bulk carriers.

Additional Equipment

Recent amendments to SOLAS Chapter XII/II, require that new bulk carriers over 150 metres in length and below, be fitted with loading instrumentation which provides information on the ship's stability.

Water ingress alarms – are required for vessels with a single cargo hold. The requirement to fit water level detectors in the lowest part of the cargo space is applicable to bulk carriers less than 80 metres in length or 100 metres in length if built before 1998, to take effect from the first renewal or intermediate survey after July 2004.

The alarms will be audible and visual to the navigation bridge and will monitor cargo spaces and other spaces forward of the collision bulkhead. The regulation does not apply to vessels with double sides up to the freeboard deck.

Additional reference: S.I. 1999 M.S. (Additional Safety Measures for Bulk Carriers) Regulations 1999 and MGN 144 (M).

FUTURE BUILDS

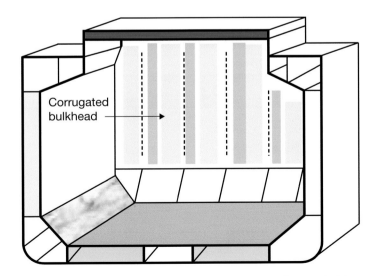

Corrugated bulkhead

FIGURE 5.3 Diamond 53, design – complete double hull in way of cargo holds.

Double Hull, Bulk Carrier Construction

The double hull types have inherent strength which allows flexible loading patterns, which will increase capacity for heavy load density cargoes such as steel coils. The design dispenses with exposed side frames in the holds and presents a flush side and hold ceiling for cargoes. Such flush features have distinct advantages for hold cleaning and cargo working options with bulk commodities.

The double design also provides a perceived safer protection against water ingress and is therefore seen as being more environmentally friendly in comparison with the single hull types. Tank arrangements permit a large water ballast capacity, in both double bottoms and side tanks, eliminating the need to input ballast into cargo spaces in the event of heavy weather.

BULK CARRIER CONSTRUCTION

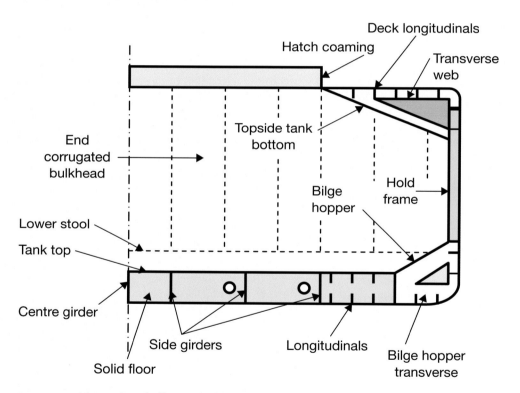

FIGURE 5.4 Athwartships half section of a bulk carrier.

Note: Framing on bulk carriers is designed as a longitudinal system in topside and double bottom tanks and as a transverse system at the cargo hold, side shell position.

Bulk Carrier Designs and Hatch Covering

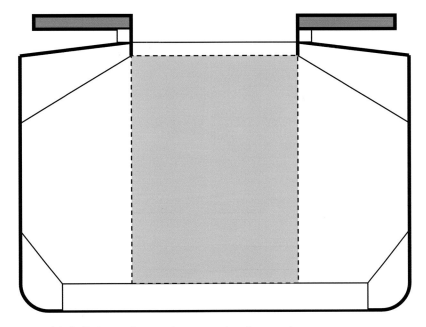

FIGURE 5.5 Double hull design (twin side moving hatch covers).

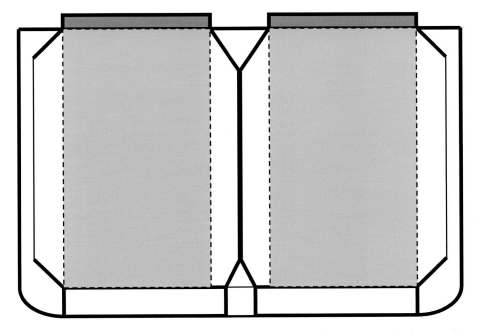

FIGURE 5.6 OJ Libaek's Optimum 2000 (capesize) bulk carrier design (twin hatch covers).

FIGURE 5.7 The bulk carrier *Berge Stahl* (IMO No. 8420804) is seen starboard side to the loading berth. The shoreside gantry cranes use 20-tonne mechanical grabs to complete the loading of iron ore.

FIGURE 5.8 The bulk carrier *Berge Nord* (IMO No. 9127150) alongside the iron ore berth Seven Islands, Canada. The upper deck shows the side moving hatch top arrangement in the closed positions.

Mechanical Grabs for Bulk Cargoes

Bulk commodities are continually being shipped to virtually all major world ports. The discharge of these various cargoes is frequently undertaken by steel grabs. These grabs come in a variety of sizes and capacities and may be operated by one of several methods:

Mechanical
Electrical
Hydraulic
Radio remote control

Some are designed for operation from a single crane wire, gantry cranes, or employing up to 4 x steel wire ropes.

Capacity of grabs in general tend to range from 5 tonnes to 60 tonnes having a space capacity averaging from 5 cu/meters to about 25 cu/meters.

The type of grab used is subject to the nature of the cargo being transferred. Such examples include clamshell bucket grabs for sand, ores, coal and powder products, while orange peel grabs/octopus types, are employed for scrap metal.

The modern designs of clamshell bucket grabs are made of steel and are often radio controlled. The jaws of the grab, if required, can also be opened automatically when the grab is touched to ground. Once sealed shut they do not allow seepage or leakage of any product contained within.

Grabs by their very nature are large and heavy objects and operators need to be attentive when delivering cargo loads to ship or shore. If not fully controlled the grab can quickly wander with a distraction to the operator or a break in the operator's concentration. Such mishaps can easily cause structural damage to ship or quayside structures

Working Bulk Cargoes

FIGURE 5.9 A mechanical grab with the tines closed, seen discharging scrap metal.

Use of Mechanical Grabs

Many types of bulk cargoes are discharged by means of 'mechanical grabs', of which there are several variations. The handling of grabs (5 to 60 tonnes) is always precarious, especially the larger capacity grabs, because they are difficult to exactly control and may cause structural damage to the vessel.

Cargoes tend to be loaded by chute, tipping or pouring, especially the grain type cargoes, ores and coals. However, discharging of ores, bulk solids and coal cargoes tend to employ grabs for discharge purposes. Bucket grabs come in various sizes ranging from 5 to 60 tonnes, the more popular range being in the 5 to 25 tonne bracket. (Manual labour for bulk commodity discharge has all but died out.)

Cargo Officers are advised that working with heavy grabs requires designated concentration by the crane drivers and even then ship damage is not unusual. A close check on the operation of grabs throughout load/discharge operations is advised and any damage to the ship's structure by contact of the grab should be reported to the Chief Officer of the vessel. Subsequent damage claims can then be made against the stevedores for relevant repair costs. Bear in mind that damage to hatch coamings may render the vessel unseaworthy if the damage prevents the closure of hatches and cannot be repaired before the time of sailing.

FIGURE 5.10 Mechanical clamshell bucket grab seen in the open position on the quayside. The crane is configured so that the controlling crane driver can open and shut the grab by means of a control wire to open and close the bucket arrangement.

Hold Preparation for Bulk Cargoes

Bulk cargoes are generally loaded in designated 'Bulk Carrier' vessels, but they can be equally transported in general cargo ships alongside other commodities. However, in such circumstances, specific stowage criteria and hold preparation would probably be a requirement. In virtually every case, except where perhaps the same commodity from the previous voyage is being carried, the cargo holds would need to be thoroughly cleaned and made ready to receive the next cargo.

The Designated 'Bulk Carrier'

1   The holds would be swept down and cleared of any residuals from the previous cargo.

2   All rubbish and waste matter must be removed from the cargo space, before loading of the next cargo can commence.

3   The hold bilge system would need to be inspected and checked to ensure that:

   a   the bilge suctions are operational;

   b   the bilge bays are clean and smelling sweet (not liable to cause cargo taint).

4   All hold lighting arrangements, together with relevant fittings, would be inspected and seen to be in good order.

5   The space, depending on the nature of the previous cargo and the nature of the next cargo to be carried, would probably require to be washed down with a salt water wash.

> Note: Following a wash down, the space would be expected to be allowed to dry out. Special commodities may require the cargo spaces to be surveyed prior to permission being granted to load another alternative cargo.

BULK CARGOES

Grain

Grain is defined in the IMO Grain Rules as:
   Wheat, Maize (Corn), Rye, Oats, Barley, Rice, Pulses or Seeds, whether processed or not, which when carried in bulk, has a behaviour characteristic similar to grain in that it is liable to shift transversely across a cargo space of a ship, subject to the normal seagoing motion.

The following terms are applicable to the Grain Rules:

Filled – When applied to a cargo space, this means that the space is filled and trimmed to feed as much grain into the space as possible, when trimming has taken place under the decks and hatch covers, etc.

Partly filled – This is taken to mean that level of bulk material which is less than 'filled'. The cargo would always be trimmed level with the ship in an upright condition.

Note: A ship may be limited in the number of partly filled spaces that it may be allowed.

Grain must be carried in accordance with the requirements of the aforementioned grain rules which consist of three parts, namely 'A', 'B' and 'C'.

Part A. Contains thirteen rules which refer among other items to definitions, trimming, intact stability requirements, longitudinal divisions (shifting boards), securing and the grain loading information which is to be supplied to the Master. This information is to include sufficient data to allow the Master to determine the heeling moments due to a grain shift. Thus there are tables of grain heeling moments for every compartment which is filled or partly filled, tables of maximum permissible heeling moments, details of scantlings of any temporary fittings, loading instructions in note form and a worked example for the Master's guidance.

Part B. Considers the effect on the ship's stability of a shift of grain. For the purpose of the Rules, it is assumed that in a filled compartment (defined as a compartment in which, after loading and trimming as required by the Rules, the bulk grain is at its highest level) the grain can shift into the void space which is always considered to exist at the side of hatchways and other longitudinal members of the structure or shifting boards, where the angle of repose of the grain is greater than 30°.

The average depth of these void spaces is given by the formula:

$$V_d = V_{d1} + 0.75 \ (d - 600) \ mm$$

where: V_d represents the average void depth in mm
V_{d1} is the standard void depth found from tables
d is the actual girder depth in mm.

The standard void depth depends on the distance from the hatch end or the hatch side to the boundary of the compartment.

The assumed transverse heeling moment can now be calculated by taking the product of the length, breadth and the ½ depth of the void (if it is triangular, over the full breadth) and the horizontal distance of the centroid of the void from the centroid of the 'filled' compartment.

The total heeling moment = 1.06 x Calculated transverse heeling moment for a full compartment.
or
1.12 x Calculated transverse heeling moment in a partly filled compartment.

It will have been noted that the above heeling moments are expressed in m units and so it is also termed a Volumetric Heeling Moment.

The reduction in GZ in the initial position (λ_0) is assumed to be

Total Volumetric Heeling Moment due to grain shift
Stowage factor of the grain x displacement

The reduction in GZ at 40° (λ_{40}) = 0.8 λ_0

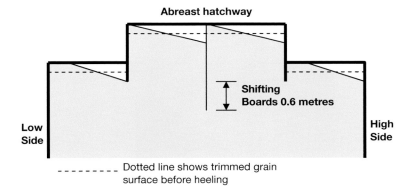

Abreast hatchway

Low
Side

High
Side

Shifting
Boards 0.6 metres

- - - - - - - - - Dotted line shows trimmed grain
surface before heeling

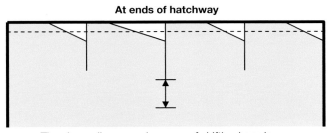

At ends of hatchway

The above diagrams show use of shifting boards
and assumed formation of voids if heeled 15°

FIGURE 5.11 The above diagrams show use of shifting boards and assumed formation of voids + heeled 15°.

Superimposing the above reductions in GZ on the vessel's curve of statical stability will give a 'heeling arm' curve (straight line). The angle at which the two curves cross is the angle of heel due to the shift of grain and this angle must not exceed 12°. Also the initial GM (after correction for free surface for liquid in tanks) must not be less than 0.30 metres.

The residual area between the original curve of righting levers and the heeling arms up to 40°, or such smaller angle at which openings in the hull, superstructures or deckhouse cannot be closed watertight when immersed (this is called 0_f – the angle of flooding – at which progressive flooding commences) must not be less than 0.075 metre-radians.

If the vessel has no document of authorisation from the contracting governments, she can still be permitted to load grain if all filled compartments are fitted with centre line divisions extending to the greater of 1/8th of the maximum breadth of the compartment or 2.4 metres. The hatches of filled compartments must be closed with the covers in place. The grain surfaces in partly filled compartments must be trimmed level and secured and she must have a GM which is to be the greater of 0.3 metres or that found from the formula I of the rules.

Part C. Concerned with the strength and fitting of shifting boards, shores, stays, and the manner in which heeling moments may be reduced by the saucering of grain. The handling of bulk and the securing of hatches of filled compartments and the securing of grain in a partly filled compartment are also detailed.

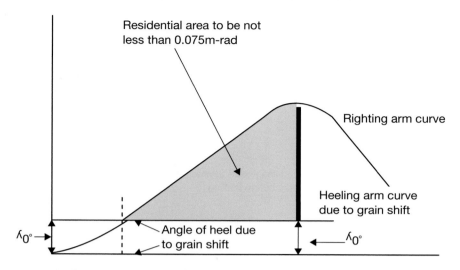

FIGURE 5.12 Heeling arm curve.

When shifting boards are fitted in order to reduce the volumetric heeling moment, they are to be of a certain minimum strength with a 15 mm housing on bulkheads and are supported by uprights spaced according to the thickness of the shifting boards (e.g. 50 mm thick shifting boards would require a maximum spacing of 2.5 m between uprights). The shores will be heeled on the permanent structure of the ship and be as near horizontal as practical but in no case more than 45° to the horizontal. Steel wire rope stays set up horizontally may be fitted in place of wooden shores but the wire must be of a size to support a load in the stay support of 500 kg/m².

The shifting boards will extend from deck to deck in a filled tween deck compartment while in a filled hold they should extend to at least 0.6 m below the grain surface after it has been assumed to shift through an angle of 15°.

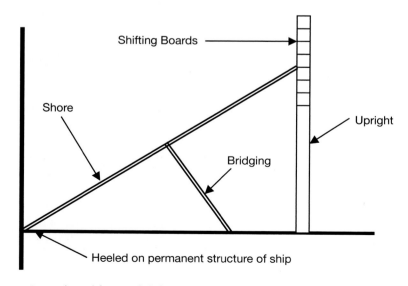

FIGURE 5.13 Strength and fitting of shifting boards.

In a partly filled compartment the shifting boards can be expected to extend from at least 1/8th of the maximum breadth of the compartment above the surface of the levelled grain to the same distance below.

A further method of reducing the heeling moment in a filled compartment is to 'saucer' the bulk in the square of the hatch and to fill the saucer with bagged grain or other suitable cargo laid on separation cloths spread over the bulk grain. The depth of the saucer on a vessel over 18.3 moulded breadth will be not less than 1.8 m. Bulk grain may be used to fill a saucer provided that it is 'bundled', which is to say that after lining the saucer with acceptable material, athwartships lashings (75 mm polypropylene or equivalent) are placed on the lining material not more than 2.4 m apart and of sufficient length to draw tight over the surface of the grain in the saucer. Dunnage 25 mm in thickness and between 150 and 300 mm wide is laid longitudinally over the lashings. The saucer is now filled with bulk grain and the lashings drawn tight over the top of the bulk in the saucer.

In a partly filled compartment, where account is not taken of adverse heeling moments due to grain shift, the surface of the bulk grain is to be trimmed level before being overstowed with bagged grain or other cargo exerting at least the same pressure to a height of not less than 1/16th of the maximum breadth of the free grain surface or 1.2 m, whichever is the greater. The bagged grain or other suitable cargo will be stowed on a separation cloth placed over the bulk grain, or a platform constructed by 25 mm boards laid over wooden bearers not more than 1.2 m apart may be used instead of separation cloths.

Note: Where lashings and bottle screw securing's are employed, these must be regularly inspected and reset taut during the voyage.

FIGURE 5.14 Bulk carrier *Alpha Afovos* (IMO No. 9221853) lies port side to the grain silos in Barcelona. The grain elevators are seen deployed into the ships hold effecting discharge.

Measures to Reduce the Volumetric Heeling Moment of 'Filled' and 'Partly Filled' Cargo Compartments

By use of longitudinal divisions – These are required to be grain tight and of an approved scantling.

FIGURE 5.15 A tractor engaged in the tween deck of a vessel discharging cereals to ensure that suction of the grain elevator has access to all residuals of the cargo product.

By means of a saucer and bundling bulk – A saucer shape is constructed of bulk bundles in the hatch square of a filled compartment. The depth of the saucer is established between 1.3 and 1.4 metres depth dependent on the ship's beam, below the deck line.

By over stowing in a partly filled compartment – Achieved by trimming the surface level flat and covering with a separation cloth then tightly stowing bagged grain to a depth of 1/16th of the depth of the free grain stow.

To Ensure Adequate Stability

The angle of heeling of the vessel which arises from the assumed 'shift of grain' must not exceed 12°.

When allowing for the assumed shift of grain, the dynamical stability remaining – that is, the residual resistance to rolling on the listed side – must be adequate.

The initial metacentric height (GM), making full allowance for the free surface effect of all partially filled tanks, must be maintained at 0.3 metres or more.

The ship is to be upright at the time of proceeding to sea.

Document of Authorisation

In order to load grain, a vessel must have a Document of Authorisation or an appropriate 'Exemption Certificate'. The authorisation means that the vessel has been surveyed and correct grain loading information has been supplied to the ship for use by the Deck Officer responsible.

The Document of Authorisation serves as evidence that the ship is capable of complying with the requirements of the code. It must be incorporated into the 'Grain Loading Manual' and contains information that enables the Master to meet the stability requirements of the Code.

Grain Awareness

When a grain cargo is loaded, compartments will contain void spaces below the crown of the hatch top. During the voyage the grain will 'settle' and these void spaces would be accentuated. In the event that the grain shifts, it will move into these void spaces to one side or another, generating an adverse list to the vessel and directly affecting the stability of the ship by reducing the resistance to roll and adversely affecting the 'range of stability'.

Measures to reduce the possibility of the grain shifting include the rigging of longitudinal shifting boards and over stowing the bulk cargo with bagged grain.

FIGURE 5.16 The suction is extended from the grain elevator into the cargo hold space to extract the cargo into the grain silo.

Cargo Hold – Longitudinal Separation

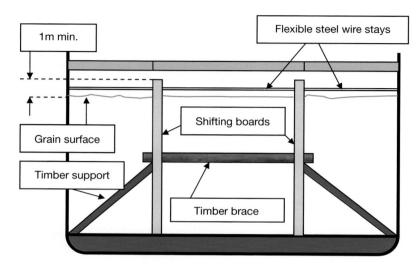

FIGURE 5.17 Longitudinal separation.

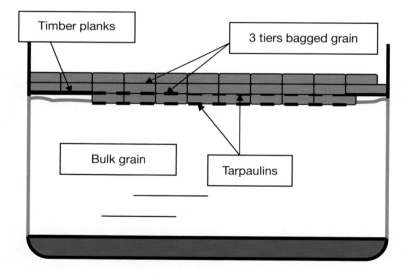

FIGURE 5.18 Over stowing bulk cargo with bagged cargo.

Trimming of Bulk Cargoes

Many bulk cargoes are trimmed (levelled) at the loading port to provide a stable stowage for when the ship is at sea. However, trimming also takes place during the period of discharge to ensure that the total volume of cargo is landed.

Permissible Grain Heeling Moment Tables

The purpose of the tables is to allow the ship's Master to ascertain whether or not a particular grain stowage condition will achieve the required stability criteria.

The obtained values can then be applied to obtain the approximate angle of heel that would result from a possible shift of the grain cargo.

$$\text{Actual Grain Heeling Moment} = \frac{\text{Total Volumetric Heeling Moment}}{\text{Stowage rate (S/F)}}$$

S/F (Bulk) = 1.20 / 1.67 m³/tonne

$$\text{Approximate angle of heel} = 12° \times \frac{\text{Actual Heeling Moment}}{\text{Permissible Heeling Moment}}$$

For a ship to be authorised to carry grain, the surveyor will have made calculations for sample cargoes to show that adequate stability for the ship exists. A grain loading information listing should be made in which the surveyor would record all the dimensions of the carriage compartments and then these would be converted into potential heeling moments for when the space is filled or partly filled.

The deck officer would be expected to make his own calculations before the intended voyage to take account of the type of grain being carried and its stowage factor. Account must also be taken for the condition of the ship at all stages of the voyage, to ensure adequate stability throughout.

FIGURE 5.19 The bulk carrier *Konkar Lydia* (IMO No. 8022444) seen here twin side opening hatch tops open lying port side to the grain silo berth in Barcelona, Spain. The tug in attendance provides size comparison of the hatch top covers.

FIGURE 5.20 The bulk carrier *Konkar Lydia* (Grt 37,340) seen in the loaded condition with the grain elevation suckers poised above the hatches prior to commencing discharge.

COAL – LOADING, CARRIAGE AND DISCHARGE

Categories of Coal

Coal – Any coal, including sized grades, small coal, coal duff, coal slurry or anthracite.

Coal duff – Coal with an upper size of 7 mm.

Coal slurry – Coal with particles generally under 1 mm minimum size.

Coke – Solid residue from the distillation of coal or petroleum.

Small coal – Sufficient particle material below 7 mm to exhibit a flow state when saturated with water.

The Characteristics of Coal

Coal cargoes are liable to spontaneous heating, especially when sufficient oxygen is available to generate combustion. The amount of heating which takes place will depend on the type of coal being carried and the ability to disperse that heat with effective ventilation methods. Unfortunately, ventilation can work against the safe carriage because of supplying unwanted oxygen while at the same time dispersing the heat concentrations. It is recommended that surface ventilation only is applied to coal cargoes. This can be applied by raising the hatch tops (weather permitting) to allow surface air and released gases to go to atmosphere and not be allowed to build up inside the cargo compartment.

Freshly mined coal absorbs oxygen which, with extrinsic moisture, forms peroxides. These in turn break down to form carbon monoxide and carbon dioxide. Heat is produced and this exothermic reaction causes further oxidation and further heat. If this heat is not dissipated, ignition will occur, i.e. spontaneous combustion. Large coal gives a good ventilation path for air flow towards surface ventilation methods, while small coal tends to retain the oxygen content and is more likely to generate spontaneous combustion.

Preparation of the holds should include the overall cleaning of the hold prior to loading, the testing of the bilge suctions and sealing the bilge bays to prevent coal dust clogging bilge bays. Spar ceiling (cargo battens) should also be removed as these would have a tendency to harbour oxygen pockets deep into the heart of the stow. Hold thermometers should be rigged at three different levels, to ensure tight monitoring of the temperatures in the compartments loaded with coal. Critical temperatures in coal vary, but heating will be accelerated in some varieties of coal from as low as 38°C (100°F). Such temperatures would create a need to keep external hull and deck surfaces as cool as possible. In the tropics, it may be appropriate to cover decks to lessen the internal heating in the compartment.

Coal Fires

Most coal fires occur at about the tween deck level, which is an area that requires more attention to temperature monitoring and ventilation.

Surface ventilation to holds should be concerned with the removal of methane gas for the first five days of the voyage, thereafter the ventilators to the lower holds should be plugged with an exception for about six hours every two days. Gas from the holds or tween deck regions may find its way into trunk sections, shaft tunnels, chain lockers, peaks and casings unless bulkheads can be maintained in a gas-tight condition.

A strict policy of no naked lights and no smoking should be followed and crew should not engage in chipping or painting below decks.

The majority of coal fires are caused by spontaneous combustion. Poor hatch cleaning prior to loading and a lack of temperature monitoring are often directly linked to the cause. In the event of a coal fire at sea, it should be realised that they are extremely hot fires and if tackled with water would generate copious amounts of steam. Unless this can be vented, the compartments could become pressurised.

If tackled from sea, it is recommended that hatches are battened down and all ventilation to the compartment sealed with the view to starve the fire of oxygen. A Port of Refuge should be sort, where the authorities can be informed to receive the vessel and dig the fire out by grabs while fire fighters are stood by to tackle the blaze once exposed.

Boundary Cooling on adjacent areas to control the spread of fire is recommended in virtually all cases, but it should be realised that water on hot surfaces will generate steam which could restrict personnel access, visibility and movement. In any event personnel so engaged would need Personnel Protective Equipment (PPE).

Loading Coal

Coal is loaded by either tipping or conveyor belt bucket system. It is recommended that the first few truck loads are lowered to the holds – this reduces breakage, as does a control rate of the chutes. Loading may take place from a single loading dispenser and as such it may become necessary to shift the ship to permit all compartments to be loaded. A loading plan to prevent undue stresses and minimum ship movements would normally be devised. Coal will need to be trimmed as its 'angle of repose' is quite high, especially for large coal.

Small coal like 'mud coal', 'slurry' or 'duff' is liable to shift, but shifting is unlikely in large coal.

Reference should be made to the (IMSBC) Code for Bulk Cargoes prior to loading any of the coal types. Information on dry bulk cargoes is given under the heading of 'Ores and Similar Cargoes' and information on wet bulk cargoes is given under the heading of 'Ore Concentrates'.

> **Example of Loading Coal** at Newport News USA: 10,000 tonnes of coal poured to a five hatch, general cargo vessel completed the full load inside a ten-hour period.

Coke

Coke and similar substances such as 'Coalite' have had their gas and benzole removed and they do not heat spontaneously. No special precautions are necessary other than to ensure that the coke is cold before loading. If hot coke is loaded this may generate a fire.

The Precautions of Loading Coal

The IMO divides coal into several categories:

Category A – No risk

Category B – Flammable gas risk

Category C – Spontaneous heating risk

Category D – Both risks

Although precautions are given to each category, the following general precautions are recommended.

1 Gas tight bulkheads and decks.

2 Spar ceiling (cargo battens) removed.

3 Measures taken to prevent gas accumulating in adjacent compartments.

4 Intrinsically safe electrical equipment inside compartments.

5 Cargo stowed away from high temperature areas and machinery bulkheads.

6 Gas detection equipment on board.

7 Trim cargo level to gain maximum benefits from surface ventilation.

8 Cargo/hatch temperatures monitored at regular intervals and logged.

9 No naked flame or sparking equipment in or around cargo hatches.

10 No welding or smoking permitted in the area of cargo hatches.

11 Full precautions taken for entry into enclosed compartments carrying coal.

12 Suitable surface ventilation procedures adopted as and when weather permits.

Note: Certain coal cargoes of small particle content are liable to shift if wet and experience liquefaction hazards. Reference to the International Maritime Solid Bulk Cargoes (IMSBC) Code should be made and appropriate precautions taken.

IRON AND STEEL CARGOES

Steelwork is carried in various forms, notably as pig iron, steel billets, round bars, pipes, castings, railway iron, 'H' girders, steel coils, scrap metal or iron and steel swarf.

Other examples are seen in the form of bulldozers, locomotives, armoured tank vehicles and designed steel constructions.

It is without doubt one of the most dangerous of cargoes worked and carried at sea. Recommendations for stowage have been made by various MSNs and MGNs in the past and yet it is still prone to 'shifting' in a rough sea condition.

Pig iron – If pig iron or billets are taken, they should be levelled and large quantities should not be carried in tween deck spaces. A preferred stow is to level in lower hold spaces and overstow by other suitable cargoes.

If it has to be carried in tween deck spaces, the maximum height to which it can be stowed should not exceed 0.22 x the height of the tween deck space.

Pig iron should be trimmed and stowed level in both tween deck and hold spaces in either a side-to-side or fore-and-aft stow. If it is not effectively overstowed it should be stowed in robust 'bins' with suitable shifting boards to prevent cargo movement. It is recommended that gloss finished pig iron is always stowed on wood ceiling or dunnage, to reduce steel-to-steel friction.

Round bars and pipes – These should be stowed preferably in the lower hold compartments and levelled off. Securing should be in the form of strong cross wires over the top of the stow and secure 'toms' at the sides. Suitable cargoes can overstow this type of steelwork.

Railway iron, 'H' girders, and long steel on the round – These should be stowed in a fore and aft direction and packed as solidly as possible. If left exposed and not overstowed, chain lashings should be secured to prevent cargo shift.

Iron and steel swarf – This may heat to dangerous levels while in transit, if the swarf is wet and contaminated with cutting oils. The carriage of swarf requires that surface temperatures of the cargo are monitored at regular intervals during the loading process and while on the voyage. If during loading the temperature of an area is noted as 48°C (120°F), loading should be temporarily suspended until a distinct fall is observed. In the event that a temperature of 38°C is observed on passage, gentle raking of the swarf surface area in the region of the high temperature to a depth of about 0.3 metres should cause the temperature to lower. If a temperature of 65°C is noted the ship is recommended to make for the nearest port.

Scrap metal – Similar problems to other steel cargoes, in that it is very heavy. It is generally loaded by elevator/conveyor or grabs and usually discharged by mechanical grabs. When loading, the first few loads are often lowered into the hold to prevent the possibility of excessive damage to ship's structures.

Scrap metal tends to come in all shapes and sizes. As such, where mechanical grabs are engaged, metal pieces frequently become dislodged from the grab when in transit from the hold to the shore while discharging or loading, respectively. Deck Officers should ensure that the working area is cordoned off and personnel on the deck area should wear hard hats and observe cargo operations from a safe distance.

Steel coils – These are stowed on the round and are frequently carried in the cargo holds of bulk carriers (see Chapter 3: General Cargoes). The overall stow is secured by steel wire and bottle screw lashings. The sides of the stow are generally chocked tight against the ship's side, if broken stowage is a feature of the loaded cargo. The first tier of coils at deck level are individually wedged at the deck, against sideways movement.

Steel coils are classed as a heavy cargo, and would be levelled to no more than two tiers in height. Individually a coil may weigh up to 15 tonnes, and they are frequently treated as 'heavy lifts'. They are prone to shifting, being stowed on the round, if the vessel encounters rough weather. Passage plans should bear this in mind and chart a Port of Refuge in case such a contingency is required.

Ores – Ore cargoes are mostly of a low stowage factor, which means that when a full cargo of ore is loaded, there will be a large volume of the hold left unused. A low stowage factor also lends itself to a 'stiff ship', unless some of the cargo can be loaded in the higher regions of the vessel.

Ore should be trimmed if possible and at the very least, the top of the heap should be knocked off. Modern bulk carrier hold design compensates in some way towards a cargo which is likely to shift. Other vessel designs have been developed as designated ore carriers, and have effective upper ballast compartments to raise the vessel's centre of gravity when carrying dedicated heavy ore cargoes.

In the event that an ore cargo is only a 'part cargo' it should be realised that some ores have a high moisture content which does not always lend to over stowing.

There are 4 types of iron ore deposit: Massive hematite, (most commonly mined), Magnetite, Titanomagnetite and Pisolitic ironstone.

Raw iron ore is separated into lumps and fines from the mining, crushing and screening, for use in the iron and steel industry. Example ore cargoes: bauxite, chrome, iron, lead, manganese.

General Information on the Loading/Discharge of Steel Cargoes

Steel cargoes in any form are probably one of, if not the most dangerous of cargoes.

Steel is shipped in many forms, from railway lines to ingots, from bulk scrap to bulldozers. It is invariably always heavy and very often difficult to control because of its size. It is a regular cargo for many ships and has been known to cause many problems by way of stability, or adverse effects to the magnetic compass. If steel shifts at sea, due to bad weather, it is unlikely that the crew would have the skills or the facilities to rectify the situation and the vessel would probably need to seek a Port of Refuge with the view to corrective stowage. Steel coils on the round are particularly notorious for moving in bad weather.

An active Cargo Officer during the loading period can ensure that correct stowage is achieved and, even more important, that correct securing is put in place. Relevant numbers of chain lashings and strong timber bearers go well with steel loads but are often required before the load becomes overstowed by light goods.

Masters should monitor progress during loading periods without being seen to interfere with the Cargo Officer's duties. Specific attention should be given to the use of rigging gangs being employed as and when required. An awareness of the needs of industrial relations without sacrificing the safety requirements can be a delicate balance when a load needs 20 securings and dock labour only wants to secure with 10.

Damage to the vessel when loading or discharging heavy steelwork is not unusual. Heavy lifts by way of bulldozers or locomotives require advance planning and a slow operation. They are awkward to manoeuvre because of weight and extended dimensions.

Heavy rig lifting gear operated by ship or shore authorities, even when taking all precautions, very often results in damaged hatch coamings or buckled deck plates. The possibility of damage to the cargo itself is also a likely occurrence.

'H' or 'T' section steel girders are difficult to control because of length and are normally loaded on the diagonal into a hatchway. Slinging is normally by long leg chains but high winds when loading can cause excessive oscillations of the load, especially with deep sections. Steadying lines of adequate size should be employed before lifting. High winds also pose problems for the lighter steel boilers. These are large but comparatively light, being hollow. Size and shape coupled with strong winds tend to cause slewing on the load in way of the hatch coaming.

Steel in any form will always be shipped and it is in the interests of all concerned to ensure safe handling and stowage. Masters tend to be wary of the stability needs and load in lower holds rather than tween decks, depending on circumstances and the needs of other cargoes. However, the need for vigilance when securing remains a high priority towards voyage safety.

BULK CARGO EXAMPLES

Concentrates are partially washed or concentrated ores. These cargoes are usually powdery in form and liable to have a high moisture content and subsequently, under certain conditions, have a tendency to behave almost as a liquid. Special stowage conditions prevail, and sampling must take place to ascertain the transportable moisture limit as provided by the IMSBC Code. They are extremely liable to shifting, and care should be taken when loading. Some cargoes may appear to be in a relatively dry condition when loading, but at the same time contain sufficient moisture to become fluid with the movement and vibrations of the vessel when at sea.

Nitrates are considered dangerous cargoes. Before stowing, the IMDG code on dangerous goods should be consulted.

Phosphates readily absorb water and should be kept dry. A variety of these is guano, which is collected from islands in the Pacific. Phosphates should be kept clear of foodstuffs.

Sulphur is a highly flammable cargo and all anti-fire precautions should be taken. It is also very dusty and highly corrosive. The risk of dust explosions when clearing holds after carriage is of concern. Fires occurring in sulphur cargoes are smothered by use of more sulphur. Personnel should be issued with personal protection equipment when loading or discharging a sulphur cargo – i.e. masks and goggles.

Nuts tend to have a high oil content and they are liable to heat and deteriorate. They should be kept dry. Precautions should be taken to prevent shifting, as per the Grain Rules.

Copra is dried coconut flesh, mainly from Malaysia. It is liable to spontaneous heating and is highly inflammable. It is suggested that cargo thermometers are rigged to monitor temperatures in the bulk. Tight anti-fire regulations should prevail around the cargo spaces, to include spark arrester gauze in place on ventilator apertures. The cargo should be kept dry and kept clear of surfaces that are liable to 'sweat'. Matting is recommended to cover the ship's steelwork for this purpose.

Salt has a high moisture content which is likely to evaporate and dry goods should not be stowed in close proximity. Prior to loading, the spaces should be clean and dry. The steelwork may be whitewashed and separation cloths may be used to keep salt off the ship's structures.

Sugar – Vessels have been specifically built for the bulk sugar trade. They are of a similar construction to those of the bulk ore carrier. This is not to say that bulk sugar cannot be carried in any other general type cargo vessel. In any event the compartments should be thoroughly cleaned out and the bilge bays made sugar-tight.

Bulk sugar must be kept dry. If water is allowed to enter by any means it would solidify the cargo and result in the product being condemned.

MAIN HAZARDS OF LOADING/SHIPPING/DISCHARGING BULK CARGOES

Dry shift of cargo – Caused by a low angle of repose and can be avoided by trimming level or the use of shifting boards.

Wet shift of cargo – Caused by liquefaction of the cargo, possibly due to moisture migration causing the cargo to act like a liquid, the moisture content of the product probably being below the transportable moisture limit.

Oxidation – The removal of oxygen from the cargo compartment by the type and nature of the cargo. Ventilation is required before entry into the compartment.

Flammable/explosive gas/dust – The nature of the cargo has a high risk and may be of a highly flammable nature, or give off explosive gases. Dusty cargoes also run the risk of a dust explosion in the atmosphere inside the compartment.

Toxic gas or dust – Identified toxic effects from products may well require personnel to wear protective clothing and masking/breathing equipment when in proximity to the product.

Corrosive elements (e.g. sulphur) – Personnel will require protective clothing and eye protective wear. Some products have a high fire risk.

Spontaneous combustion – A self-heating cargo which needs to be monitored by the use of cargo and hatch thermometers throughout the period of the voyage. It should be stowed clear of machinery space bulkheads and provided with recommended ventilation where appropriate – e.g. coal: surface ventilation; bagged fishmeal: channel ventilation inside stow.

Reaction cargoes – Products that may react with other cargoes, and as such may require separate stowage compartments.

High density cargoes – May cause structural damage to the vessel and pose stability problems from the position of stow. Could well affect bending and shear force stress effects on the hull.

Infectious cargoes (e.g. guano) – Exposed personnel would require personal protection inclusive of respirators.

Structural damage – Through excessive bending and shear forces caused by poor distribution of and/or inadequate trimming of certain cargoes, or sailing with partly filled holds or empty holds.

ASCERTAINING TOTAL CARGO LOAD

Many cargo ships, especially the large bulk carriers, will have a general idea on how much cargo has been loaded at a specific port. However, to obtain the exact displacement by carrying out a draught survey would provide a more determined figure. A draught survey would also ascertain any bending experienced throughout the ships length.

Following a draught survey, the true mean draught can be used with the ships hydrostatic tables and the deadweight scale. Bulk carriers can then determine a more exact deadweight for the loaded vessel and be reassured with total cargo values on board.

DRAUGHT SURVEY FORMAT

Port .... Surveyor..................................... Date..............

General Particulars	Hydrostatic Particulars	
Ships Name	Displacement (from table)	Df (P)
Port of Registry	Density Correction	Df (S)
Official Number	Displacement total	Df (M)
Masters Name	Lightship	FPP corr'n (t x d÷LBP)
Loading Port	Heavy fuel	Df @ FPP
Discharging Port	Diesel Fuel	Da (P)
Cargo Particulars	Lub Oils	Da (S)
Moisture Content	Dirty Oils	Da (M)
Stowage Factor	Fresh Water	APP corr'n (t x d÷LBP)
Dock Water R.D.	Ballast Water	Da @ APP
	Stores	D mid (P)
	Constant	D mid (S)
	Total Deductions	D mid (M)
	Cargo Figure	Trim (Da APP – Df FPP)
		Dm=(T.Da ~ T.Df) x ½
		Corr'n to Dm*
		True Dm

*Correction = [D mid (M) ~ Dm] x 0.75
Add correction if D mid (M) > Dm. Subtract correction if D mid (M) < Dm.

NB. To correct draughts observed or draughts at PP ~ Correction

$$= t \times d \div LBP.$$

ADD if trimmed towards the perpendicular, where 'd ' is the horizontal distance (m) between draught marks at PP.

FIGURE 5.21 Draught survey format.

Displacement @ corrected Dm	Hydrostatic saltwater scale reading	Tonnes
Trim Correction	TPC(corrected) x layer corr'n +/-	Tonnes
Nemoto's Correction	Trim² x 50 x "Z" ÷ LBP +	Tonnes
Displacement fully corrected (SW)		Tonnes
R.D. Correction	W x R.D. (DW) ÷ 1.025	Tonnes
Total Displacement (Dock Water)		Tonnes

FIGURE 5.21 (*Continued*)

"Z" = MCTC @ True Dm + 0.5m ~ MCTC @ True Dm - 0.5m

Draught surveys are usually carried out by a cargo surveyor for ships carrying bulk commodities, in order to determine the exact loaded draught. This figure can then be used to assess the vessels true deadweight, providing a specific cargo tonnage on board.

ADDITIONAL READING AND REFERENCES FOR BULK CARGOES

The International Marine Solid Bulk Cargoes Code (IMSBC Code)

Code of Practice for the Safe Loading and Unloading of Bulk Cargoes (BLU Code)

IBC Code for the International standards for safe carriage in bulk by sea of dangerous chemicals and noxious liquid substances.

IMO International Code for the safe carriage of grain in bulk.

ADDITIONAL REFERENCES FOR BRITISH REGULATIONS

MGM 108 (M) – Hull Stress Monitoring Systems

MGN 144 (M) – The Merchant Shipping (additional safety measures for Bulk Carriers) Regulations 1999. S.I.1999/1644.

MGN 511(M) – Solid Bulk Cargoes: Adoption of amendment 02–13 to the IMSBC Code.

MGN 512(M) – Solid Bulk Cargoes: Guidelines for the submission of information and completion of the format for the properties of cargoes not listed in IMSBC Code.

MGN 513(M) – Solid Bulk Cargoes: Guidelines for the developing and approving procedures for sampling, testing and controlling the moisture content for solid bulk cargoes which may liquefy.

S.I 1999 No. 336, The Merchant Shipping (Carriage of Cargoes) Regulations 1999.

CHAPTER 6

Tanker Cargoes

INTRODUCTION

At the present time modern civilisation still remains largely dependent on oil and its by-products. Vast quantities of liquid products are transported by tankers throughout the world and as such they have a high profile in the eyes of the general public. However, it should be realised from the outset that not all tankers are in the oil trade. Many transport wine or liquid chemicals, or liquid natural gas (LNG), but generally the tanker vessel is synonymous with the carriage of bulk oil or oil-based products.

Concern for the environment associated with tanker traffic has become a number one priority in the anti-pollution campaign, and rightly so. The marine industry must respect the environment and the well-being of the planet on which we all exist. To this end the MARPOL Convention has gone some way to establish standards of oil operations around the globe.

The main concern with the demands of a modern society has always been the cost of pollution scaled against society's need for oil. Those countries that have oil need to go to market to strengthen their national economy, while those that are without oil need to import it to strengthen their economy. Clearly, this is part of the endless cycle of world trade and economics.

Unfortunately, the tanker accident is not unheard of: the *Amoco Cadiz*, the *Torrey Canyon*, the *Exxon Valdez* and the *Sea Empress* are hard examples to live with. Oil tankers and their product have two main problems, one being pollution and the other being from the danger of fire.

Our seafarers must be educated not only to the public outrage that accompanies poor seamanship which generates most modern day accidents, but also to the ways that prevent such catastrophes happening in the first place. The training of all our seafarers, especially tanker personnel, is an aspect of the marine industry which must take precedence within an industry which continues to drill for oil in the deepest and most remote quarters of the earth.

Note: 'Global Warming' is impacting on many industries including Civil Aviation and shipping. Cleaner, greener fuels like 'Hydrogen' are seemingly moving to replace fossil fuels, which can expect to effect the tanker trades, in time. However, pressures to generate a cleaner environment are increasing and impacting on all the oil companies.

DOI: 10.4324/9781003407706-6

DEFINITIONS FOR USE (WITHIN THE UNDERSTANDING OF MARPOL) AND TANKER OPERATIONS (GAS AND CHEMICAL)

Administration – The government of the state under whose authority the ship is operating.

Ammonia (NH$_3$) – A pressurised liquid (toxic) gas. Used as a carrier of Hydrogen. It is colourless and with a strong sharp unpleasant smell. Human contact would experience skin irritation to eyes, nose and throat.

Associated piping – The pipeline from the suction point in a cargo tank to the shore connection used for unloading the cargo and includes all the ship's piping, pumps and filters which are in open connection with the cargo unloading line.

Bulk chemical code – The code for the construction and equipment of ships carrying dangerous chemicals in bulk. (Ships must have a Certificate of Fitness for the carriage of dangerous chemicals.)

Cargo area – That part of a ship which contains cargo spaces, slop tanks and pump rooms, cofferdams, ballast and void spaces adjacent to cargo tanks and also deck areas throughout the length and breadth of the part of the ship over such spaces.

Centre tank – Any tank inboard of a longitudinal bulkhead.

Chemical tanker – A ship constructed or adapted primarily to carry a cargo of noxious liquid substances in bulk and includes an oil tanker as defined by Annex I of MARPOL, when carrying a cargo or part cargo of noxious liquid substances in bulk. (See also Tanker)

Clean ballast – Ballast carried in a tank which since it was last used to carry cargo containing a substance in Category A, B, C or D has been thoroughly cleaned and the residues resulting therefrom have been discharged and the tank emptied in accord with Annex II of MARPOL.

Cofferdam – An isolating space between two adjacent steel bulkheads or decks. This space may be a void space or a ballast space.

Combination carrier – A ship designed to carry either oil or solid cargoes in bulk.

Continuous feeding – The process whereby waste is fed into a combustion chamber without human assistance while the incinerator is in normal operating condition with the combustion chamber operative temperature between 850°C and 1200°C.

Critical structural areas – Locations which have been identified from calculations to require monitoring or from service history of the subject ship or from similar or sister ships to be sensitive to cracking, buckling or corrosion, which would impair the structural integrity of the ship.

Crude oil – Any liquid hydrocarbon mixture occurring naturally in the earth, whether or not treated to render it suitable for transportation, and includes:

a   Crude oil from which certain distillate fractions may have been removed.

b   Crude oil to which certain distillate fractions may have been added.

Dedicated ship – A ship built or converted and specifically fitted and certified for the carriage of:

a   One named product.

b   A restricted number of products each in a tank or group of tanks such that each tank or group of tanks is certified for one named product only or compatible products not requiring cargo tank washing for change of cargo.

Domestic trade – A trade solely between ports or terminals within the Flag State, which the ship is entitled to fly, without entering the territorial waters of any other state.

Discharge – In relation to harmful substances or effluent containing such substances, means any release howsoever caused from a ship and includes any escape, disposal, spilling, leaking, pumping, emitting or emptying.

Emission – Any release of substance subject to control by Annex VI, from ships into the atmosphere or sea.

Flammability limits – The conditions defining the state of fuel oxidant mixture at which application of an adequately strong external ignition source is only just capable of producing flammability in a given test apparatus.

Flammable products – Those products identified by an 'F' in column 'F' of the table in Chapter 19 of the International Gas Code (IGC).

Flashpoint (of an oil) – The lowest temperature at which the oil will give off vapour in quantities that when mixed with air in certain proportions are sufficient to create an explosive gas.

Garbage – All kinds of victual, domestic and operational waste, excluding fresh fish and parts thereof, generated during the normal operation of the ship and liable to be disposed of continuously or periodically, except those substances which are defined or listed in other Annexes to the present Convention.

Gas carrier – A cargo ship constructed or adapted and used for the carriage in bulk of any liquefied gas or other products listed in the table in Chapter 19 (IGC) Code.

Good condition – A coating condition with only minor spot rusting.

Harmful substance – Any substance which, if introduced into the sea, is liable to create hazards to human health, to harm living resources and marine life, to damage amenities or to interfere with legitimate use of the sea, and includes any substance subject to control by the present Convention.

Hold space – The space enclosed by the ship's structure in which a cargo containment system is situated.

Holding tank – A tank used for the collection and storage of sewage.

Hydrogen (H_2) – A nontoxic gas colourless, odourless, and tasteless. Now seen as a future fuel replacement for fossil fuels. Manufacturing is now developing Green Hydrogen (NH_3) as a cheaper fuel option.

IBC Code Certificate – Refers to an international Certificate of Fitness for the Carriage of Dangerous Chemicals in Bulk, which certifies compliance with the requirements of the IBC Code.

IGC Code – Refers to the International Code for the Construction and Equipment of ships carrying liquefied gases in bulk.

Ignition point (of an oil) – This is defined by the temperature to which an oil must be raised before its surface layers will ignite and continue to burn.

Incident – Any event involving the actual or probable discharge into the sea of a harmful substance, or effluents containing such a substance.

Instantaneous rate of discharge of oil content – The rate of discharge of oil in litres per hour at any instant divided by the speed of the ship in knots at the same instant.

International trade – A trade which is not a domestic trade as defined above.

Liquid substances – Those having a vapour pressure not exceeding 2.8 kPa/cm² when at a temperature of 37.8°C.

MARVS – The maximum allowable relief valve setting of a cargo tank.

Miscible – Soluble with water in all proportions at wash water temperatures.

NLS Certificate – An international Pollution Prevention Certificate for the Carriage of Noxious Liquid Substances in Bulk, which certifies compliance with Annex II of MARPOL.

Noxious Liquid Substance – Any substance referred to in Appendix II of Annex II of MARPOL, or provisionally assessed under the provisions of Regulation 3(4) as falling into Category A, B, C or D.

NO$_x$ Technical Code – The Technical Code on Control of Emission of Nitrogen Oxides from Marine Diesel Engines, adopted by the Conference, Resolution 2, as may be amended by the organisation.

Oil – Petroleum in any form, including crude oil, fuel oil, sludge oil refuse and refined products (other than petrochemicals which are subject to the provisions of Annex II).

Oil fuel unit – The equipment used for the preparation of oil fuel for delivery to an oil-fired boiler, or equipment used for the preparation for delivery of heated oil to an internal combustion engine, and includes any oil pressure pumps, filters and heaters with oil at a pressure of not more than 1.8 bar gauge.

Oily mixture – A mixture with any oil content.

Oil tanker – A ship constructed or adapted primarily to carry oil in bulk in its cargo spaces, including combination carriers and any 'Chemical Tanker' as defined by Annex II, when it is carrying a cargo or part cargo of oil in bulk.

Organisation – The International Maritime Organization (IMO), formerly known as the Inter-Governmental Maritime Consultative Organization (IMCO).

Permissible Exposure Limit (PEL) – An exposure limit which is published and enforced by the Occupational Safety and Health Administration (OSHA) as a legal standard. It may be either time-weighted-average (TWA) exposure limit (8 hours), or a 15-minute short term exposure limit (STEL), or a ceiling (C).

Primary barrier – The inner element designed to contain the cargo when the cargo containment system includes two boundaries.

Product carrier – An oil tanker engaged in the trade of carrying oil other than crude oil.

Residue – Any noxious liquid substance which remains for disposal.

Residue/water mixture – Residue in which water has been added for any purpose (e.g. tank cleaning, ballasting, bilge slops).

Secondary barrier – The liquid-resisting outer element of a cargo containment system designated to afford temporary containment of any envisaged leakage of liquid cargo through the primary barrier and to prevent the towering of temperature of the ship's structure to an unsafe level.

Segregated ballast – Ballast water introduced into a tank which is completely separated from the cargo oil and fuel oil system and which is permanently allocated to the carriage of ballast or to the carriage of ballast or cargoes other than oil or noxious substances.

Sewage –

a   Drainage and other wastes from any form of toilet, urinals and WC scuppers.

b   Drainage from medical premises (dispensary, sick bay etc.) via washbasins, wash tubs and scuppers located in such premises.

c   Drainage from spaces containing living animals.

d   Other waste waters when mixed with drainage as listed above.

Ship – A vessel of any type whatsoever operating in the marine environment, including hydrofoil boats, air cushion vehicles, submersibles, floating craft and fixed or floating platforms.

Shipboard incinerator – A shipboard facility designed for the primary purpose of incineration.

Slop tank – A tank specifically designated for the collection of tank draining's, tank washings and other oily mixtures.

Sludge oil – Sludge from the fuel or lubricating oil separators, waste lubricating oil from main or auxiliary machinery, or waste oil from bilge water separators, oil filtering equipment or drip trays.

SO_x (oxides of Sulphur) emission control area – An area where the adoption of special mandatory measures for SO_x emissions from ships is required to prevent, reduce and control air pollution from SO_x and its attendant adverse impacts on land and sea areas. SO_x emission control areas shall include those listed in Regulation 14 of Annex VI.

Special area – A sea area where, for recognised technical reasons in relation to its oceanographically and ecological condition and to the particular character of its traffic, the adoption of special mandatory methods for the prevention of sea pollution by oil is required. Special areas include: the Mediterranean Sea, the Baltic Sea, the Black Sea, the Red Sea, the Gulf Area, the Gulf of Aden, the North Sea, the English Channel and its approaches, the Wider Caribbean Region and Antarctica.

Substantial corrosion – An extent of corrosion such that the assessment of the corrosion pattern indicates wastage in excess of 75% of the allowable margins, but within acceptable limits.

Suspect areas – Locations showing substantial corrosion and/or considered by the attending surveyor to be prone to rapid wastage.

Tank – An enclosed space which is formed by the permanent structure of the ship and which is designed for the carriage of liquid in bulk.

Tank cover – The protective structure intended to protect the cargo containment system against damage where it protrudes through the weather deck or to ensure the continuity and integrity of the deck structure.

Tank dome – The upward extension of a position of a cargo tank. In the case of below deck cargo containment system the tank dome protrudes through the weather deck or through a tank covering.

Tanker – An oil tanker as defined by the regulation 1 (4) of Annex 1, or a chemical tanker as defined in regulation 1 (1) of Annex II of the present Convention (see Figs 6.3, 6.4).

Threshold Limit Value (TLV) – Reference to the British Occupational Hygiene Society (BOHS) for level concentrations. (US: Airborne concentrations of substances devised by the American Conference of Government Industrial Hygienists (ACGIH)). Representative of conditions under which it is believed that nearly all workers may be exposed day after day with no adverse effects. There are three different types of TLV: Time weighted average, Short-term exposure limit and Ceiling.

Note: TLVs are advisory exposure guidelines, not legal standards and are based on evidence from industrial experience and research studies.

Time Weighted Average (TWA) – The average time over a given work period (e.g. 8-hour working day) of a person's exposure to a chemical or an agent. The average is determined by sampling for the containment throughout the time period and represented by (TLV – TWA).

Toxic products – Those products identified by a 'T' in column 'F' in the table of Chapter 19 (IGC) Code.

Ullage – The measured distance between the surface of the liquid in a tank and the underside decking of the tank.

Vapour pressure – The equilibrium pressure of the saturated vapour above the liquid expressed in bars absolute at a specified temperature.

Void space – An enclosed space in the cargo area external to a cargo containment system, other than a hold space, ballast space, fuel oil tank, cargo pump or compressor room, or any space in normal use by personnel.

Volatile liquid – A liquid which has a tendency to evaporate quickly and has a flash point of less than 60°C.

Wing tank – Any tank which is adjacent to the side shell plating.

FIGURE 6.1 The view of the Single Buoy Mooring (SBM) through the pipe lead of a tanker as it approaches to pick up the SBM securing. The floating pipeline is seen on the surface leading from the buoy float.

FIGURE 6.2 Tanker moorings and floating pipeline seen prior to liquid cargo transfer taking place from the Floating Storage Unit (FSU) to the tanker.

TANKER STRUCTURE

The size and sophistication of the modern tanker has changed considerably over the decades. World economics have influenced the capacity, while legislation has changed all future construction into the double hull category.

FIGURE 6.3 The *Jahre Viking* at 564,000 dwt (IMO No. 7381154) was the largest man-made transport in the world. It is seen manoeuvring with tugs off the Dubai Dry Dock. This vessel has been converted to a floating oil storage unit to prolong its active life.

The Design of the Oil Tanker

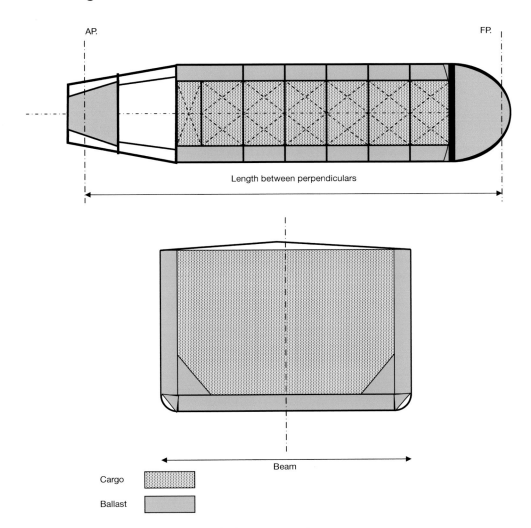

FIGURE 6.4 Double hull design example for 150,000 dwt.

Double Hull Tanker Construction

Tankers were generally constructed with either centre tanks or wing tanks dividing the vessel into three athwartships sections by two longitudinal bulkheads, individual tanks being segregated by transverse bulkheads. Modern construction which integrates the double hull has meant that construction designs have changed and twin tanks are now positioned to either side of a centre line bulkhead.

The maximum length of an oil tank is 20% L (L represents the ship's Length) and there is at least one wash bulkhead if the length of the tank exceeds 10% L or 15m.

It should be appreciated that in a large tanker of 300 m length and 30 m beam and equivalent depth, each tank would have a capacity for over 20,000 tonnes of oil.

Tanks are usually numbered from forward to aft with pump rooms usually situated aft so that power can be easily linked direct from the engine room. Pipeline systems provide flexibility in loading/discharging, interconnecting the tanks to the pumping arrangement.

FIGURE 6.5 Athwartships cross-section of the modern double hull tanker seen at a late construction stage prior to assembly.

Equipment Regulation Requirements

Tankers now require:

Cargo Tank Pressure Monitoring systems, required under SOLAS II–2, Regulation 59/ IBC Code, Chapter 8.3.3 to be fitted after the first dry docking after 1 July, between 1998 and 2002. New build vessels would be similarly equipped.

Cargo Pump Bearing Temperature Monitoring systems must be fitted under SOLAS II–2, Regulations 4 and 5.10.1 at the next dry docking after 1 July 2002.

Cargo Pump Gas Detection/Bilge Alarm systems are now required under SOLAS II–2, Regulations 4 and 5.10.3/5.10.4 to be fitted at the next dry docking after 1 July 2002.

High Level and Overfill Alarm system, is now required under USCG Regulation 39.

Emergency Escape Breathing Devices (EEBDs) are now required under SOLAS II–2, Regulation 13.3.4 by the first survey after 1 July 2002.

Emergency Towing Arrangements are now required for all tankers over 20,000 grt. This requirement is effective from January 2010 and must be fitted both in fore and aft positions and be capable of rapid deployment.

FIGURE 6.6 The *Seoul Spirit* (IMO No. 9248409) VLCC tanker seen in the Lisnave shipyard and dry docking complex in Portugal following hull coating and general maintenance.

TANKER PIPELINES

There are three basic types of pipeline system:

1 Direct system.
2 Ring main system.
3 Free flow system.

Each system has their uses and is designed to fulfil a need in a particular type of vessel.

Direct System

This is the simplest type of pipeline system which uses fewer valves than the others. It takes oil directly from the tank to the pump and so reduces friction. This has an effect of increasing the rate of discharge and at the same time improves the tank suction. It is cheaper to install and maintain than the ring main system because there is less pipeline length set, with fewer valves, so less likelihood of malfunction. However, the layout is not as versatile as a ring main system and problems in the event of faulty valves or leaking pipelines could prove more difficult to circumvent. Also the washing is more difficult since there is no circular system and the washings must be flushed into the tanks.

The advantages are:

1   It is easy to operate and less training of personnel is required.
2   As there are fewer valves it takes less time to set up the valve system before commencing a cargo operation.
3   Contamination is unlikely, as it is easy to isolate each section.

The disadvantages are:

1   It is a very inflexible system, which makes it difficult to plan for a multi-port discharge.
2   Block stowage has to be used, which makes it difficult to control trim.
3   Carrying more than three parcels concurrently can be difficult.

Direct Line System

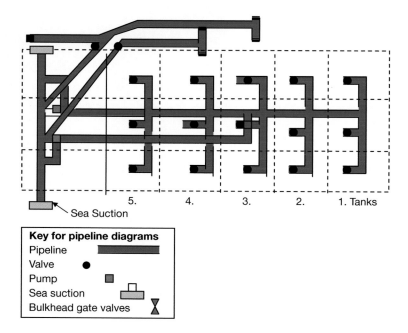

FIGURE 6.7 Direct line system. Used mainly on crude and black oil tankers where separation of grade oils is not as important.

Ring Main System

This is basically a ring from the pump room around the ship, with crossover lines at each set of tanks. There are various designs usually involving more than one ring. It is extensively employed on Product Tankers where the system allows many grades of cargo to be carried without contamination. This is a highly versatile system which permits several different combinations of pump and line for any particular tank.

The advantages are:

1   Cargoes can be more easily split into smaller units and placed in various parts of the ship.
2   Line washing is more complete.
3   A greater number of different parcels of cargo can be carried.
4   Trim and stress can be more easily controlled.

The disadvantages are:

1   Because of the more complicated pipeline and valve layout, better training in cargo separation is required.
2   Contamination is far more likely if valves are incorrectly set.
3   Fairly low pumping rates are achieved.
4   Costs of installation and maintenance are higher because of more pipeline and an increased number of valves.

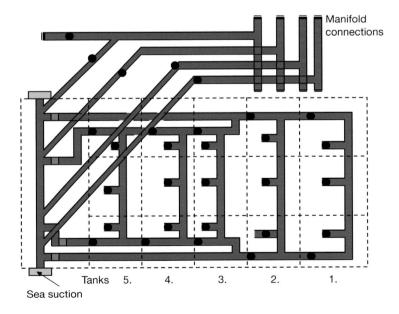

FIGURE 6.8 Ring main system pump room aft.

The Free Flow System

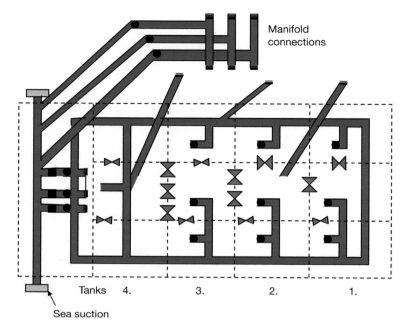

FIGURE 6.9 The free flow system.

The Free Flow System employs sluice valves in the tank bulkheads rather than pipelines. With a stern trim this system can discharge all the cargo from the aftermost tank via direct lines to the pump room. The result is that a very high speed of discharge can be achieved and, as such, the system is suitable for large crude carriers with a single grade cargo. Tank drainage is also very efficient since the bulkhead valves allow the oil to flow aft easily. There are fewer tanks with this system and it has increased numbers of sluice valves the farther aft you go. The increased number of sluices is a feature to handle the increased volume being allowed to pass from one tank to another.

The main advantage is that a very high rate of discharge is possible with few pipelines and limited losses to friction. The main disadvantage is that overflows are possible if the cargo levels in all tanks are not carefully monitored.

TANKER DECK ARRANGEMENT

FIGURE 6.10 Typical example of the pipeline and ventilation arrangements (fore deck) of a medium-sized oil tanker seen in the seagoing environment.

Manifold and Pipeline Connections

Most tanker vessels now have designated lifting gear in the form of derricks or deck cranes to carry out the hoisting and lifting of pipelines to respective connections. These derricks or cranes are usually positioned in way of the 'manifolds' to compliment the pick-up and connection. Oil pipe lines have increased in size and weight since the days of the early small tank vessel. They are considered heavy duty and difficult to manhandle when loading and securing to the ships manifold connections.

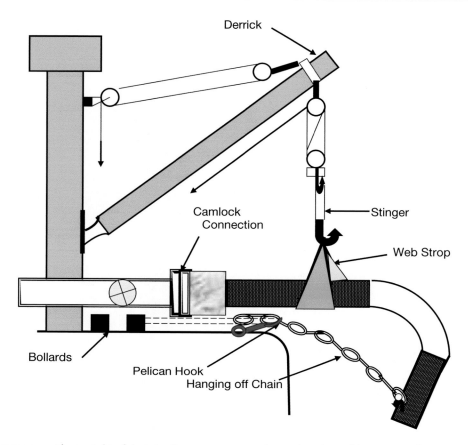

FIGURE 6.11 The weight of the pipeline connection is hoisted on board by means of a 'Hanging Off Chain'. A heavy duty wire, known as a 'stinger' accepts the end weight in conjunction with web strops.

FIGURE 6.12 Upper deck of an oil tanker showing the manifold connections, seen inside the Lisnave dry dock complex.

FIGURE 6.13 Deck pipeline example on the upper deck of an oil tanker.

Manifold and Pipeline Connection

FIGURE 6.14 Manifold and pipeline connection. Typical 23-inch oil pipe connection.

MEASUREMENT OF LIQUID CARGOES

The volume of oil in a tank is ascertained by measuring the distance from a fixed point on the deck to the surface of the oil. The distance is known as the 'ullage' and is usually measured by means of a plastic tape. A set of tables is supplied to every ship, which indicate for each cargo compartment the volume of liquid corresponding to a range of ullage measurements. The ullage opening is usually set as near as possible to the centre of the tank so that for a fixed volume of oil, the ullage is not appreciably affected by conditions of trim and list. If a favourable siting is not possible then the effects of trim and list should be allowed for.

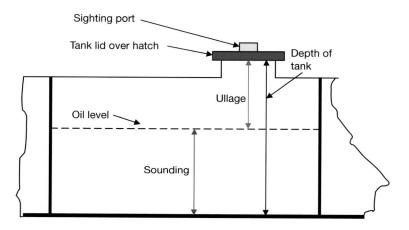

FIGURE 6.15 Measurement of liquid cargoes.

The important measure of oil is weight and this must be calculated from the volume of oil in each tank. Weight in tonnes is quickly found by multiplying the volume of oil in cubic metres by the relative density of the oil. This density is a fraction and may be taken out of petroleum tables when the relative density of the oil is known.

Example

To find the weight of 125 m³ of oil at a Relative Density of 0.98:

Density of oil = 0.98 t/m³
Weight of oil = volume x density
              = 125 m³ x 0.98 tonnes
              = 122.5 tonnes.

Oil expands when heated and its Relative Density (RD) therefore decreases with a rise in temperature. In order that the weight may be calculated accurately, it is important that when ullages are taken the RD of the oil should also be known. This may be measured directly, by means of a hydrometer.

The RD of a particular oil may be calculated if the temperature of the oil is taken. The change of RD due to a change of one degree in temperature is known as the RD coefficient. This lies between 0.0003 and 0.0005 for most grades of oil and may be used to calculate the RD of an oil at any measured temperature, if the RD at some standard temperature is known.

Examples

A certain oil has a RD of 0.75 @ 16°C

Its expansion coefficient is 0.00027/°C
Calculate its RD at 26°C

Temperature difference = 26°C – 16°C = 10°C
Change in RD   = 10 x 0.00027
         = 0.0027
RD @ 16°C   = 0.75
RD @ 26°C   = 0.7473

An oil has a RD 0.75 @ 60°F
Its expansion coefficient is 0.00048/°F
Calculate its RD at 80°F

Temperature difference = 80°F – 60°F = 20°F
Change in RD   = 20 x 0.00048
         = 0.0096
RD @ 60°F   = 0.75
RD @ 80°F   = 0.7404

TANK MEASUREMENT AND ULLAGING

Use of the Whessoe Tank Gauge

The function of the gauge is to register the ullage of the tank at any given time, in particular when the liquid level in the tank is changing during the loading and discharge periods. The gauge is designed to record the readings not only at the top deck level of the tank but also remotely at a central cargo control room. A transmitter is fitted on the head of the gauge for just this purpose.

The unit is totally enclosed and various models manufactured are suitable for use aboard not only oil tankers, but also chemical and gas carriers.

Inside the gauge housing is a calibrated ullage tape, perforated to pass over a sprocket wheel and guided to a spring-loaded tape-drum. The tape extends into the tank and is secured to a float of critical weight. As the liquid rises or falls, the tape is drawn into or extracted out from the drum at the gauge head.

The tape-drum, being springloaded, provides a constant tension on the tape, regardless of the amount of tape paid out. A counter window for display is fitted into the gauge head, allowing the ullage to be read on site at the top of the tank.

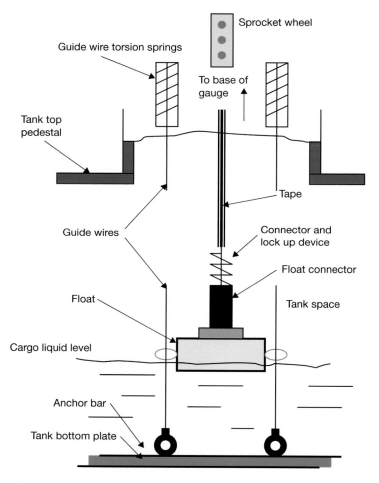

FIGURE 6.16 Whessoe tank gauge, working principle.

Tank Measurement – Radar System

This is a totally enclosed measuring system which can only be employed if the tank is fully inerted. Systems are generally fitted with oxygen sensor and temperature sensor switches, so if the atmosphere in the tank is hot or flammable the radar will not function.

The main unit of the system is fitted on the deck with an inserted cable tube into the tank holding a transducer. A cable then carries the signal to a control unit in the cargo control room, where the signal is converted to give a digital read-out for each tank monitored.

The transducer would be fitted as close to the centre of the tank area as was possible. Such siting tends to eliminate errors due to trim and list.

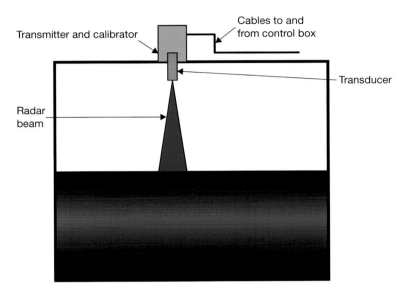

FIGURE 6.17 The transducer would be fitted as close to the centre of the tank area as physically possible. Such siting tends to eliminate errors due to trim and list.

LOADING

Loading of tankers takes place at jetties or from Floating Storage Units (FSUs) or from Single Point Moorings (SPMs). FSUs are often fitted with booms carrying oil-bearing pipes ready to be connected. These pipes and fittings are usually insulated to prevent stray currents flowing, as from corrosion prevention systems employed on both ships and jetties. The flow of current in itself should not be a problem, but it may give rise to a spark when making or breaking connections to the manifold. For this reason, these sections are tested regularly for efficient insulation. Lines are often bonded to reduce static electricity effects which could also give rise to an unwanted source of ignition from fast pumping of liquids.

These points are highlighted to illustrate that a high degree of awareness is required in all tanker operations whether loading, discharging, or gas-freeing. Fire precautions are paramount because the risk of fire aboard the tanker is a real hazard and stringent fire precautions must be adopted throughout cargo operations of every kind.

LOADING PROCEDURAL CHECK LIST

Company policy on loading procedures vary and Cargo Officers should adhere to the company procedures and take additional reference from the International Safety Guide for Oil Tankers and Terminals (ISGOTT).

1   Complete and sign the ship/shore checklist.
2   Establish an agreed communication network.
3   Agree the loading plan by both parties and confirm in writing.
4   Loading and topping off rates agreed.
5   Emergency stop procedures and signals agreed.
6   All affected tanks, lines, hoses inspected prior to commencing operations.
7   Overboard valves sealed.
8   All tanks and lines fully inerted.
9   IG system shut down.
10 Pump room isolated and shut down.
11 Ship's lines set for loading.
12 Offside manifolds shut and blanked off.
13 All fire-fighting and SOPEP equipment in place.
14 Notice of readiness accepted.
15 First set of tanks and manifold valves open.
16 Commence loading at a slow rate.
17 Check and monitor the first tanks to ensure cargo is being received.
18 Carry out line sample.
19 Check all around the vessel and overside for leaks.
20 Increase loading rate to full.
21 Check ullages at half-hourly intervals and monitor flow rate to confirm with shoreside figures.
22 Check valves operate into next set of tanks prior to changeover.
23 Reduce loading rate when topping off final tank.
24 Order stop in ample time to achieve the planned ullage/line draining.
25 When the cargo flow has completely stopped, close all valves.
26 After settling time, take ullages, temperatures and samples.
27 Ensure all log book entries are completed.
28 Cause an entry to be made into the Oil Record Book.

Note: The loading plan devised by Chief Officers and shoreside authorities would take account of the ship's stability and the possibility of stresses being incurred during all stages of the loading procedure.

LOADING CAPACITY

The amount of cargo a tanker can lift will depend upon the vessel's deadweight when the vessel is floating at her designated loadline. The amount of bunkers, fresh water and stores would be deducted to give the total weight of cargo on board. The order of loading tanks is of high priority in order to avoid excessive stresses occurring. Visible damage might not be an immediate result of a poor loading sequence but subsequent damage may be caused later, when in a seaway, which could be attributable to excessive stresses during loading periods.

Nowadays vessels are equipped with designated 'loadicators' or computer software programmes to establish effective loading plans and show shear forces and bending moments throughout the ship's length. Such aids are beneficial to ships' officers in illustrating immediate problems and permitting ample time to effect corrective action.

Although a high rate of loading is usually desirable, this in itself generates a need for tight ship-keeping. Moorings will need to be tended regularly and an efficient gangway watch should be maintained. Communications throughout the loading period should be effective and continuous with shoreside authorities, with adequate notice being given to the pumping station prior to 'topping off'.

CARE DURING TRANSIT

It would be normal practice that through the period of the voyage regular checks are made on the tank ullage values and the temperatures of all tanks. Empty tanks and cofferdams together with pump rooms should be sounded daily to ensure no leakage is apparent. Generally oil is loaded at a higher temperature than that which will be experienced at sea; as such, it would be expected that the oil will cool and the ullage will increase for the first part of the voyage.

Viscous oils like fuel oil or heavy lubricating oil would normally be expected to be heated for several days before arrival at the port of discharge. Heating will decrease the viscosity and a higher rate of discharge can be anticipated. Overheating should be avoided as this could affect the character of the product and may strain the structure of the vessel.

Tanks are vented by exhaust ventilators above deck level via masts and Samson posts. Volatile cargoes such as gasoline are vented via pressure relief valves which only operate when the tank pressure difference to atmosphere exceeds 0.14 kg/cm^2. This prevents an excessive loss of cargo due to evaporation. Evaporation of cargo can also be reduced in hot weather by spraying the upper decks cool with water.

DISCHARGING

Flexible hoses are connected to the ship's manifold, as at the loading port, and the ship to shore checklist would be completed. Good communication between the ship and the shore authority is essential. All overboard discharges should be checked and if all valves are correct, discharge would be commenced at an initial slow rate. This slow rate is commenced to ensure that if a sudden rise in back pressure is experienced in the line, the discharge can be stopped quickly. Such an experience would probably indicate that the receiving lines ashore are not clear.

Back pressure should be continually monitored during discharge operations and the ship, using ship's pumps, should be ready to stop pumping at short notice on a signal from the terminal. The waterline around the ship should also be kept under regular surveillance in the event of leakage occurring.

As with loading operations, the deck scuppers should all be sealed and SOPEP and fire-fighting equipment should be kept readily available throughout the operation.

BALLASTING

In order that no oil is allowed to escape into the sea when engaging in ballast operations, the pumps should be started before the sea valves are opened. If it is intended to ballast by gravity, it is still preferable to pump for the first ten minutes or so to ensure that no oil leaks out.

Care should also be taken when topping up ballast tanks, since any water overflow could be contaminated with oil. Any gas forced out of tanks during ballast operations constitutes a fire risk as equally dangerous as when loading.

All ballast operations should be recorded in the Ballast Management Record Book and any transfers of oil content should be recorded in the Oil Record Book. Log books should take account of all tank operations regarding loading, discharging, ballasting or cleaning.

TANK CLEANING METHODS

There are generally three methods of cleaning tanks:

a   Bottom flushing with water, petroleum product or chemical solvents.
b   Water washing (hot or cold) employing tank-washing machinery.
c   Crude oil washing (COW).

a)   **Bottom washing** – Bottom flushing is usually carried out to rid the tank bottoms of previous cargo prior to loading a different, but compatible, grade of cargo. It can be effective when carrying refined products in small quantities. Bottom washing with acceptable solvents is sometimes conducted, especially where a tanker is to take, say, paraffin (kerosene) products after carrying leaded gasolines.

   It should be realised that bottom washing will not remove heavy wax sediments from the bottom of tanks and is used purely as a means of removing the traces of previous cargo.

b)   **Portable or fixed washing machines** – Using a high-pressure pump and heater, sea water, via a tank-cleaning deck line, is used to wash the tank thoroughly. The dirty slop water is then stripped back to the slop tank where it is heated to separate the oil from the water. This is considered an essential method when changing trades from carrying crude to the white oil cargoes, or when the tank is required for clean ballast, or if it is to be gas-freed.

c) **Crude oil washing** – A procedure that is conducted during the discharge which has positive advantages over water washing methods. All crude oil carriers over 40,000 dwt tonnes must now be fitted and use a crude oil washing facility.

The method employs a high-pressure jet of crude oil from fixed tank-cleaning equipment. The jet is directed at the structures of the tank and ensures that no slops remain on board after discharge, every last drop of cargo going ashore. The advantages are that tank cleaning at sea is avoided, with less likelihood of accidental pollution, less tank corrosion is experienced than from water washing, increased carrying capacity for the next cargo, full tank drainage is achieved and time saved in gas-freeing for dry dock periods.

Some disadvantages of the system include that crew workload is increased at the port of discharge and discharge time is increased. It has a high installation cost and maintenance costs are also increased, while crew need special training in operational aspects.

ASPECTS OF CRUDE OIL WASHING (COW)

The operational principle of the COW system is to use dry crude from a full tank to wash the tanks being discharged. Crude containing water droplets from the bottom of a tank should not be used for washing purposes as this may introduce water droplets that have become electrostatically charged and produce an unnecessary source of ignition in the tank atmosphere.

To this end, any tank designated for use as crude oil washing should be first de-bottomed into the slop tank or bled ashore with the discharge pump.

One of the main cargo pumps is used to supply the crude oil washing line with pressurised crude for washing operations. The line, along the deck, will carry branch lines to all the fixed machines. Large VLCC vessels may have up to six (6) machines per tank.

Safety in Operation

Tanks must be fully inerted prior to commencing washing operations and the heater in the tank-washing system must be isolated by blanks. The line would need to be pressurised and tested for leaks prior to commencing washing.

Operation – Stage One

The limits to cover the top of the cycle would need to be adjusted to be pointing upwards. Where portable drive units are employed, these would have to be initially fitted and limits set accordingly.

Operation - Stage Two

The second stage starts when one-third of the tank is discharged and the washing jet will only be allowed to travel down to a point where the jet strikes the bulkhead just above the level of the oil in the tank. At this stage the machine completes 1½ cycles and must therefore be adjusted up again, before the start of the next stage.

Operation - Stage Three

The third stage is where the machine washes from where two-thirds of the tank has been discharged and between one- and two-thirds of the tanks structure is washed.

Operation - Last Stage

The final stage washes the last third and the bottom of the tank with the jet pointing in the downward position.

Crude Oil Washing Cycles

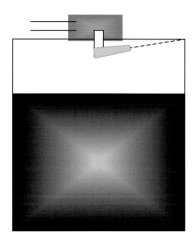

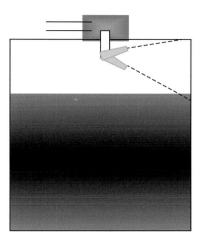

1st Cycle Stage One: nozzle elevated for upper level wash. Cargo discharged, to wash upper third of tank.

2nd Cycle Stage Two: one-third of nozzle programmed.

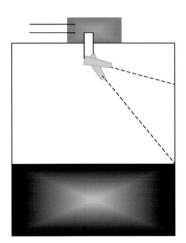

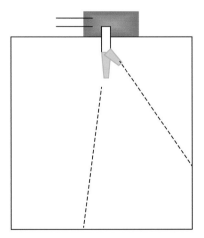

3rd Cycle Stage Three: two-thirds of the cargo is discharged. Nozzle programmed to wash mid-levels of the tank with the end of discharge.

4th Cycle, last stage: machine programmed so that the lower levels and the last washing cycle coincide.

FIGURE 6.18 The four cycles of Crude Oil Washing (COW).

CRUDE OIL WASHING – PREPARATION AND ACTIVITIES

Prior to arrival at the port of discharge:

1 Has the terminal been notified?
2 Is oxygen analysing equipment tested and working satisfactorily?
3 Are tanks pressurised with good quality inert gas (maximum 8% oxygen)?
4 Is the tank-washing pipeline isolated from water heater and engine room?
5 Are all the hydrant valves on the tank-washing line securely shut?
6 Have all tank-cleaning lines been pressurised and leakages made good?

In port:

1 Is the quality of the inert gas in the tanks satisfactory (8% oxygen or less)?
2 Is the pressure on the inert gas satisfactory?
3 Have all discharge procedures been followed and ship to shore checklist completed?

Before washing:

1 Are valves open to machines on selected tanks for washing?
2 Are responsible persons positioned around the deck to watch for leaks?
3 Are tank ullage gauge floats lifted on respective tanks to be washed?
4 Is the inert gas system in operation?
5 Are all tanks closed to the outside atmosphere?
6 Have tanks positive inert gas pressure?

During washing:

1 Are all lines oil-tight?
2 Are tank-washing machines functioning correctly?
3 Is the inert gas in the tanks being retained at a satisfactory quality?
4 Is positive pressure available on the inert gas system?

After washing:

1 Are all the valves between the discharge line and the tank-washing line shut down?
2 Has the tank-washing main pressure been equalised and the line drained?
3 Are all tank-washing machine valves shut?

After departure:

1   Have any tanks due for inspection been purged to below the Critical Dilution Level, prior to introducing fresh air?
2   Has oil been drained from the tank-washing lines before opening hydrants to the deck?

THE INERT GAS SYSTEM

Tanker vessels have an inherent danger from fire and/or explosion and it is desirable that the atmosphere above an oil cargo or in an empty tank is such that it will not support combustion. The recognised method of achieving this status is to keep these spaces filled with an inert gas. Such a system serves two main functions:

1   Use of inert gas inhibits fire or explosion risk.
2   It inhibits corrosion inside cargo tanks.

As inert gas is used to control the atmosphere within the tanks, it is useful to know exactly what composition the gases are, not only from a safety point of view but to realise what effect such an atmosphere would have on the construction of the tanks.

Boiler flu gas consists of the following mix (assuming a well-adjusted boiler):

TABLE 6.1

Component	% of Inert Gas
Nitrogen	83%
Carbon Dioxide	13%
Carbon Monoxide	0.3%
Oxygen	3.5%
Sulphur Dioxide	0.005%
Nitrogen Oxides	Traces
Water vapour	Traces
Ash	Traces
Soot	Traces

Flu gases leave the boiler at about 300°C, contaminated with carbon deposits and sulphurous acid gas. The gas then passes through a scrubber which washes out the impurities and reduces the temperature to within 1°C of the ambient sea temperature.

The clean cooled gas is now moisture-laden and passes through a demister where it is dried. It is then fan-assisted on passage towards the cargo tanks, passing through a deck water seal and then over the top of an oil seal to enter at the top of the tank. It is allowed to circulate and is purged through a pipe which extends from the deck to the bottom of the tank.

There is a sampling cock near the deck water seal for monitoring the quality of the inert gas. Individual tank quality is tested by opening the purge pipe cover and inserting a sample probe.

Excess pressure in the cargo tanks is vented through a pressure vacuum valve (P/V valve) set at 2 psi, which is then led to a mast riser fitted with a gauze screen. The excess is then vented to atmosphere as far from the deck as practicable.

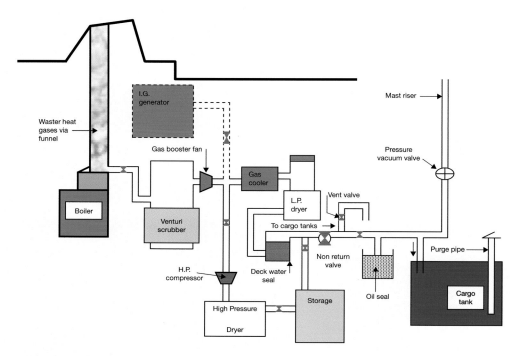

FIGURE 6.19 The inert gas system.

Requirements for Inert Gas Systems

Additional reference should be made to the Revised Guidelines for Inert Gas Systems adopted by the Maritime Safety Committee, June 1983 (MSC/Circ. 353). In the case of Chemical Tankers, reference Resolution A.567 (14) and A.473 (XIII).

1   Tankers of 20,000 tonnes deadweight and above, engaged in carrying crude oil, must be fitted with an Inert Gas System.

    a   Venting systems in cargo tanks must be designed to operate to ensure that neither pressure nor vacuum inside the tanks will exceed design parameters, for volumes of vapour, air or inert gas mixtures.

    b   Venting of small volumes of vapour, air or inert gas mixtures, caused by thermal variations affecting the cargo tank, must pass through pressure vacuum valves.

Large volumes caused by cargo loading, ballasting or during discharge must not be allowed to exceed design parameters.

A secondary means of allowing full flow relief of vapour, air or inert gas mixtures, to avoid excess pressure build-up, must be incorporated, with a pressure sensing monitoring arrangement. This equipment must also provide an alarm facility activated by over-pressure.

2 Tankers with double hull spaces and double bottom spaces shall be fitted with connections for air and suitable connections for the supply of inert gas (IG). Where hull spaces are fitted to the IG permanent distribution system, means must be provided to prevent hydrocarbon gases from cargo tanks entering double hull spaces.

(Where spaces are not permanently connected to the IG system, appropriate means must be provided to allow connection to the IG main.)

3 Suitable portable instruments and/or gas sampling pipes for measuring flammable vapour concentrations and oxygen must be provided to assess double hull spaces.

4 All tankers operating with a Crude Oil Washing (COW) system must be fitted with an inert gas system.

5 All tankers fitted with an inert gas system shall be provided with a closed ullage system.

6 The IG system must be capable of inerting empty cargo tanks by reducing the oxygen content to a level which will not support combustion. It must also maintain the atmosphere inside the tank with an oxygen content of less than 8% by volume and at a positive pressure at all times in port or at sea, except when necessary to gas-free.

7 The system must be capable of delivering gas to the cargo tanks at a rate of 125% of the maximum rate of discharge capacity of the ship, expressed as a volume.

8 The system should be capable of delivering IG with an oxygen content of not more than 5% by volume in the IG supply main to cargo tanks.

9 Flu gas isolating valves must be fitted to the IG main, between the boiler uptakes and the flue gas scrubber. Soot blowers will be arranged so as to be denied operation when the corresponding flue gas valve is open.

10 The 'scrubber' and 'blowers' must be arranged and located aft of all cargo tanks, cargo pump rooms and cofferdams separating these spaces from machinery spaces of Category 'A'.

11 Two fuel pumps or one with sufficient spares shall be fitted to the inert gas generator.

12 Suitable shut-offs must be provided to each suction and discharge connection of the blowers. If blowers are to be used for gas-freeing they must have blanking arrangements.

13 An additional water seal or other effective means of preventing gas leakage shall be fitted between the flue gas isolating valves and scrubber, or incorporated in the gas entry to the scrubber, for the purpose of permitting safe maintenance procedures.

14 A gas regulating valve must be fitted in the IG supply main, which is automatically controlled to close at predetermined limits. (This valve must be located at the forward bulkhead of the foremost gas-safe space.)

15 At least two non-return devices, one of which will be a water seal, must be fitted to the IG supply main. These devices should be located in the cargo area, on deck.

16 The water seal must be protected from freezing and prevent backflow of hydrocarbon vapours.

17 The second device must be fitted forward of the deck water seal and be of a non-return valve type or equivalent, fitted with positive means of closing.

18 Branch piping of the system to supply IG to respective tanks must be fitted with stop valves or equivalent means of control, for isolating a tank.

19 Arrangements must be provided to connect the system to an external supply of IG.

20 Meters which indicate the pressure in slop tanks when isolated from the IG main supply must be fitted in the navigation bridge of combination carriers. Meters must also be situated in machinery control rooms for the pressure and oxygen content of IG supplied. (Where a cargo control room is a feature, these meters would be fitted in such rooms.)

21 Automatic shutdown of inert gas blowers and the gas regulating valve shall be arranged on pre-determined limits.

22 Alarms shall be fitted to the system and indicated in the machinery space and the cargo control room. These alarms monitor the following:

   a   Low water pressure or low water flow rate to the flu gas scrubber.

   b   High water level in the flu gas scrubber.

   c   High gas temperature.

   d   Failure of the IG blowers.

   e   Oxygen content in excess of 8% by volume.

   f   Failure of the power supply to the automatic control system, regulating valve and sensing/monitoring devices.

   g   Low water level in the deck water seal.

   h   Gas pressure less than 100 mm water gauge level.

   i   High gas pressure.

   j   Insufficient fuel oil supply to the IG generator.

   k   Power failure to the IG generator.

   l   Power failure to the automatic control of the IG generator.

HAZARDS WITH INERT GAS SYSTEMS

The inert gas system aboard any vessel has two inherent hazards:

1 If the cooling water in the scrubber should fail, then uncooled gas at 300°C would pass directly to the cargo tank. This is prevented by the fitting of two water sensors in the base of the scrubber, which if allowed to become uncovered would generate an alarm signal which shuts the system down and vents the gas to atmosphere. In the event that both sensors failed, two thermometer probes at the outlet of the scrubber would sense an unacceptable rise in temperature and initiate the same shut-down procedure.

2   If there was a failure in the pressure vacuum (P/V) valve at the same time as a rise in the pressure within the cargo tank, it would result in pressure working backwards towards the boiler with a possible risk of explosion. This is prevented by the water in the deck seal forming a plug in the IG line, until a sufficient head is generated to blow out the oil seal and the excess pressure vents to the deck. The pressure of water in the water seal is essential, therefore the two water sensors would sense its absence and shut down the plant as previously stated.

Inert gas pressure should be maintained at a positive pressure at all times, to avoid air being forced into the cargo spaces. Such a positive pressure is also exerted onto the surface of the oil cargo and assists to push the oil along the suction line towards the cargo pump, and in so doing assists the draining of the tanks. Any excess pressures in the cargo tanks are vented through the P/V valve.

INERT GAS – VOYAGE CYCLE

Phase 1: Vessel departs dry dock with all tanks vented to atmosphere and partially ballasted. The IG plant is started, empty tanks and ullage spaces purged to atmosphere until oxygen levels are acceptable. IG quality is monitored and maintained throughout the ballast voyage.

Phase 2: Prior to arrival at the loading port the IG plant would be started and ballast reduced to about 25% of the ship's deadweight, ballast being replaced by inert gas. After berthing, the remainder of the sea water ballast would be discharged and replaced by IG. The IG plant would then be shut down, the deck isolation valve would be closed and the mast riser opened, prior to commencing loading. Inert gas would be displaced through mast risers. On completion of loading the IG would be topped up to a working pressure which would be maintained through the loaded voyage. (This would be expected to reduce evaporation and prevent oxygen access.)

Phase 3: On arrival at the port of discharge, the IG plant would be set to maximum output with discharge pumps at maximum output. The IG pressure should be carefully monitored and if it approaches a negative, the rate of discharge of the cargo should be reduced. The mast riser should never be opened during the discharge period to relieve the vacuum.

Phase 4: On completion of discharge, the IG system is shut down. If and when ballasting takes place, the IG and hydrocarbons would be vented to atmosphere.

Phase 5: On departure from the discharge port all tanks must be drained to the internal slop tank, then purged with IG to reduce the hydrocarbon levels to below 2%.

Phase 6: Tank cleaning can now be permitted with IG in fully inerted tanks. This weakens the hydrocarbon level and the positive pressure prevents pumps draining or drawing atmosphere into the tanks.

Phase 7: When all the vessel's tanks have been washed and ballast changed, it may be necessary to carry out tank inspections. If this is the case, all tanks would then have to be purged with IG to remove all traces of hydrocarbon gas before venting by fans. All tanks would then be tested with explosi-meter and oxygen analyser. (Full procedure for enclosed space entry must be observed before internal inspection.)

Advantages and Disadvantages of the Inert Gas System

Advantages

1 A safe tank atmosphere is achieved which is non-explosive.
2 Allows high-pressure tank washing and reduces tank cleaning time.
3 Allows crude oil washing.
4 Reduces corrosion in tanks – with an efficient scrubber in the system.
5 Improves stripping efficiency and reduces discharge time.
6 Aids the safe gas-freeing of tanks.
7 Economical to operate.
8 Forms a readily available extinguishing agent for other spaces.
9 Reduces the loss of cargo through evaporation.
10 Complies with legislation and reduces insurance premiums.

Disadvantages

1 Additional costs for installation.
2 Maintenance costs are incurred.
3 Low visibility inside tanks.
4 With low oxygen content, tank access is denied.
5 Could lead to contamination of high-grade products.
6 Moisture and sulphur content corrodes equipment.
7 An established reverse route for cargo to enter the engine room.
8 Oxygen content must be monitored and alarm sensed at all times.
9 Instrumentation failure could affect fail-safe devices, putting the ship at risk through the IG system.
10 An additional gas generator is required in the system in the absence of waste heat products from boiler flue gases.

Note: Instrumentation of the system to cover: Inert gas temperature pressure read outs and recorders. Alarms for: blower failure, high oxygen content, high and low gas pressure, high gas temperature, low sea water pressure and low level alarm in the scrubber and the deck water seal, respectively.

Deck Water Seal Operation

The water level in the seal is maintained by constant running of the sea water pump and a gooseneck drain system. Under normal IG pressure the inert gas will bubble through the liquid from the bottom of the IG inlet pipe and exit under normal operating pressure.

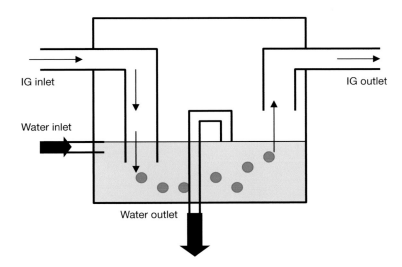

FIGURE 6.20 Operation of deck water.

In the event a back pressure did develop and the water surface experienced increased pressure, this would force the water level up the IG inlet pipe, sealing this pipe entrance and preventing hydrocarbons entering the scrubber.

Tank Atmosphere

The Cargo Officer will need to be able to assess the condition of the atmosphere inside the tank on numerous occasions. To this end various monitoring equipment is available to carry out gas detection and oxygen content. The officer should be familiar with the type of equipment aboard his/her own vessel and have a degree of understanding how such instrumentation operates.

Gas Detection

It should be understood from the outset that many accidents and loss of life have occurred through lack of knowledge of gas detection methods and the correct practice concerning this topic. The explosi-meter, of which there are several trade names available, is used for detecting the presence of flammable gas and/or air mixture.

Monitoring of Tank Operations

The modern tanker will be equipped with state-of-the-art monitoring equipment in the way of data logging of all facets of cargo operations. Sensing devices on quantity, temperature, pressure levels, etc. will be associated to all cargo spaces and protected by visual and audible alarm systems. Mimic diagrams of cargo movement and cargo quality will normally be continuously available to watch-keepers in analogue and digital presentation. Automatic and manual pump and valve control is usually a product of such systems.

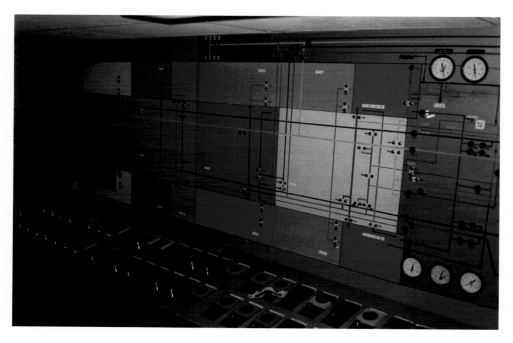

FIGURE 6.21 Mimic diagram presentation for tanker pipeline system seen in the control room of a tanker. Similar diagrams would present ballast and fresh water conditions also.

THE EXPLOSI-METER (COMBUSTIBLE GAS DETECTOR)

This is an instrument which is specifically designed for measuring the Lower Flammable Limit (LFL). It will only function correctly if the filament has an explosive mixture in contact with it. It is contained in a hand-held size box with a battery power supply.

When in use, the sample tube is lowered into the tank and a sample of the atmosphere is drawn up into the instrument by several depressions of the rubber aspirator bulb. If the sample contains an explosive mixture, the resistance of the catalytic filament will change due to the generated heat. An imbalance of the wheatstone bridge is detected by the ohmmeter which tells the operator that hydrocarbon gas is present in the tank in sufficient quantity to support combustion.

Note: Combustibles in the sample are burned on the heated filament which raises its temperature and increases the resistance in proportion to the concentration of combustibles in the sample. This then causes the imbalance in the wheatstone bridge.

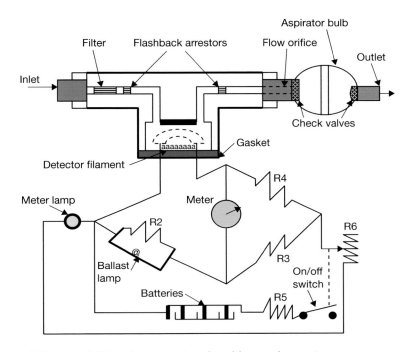

FIGURE 6.22 MSA model 2E explosi-meter (combustible gas detector).

However, it should be realised that a zero reading does not necessarily indicate that there is no hydrocarbon gas present, nor does it mean that no oxygen is present. All it signifies is that the sample taken is either too rich or too lean to support combustion. Care must be taken when testing the atmosphere in enclosed spaces to give consideration for the relative vapour density where mixtures of gases are encountered. A test at one particular level in a tank should be realised as not necessarily being an equivalent reading for other different levels in the same tank.

The electrical bridge circuit of the instrument is designed so that its balance is established at the proper operating temperature of the detecting filament. The circuit balance and detector current are adjusted simultaneously by adjustment of the rheostat. The proper relationship between these two factors is maintained by a special ballast lamp in the circuit.

The graduations on the meter are a percentage of the lower explosive limit (LEL) reading between 0 and 100%. A deflection of the meter between 0 and 100% shows how close the atmosphere being tested approaches the minimum concentration required for explosion. When a test is made with the explosi-meter and a deflection to the extreme right-hand side of the scale is noted and remains there, then the atmosphere under test is explosive.

Limitations of Explosi-meters

The explosi-meter has been designed to detect the presence of flammable gases and vapours. The instrument will indicate in a general way whether or not the atmosphere is dangerous from a flammability point of view. It is important to realise that such information obtained from the instrument is appraised by a person skilled in the interpretation of the reading, bearing in mind the environment. For example, the atmosphere sample which is indicated as being non-hazardous from the standpoint of fire and explosion may, if inhaled, be toxic to workers who are exposed to that same atmosphere.

Additionally a tank which is deemed safe before work is commenced may be rendered unsafe by future ongoing operations, e.g. stirring or handling bottom sludge. This would indicate the need for regular testing practices to be in place in questionable spaces while work is in progress.

Explosi-meter Special Uses

Where the explosi-meter is employed to test an atmosphere which is associated with high boiling point solvents, it should be borne in mind that the accuracy of the reading may be questionable. The space may be at a higher temperature than the instrument and therefore it must be anticipated that some condensation of combustible vapours would be in the sampling line. As a consequence the instrument could read less than the true vapour concentration.

A way around this would possibly be to warm the sampling line and the instrument unit to an equivalent temperature to that of the space being tested.

Note: Under no circumstances should such instruments be heated over 65°C (150°F).

Furthermore, some types of instrument are designed to measure combustible vapours in air. They are not capable of measuring the percentage vapours in a steam or inert atmosphere, due to the absence of oxygen necessary to cause combustion.

Care in Use

When sampling over liquids care should be taken that the sampling tube does not come into contact with the liquid itself. A probe tube can be used in tests of this character, to prevent liquid being drawn into the sampling tube.

Drager Instrument

This is an instrument which draws a gas or vapour through an appropriate glass testing tube. Each tube is treated with a chemical that will react with a particular gas, causing discolouration progressively down the length of the tube. When measured against a scale, the parts per million (ppm) can be ascertained. The instrument is used extensively on the chemical carrier trades, though it does have tubes for use with hydrocarbons, which makes it suitable for use on tankers.

Alarm System Detector

An instrument which is taken into a supposedly gas-free compartment and used while work is ongoing. If gas is released or disturbed in the workplace a sensitive element on the instrument triggers an audible and visual alarm. Once the alarm has been activated, personnel would be expected to evacuate the compartment immediately.

Oxygen Analyser

In order for an atmosphere to support human life it must have an oxygen content of 21%. The oxygen analyser is an instrument that measures the oxygen content of an atmosphere to establish whether entry is possible, but it is also employed for inerted spaces which must be retained under 5% oxygen to effect a safe atmosphere within the tank.

The oxygen sensor will be either an electromagnetic heated filament or an electro-chemical resistor cell. The instrument is designed to measure the oxygen content only and

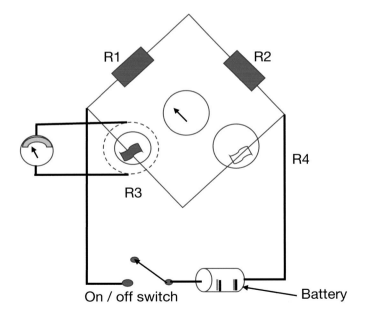

FIGURE 6.23 Oxygen analyser – circuitry principle.

will not detect the presence of any other gases. As shown in figure 6.23, the resistor filaments R3 and R4 are of equal rating. The resistor filament R3 is surrounded by a magnetic field. The atmosphere sample is drawn past the filament and, depending on the amount of oxygen in the sample, will depend on the permitted current flow through the coil and meter.

Oxygen Analyser – Circuitry Principle

Oxygen analysers are portable instruments which withdraw a sample of the atmosphere for testing through a sampling hose by means of a rubber aspirator bulb. The principle of operation is a self-generating electrolytic cell in which the electric current is directly proportional to the percentage oxygen in a salt solution connecting to the electrodes. The electrodes are connected to a micro-ammeter, so that the current read by the meter can be calibrated to indicate directly the percentage oxygen of the sample.

There are variations and different types of instrument available. Manufacturer's instructions and manuals for use and maintenance should therefore be followed when these instruments are employed.

Chemical Reaction Measuring Device

Gas detection can also be achieved by using a test sample of the atmosphere to pass over a chemical-impregnated paper or crystal compound. The chemicals subsequently react with specific gases on contact. The amount of discolouration occurring in the crystals or on the paper can then be compared against a scale to provide the amount of gas within the sample.

The operation uses a bellows to draw through 100 cm^3 of sample gas and a variety of tubes can be used to indicate specific gases. Example gases indicated are likely to be, but not limited to: carbon monoxide, hydrogen sulphide, hydrocarbon, radon, and nitrous oxide.

A popular instrument is the 'Drager Tube System' for gas detecting. Although well used in the industry, the system does have drawbacks in the fact that the tubes required for different gases have a limited shelf life. The bellows can develop leaks and they can be affected by temperature extremes. Tube insertion must also be carried out the correct way.

COASTAL AND SHUTTLE TANKER OPERATIONS

Numerous small tanker operations are engaged in coastal regions around the world and employ the services of coastal-sized craft to shuttle cargo parcels between main terminal ports, floating storage units (FSUs) and the smaller out-of-the-way ports. Restrictions are often put on direct delivery from the ocean-going vessels because of the available depth of water in the smaller enclaves and, as such, the geography imposes draught restrictions on the larger vessels. This particular drawback is also affecting the container trade; with container vessels currently being increased in overall size, the larger vessels are finding some ports are not available to them because of similar draught restrictions.

FIGURE 6.24 The *Stolt Shearwater* (IMO No. 9148958) inbound to the United Kingdom via the River Mersey.

FIGURE 6.25 The floating production, storage and offloading system (FPSO) *Seaway Falcon* (IMO No. 7409401).

EXAMPLES OF TANKER CARGOES

Ammonia carriers – Carriage is as a liquid and has significant hazards inclusive of injury risk or death as well as threats to the environment. Leaks of cargo in gas or liquid form should be avoided at all costs. If evaporation takes place the danger of fire or explosion because of high expansion is a real threat.

Bitumen – This cargo solidifies at normal temperatures and must be kept hot during transit. Ships are specifically designed for this trade, with large centre tanks and additional heating coils, the centre tanks being used for cargo and the wing tanks for ballast.

Chemicals (various) – Precautions for these cargoes as outlined above. Additional reference to the IMDG code and respective precautions pertaining to the type of commodity.

Creosote – This is a very heavy cargo and requires constant heating during the voyage.

Crude oil – Varies greatly with Relative Density and viscosity. It is not heated unless it is of a very heavy grade, as heating evaporates the lighter fractions. Crude oil has a high fire risk.

Diesel oil – An intermediate between fuel oil and gas oil. It is generally regarded as a dirty oil but its viscosity is such that it does not require heating prior to discharge.

Fuel oil – A black oil which is graded according to its weight and viscosity. It has a low fire risk and generally requires heating prior to discharge.

Gas oil – A clean oil which is used for light diesel engines as well as for making gas. A reasonable level of cleanliness is required before loading this cargo, which may be used as a transition cargo when a ship is being changed from a black oil carrier to a clean oil trade. Fire risk is low and no heating is required.

Gasoline (petrol) – Light and volatile. It has a high fire risk and may easily be contaminated if loaded into tanks which are not sufficiently clean.

Grain – May be successfully carried in tankers since when in a bulk state it has many of the qualities of a liquid. It requires very careful tank preparation and tankers would only normally enter the trade if the oil market was depressed.

Kerosene (paraffin) – A clean oil which is easily discoloured. Precautions should be taken to prevent the build-up of static. These may include a slow loading and discharge pattern being employed.

Latex – An occasional cargo carried in tankers and in ships' deep tanks. Usually has added ammonia. The tanks should be exceptionally clean and fitted with pressure relief valves. Steelwork is pre-coated in paraffin wax and heating coils in tanks should be removed. Following discharge the tanks should be washed with water to remove all traces of ammonia.

Liquefied gases – Generally carried in specifically designed vessels for the transport of Liquid Natural Gas (LNG) and Liquid Propane Gas (LPG).

Lubricating oils – These are valuable cargoes and are usually shipped in the smaller product carriers. Good separation is necessary to avoid contamination between grades. Tanks and pipelines must be free of water before loading. Some grades may require heating before discharging.

Molasses – A heavy viscous cargo which is normally carried in designated tankers specific for the trade. A comprehensive heating system is necessary and special pumps are provided to handle the thick liquid.

Propane – Similar to butane. (See 'Liquefied gases'.)

Vegetable oils – These are generally carried in small quantities in the deep tanks of cargo ships but some, such as linseed oil, may be carried in tankers. Exceptional cleanliness of the tanks is required prior to loading such a cargo.

Whale oil – Whale factory ships are basically tankers carrying fuel oil on the outward passage and whale oil when homeward bound. Careful cleaning is required before carrying

FIGURE 6.26 Conventional VLCC oil tanker *Jialong Spirit* (IMO No. 9379208) seen in a light condition after leaving the Lisnave Dry Dock complex (Portugal), having completed hull and general maintenance.

whale oil in tanks which previously carried fuel oil. (Fortunately the practice of whale hunting has been severely restricted.)

Wine – Can be carried in tankers but they are usually dedicated ships to the trade. Similar vessels sometimes engage in the carriage of fruit juices, especially orange juice. A high degree of cleanliness in the tanks is expected.

PRODUCT TANKERS

Product tankers tend to be smaller and more specialised than the large crude oil carriers and generally lay alongside specialised berths when loading and discharging, employing specialist product lines to avoid contamination of cargoes.

Bulk Liquid Chemical Carriers

Phrases and terminology associated with the chemical industry:

Adiabatic expansion – An increase in volume without a change in temperature or without any heat transfer taking place.

Anaesthetics – Chemicals that affect the nervous system and cause anaesthesia.

Aqueous – A compound within a water-based solution.

Auto-ignition – A chemical reaction of a compound causing combustion without a secondary source of ignition.

Boiling point – The temperature at which a liquid's vapour pressure is equal to the atmospheric pressure.

Catalyst – A substance that will cause a reaction with another substance or one that accelerates or decelerates a reaction.

Critical pressure – The minimum pressure which is required to liquefy a gas at its critical temperature.

Critical temperature – The maximum temperature of a gas at which it can be turned into a liquid by pressurisation.

Filling ratio – The percentage volume of a tank which can be safely filled to allow for the expansion of the product.

Freezing point – The temperature at which a substance must be at to change from a liquid to a solid state, or vice versa.

Hydrolysis – The process of splitting a compound into two parts by the agency of water, one part being combined with hydrogen, the other with hydroxyl.

Hydroscopic – The ability of a substance to absorb water or moisture from the atmosphere.

Inhibitor – A substance which when introduced to another will prevent a reaction.

Narcosis – A human state of insensibility resembling sleep or unconsciousness, from which it is difficult to arouse.

Oxidising agent – An element or compound that is capable of adding oxygen to another.

Padding – A procedure of displacing air or unwanted gases from tanks and pipelines with another compatible substance – e.g. inert gas, cargo vapour, or liquid.

Polymerisation – The process due to a chemical reaction within a substance, which is capable of changing the molecular structure within that substance – i.e. liquid to solid.

Reducing agent – An element or compound that is capable of removing oxygen from a substance.

Reid vapour pressure – The vapour pressure of a liquid as measured in a Reid apparatus at a temperature of 100°F expressed in psi/°A.

Self-reaction – The ability of a chemical to react without other influence which results in polymerisation or decomposition.

Sublimation – The process of conversion from a solid to a gas, without melting. (An indication that the flashpoint is well below the freezing point.)

Threshold Limit Value (TLV) – The value reflecting the amount of gas, vapour, mist or spray mixture to which a person may be daily subjected without suffering any adverse effects. (Usually expressed in parts per million (ppm)).

Vapour density – The weight of a specific volume of gas compared to an equal volume of air, in standard conditions of temperature and pressure.

Vapour pressure – The pressure exerted by a vapour above the surface of a liquid at a certain temperature (measured in mm of mercury).

FIGURE 6.27 The *Saargas* (IMO No. 9135781) LPG/chemical carrier seen underway and leaving the River Mersey, Liverpool, England. The raised upper covers to the tank domes are prominent above the upper deck.

BULK CHEMICAL CARGOES

The term 'liquid chemicals' within the industry is meant to express those chemicals in liquid form at an ambient temperature or which can be liquefied by heating, when carried at pressures up to 0.7kg/cm^2. Above this pressure would fall into the category of 'liquefied gases'.

The chemicals carried at sea have a variety of properties. Nearly half of the 200 chemicals commonly carried have fire or health hazards no greater than petroleum cargoes. Therefore they can be safely carried by way of ordinary product carriers, though some modification is sometimes required to avoid contamination.

Other chemical substances require quality control much more stringent than petroleum products. Contamination, however slight, cannot be allowed to occur and for this reason tanks are nearly always coated or made of special materials such as stainless steel.

Extreme care must be exercised when loading such cargoes that any substances which could cause a reaction are kept well separated. To ensure quality and safe carriage, separate pipelines, valves and separate pumps are the norm for specific cargo parcels. Also reactionary chemicals cannot be placed in adjacent tanks with only a single bulkhead separation. Neither can pipelines carrying one substance pass through a tank carrying another substance with which it may react. Chemical products which react with sea water are carried in centre tanks while the wing tanks are employed to act as cofferdams.

Chemical carriers require experienced and specialised trained personnel in order to conduct their day-to-day operations safely. They also require sophisticated cargo handling and monitoring equipment. The ships must conform in design and construction practice to the IMO Code for the Construction and Equipment of Ships Carrying Dangerous Chemicals in Bulk. The purpose of the code is to recommend suitable design criteria, safety measures and construction standards for ships carrying dangerous chemical substances. Much of the content of the code has been incorporated into the construction regulations produced by the Classification Societies.

Classification – Chemical Carriers

(Chapter references to the International Code for the Construction and Equipment of Ships Carrying Dangerous Chemicals in Bulk (IBC).)

In general, ships carrying chemicals in bulk are classified in three types:

1   **A 'Type 1' ship** is a chemical tanker intended to transport Chapter 17 products with very severe environmental and safety hazards which require maximum preventative measures to preclude an escape of such cargo.

2   **A 'Type 2' ship** is a chemical tanker intended to transport Chapter 17 products with appreciably severe environmental and safety hazards which require significant preventative measures to preclude an escape of such cargo.

3   **A 'Type 3' ship** is a chemical tanker intended to transport Chapter 17 products with sufficiently severe environmental and safety hazards which require a moderate degree of containment to increase survival capability in a damaged condition.

Many of the cargoes carried in these ships must be considered as extremely dangerous and, as such, the structure of the ship's hull is considered in the light of the potential danger which might result from damage to the transport vessel. Type 3 ships are similar to product tankers in that they have double hulls but have a greater sub-division requirement, whereas Types 1 and 2 ships must have their cargo tanks located at specific distances inboard to reduce the possibility of impact load directly onto the cargo tank.

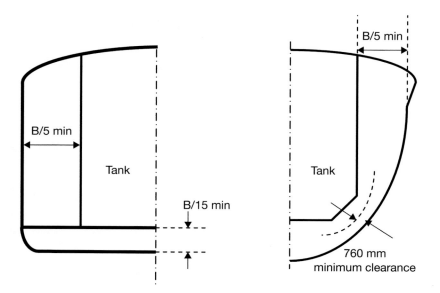

FIGURE 6.28 Type 1 ship.

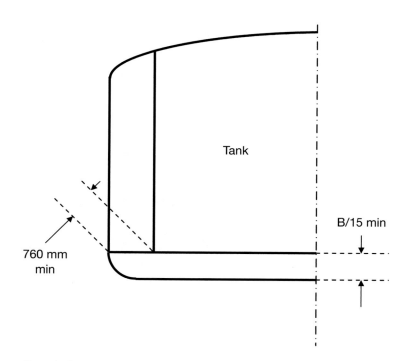

FIGURE 6.29 Type 2 ship.

Tank Plan – (all types)

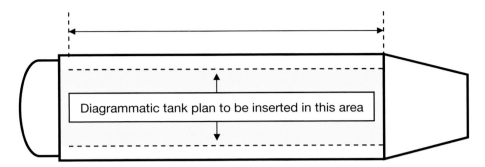

Diagrammatic tank plan to be inserted in this area

FIGURE 6.30 The tank arrangement must be attached to the International Certificate of Fitness for the Carriage of Dangerous Chemicals in Bulk.

PARCEL TANKERS – CONSTRUCTION FEATURES

Ships built specifically as parcel tankers with the intention of carrying a wide variety of cargoes will generally have some tanks of stainless steel or tanks clad in stainless steel. For reasons of construction and cost, this means having a double skin. Mild steel tanks may similarly be built with side cofferdams and a double bottom and are usually coated in either epoxy or silicate. Chemicals of high density like ethylene dibromide may have specially constructed tanks or, in some cases, only partly filled cargo tanks will be carried.

Similarly, cargoes with higher vapour pressures may generate a need for tanks to be constructed to withstand higher pressures than, say, the conventional tanker. This is particularly relevant where the boiling point of the more volatile cargoes is raised and the risk of loss is increased.

The IBC Code specifies requirements for safety equipment to monitor vapour detection, fire protection, ventilation in cargo handling spaces, gauging and tank filling. Once all criteria are met, the Marine Authority (MCA in the United Kingdom) will issue, on application, an MCA/IMO Certificate of Fitness for the Carriage of Dangerous Cargoes in Bulk.

Vapour Lines

In general each tank will have its own vapour line fitted with P/V valves but grouped tanks may have a common line. Since some vapours from specific cargoes are highly toxic or flammable, the lines are led well over accommodation and are expected to release vapour as near as possible in a vertical direction. Some vessels carry provision to return vapour expelled during the loading process to the shoreside tank. Examples are when the cargo is highly toxic or the chemicals react dangerously with air.

MAIN HAZARDS TO HUMANS ASSOCIATED WITH CHEMICALS

The substances carried in chemical tankers present certain hazards to operations of transport and to the crews of the ships. The main hazards fall into one of a combination of the following:

1   Danger to health – toxicity and irritant characteristics of the substance or vapour.

2   Water pollution aspect – human toxicity of the substance in the solution.

3   Reactionary activity with water or other chemicals.

4   Fire and/or explosion hazard.

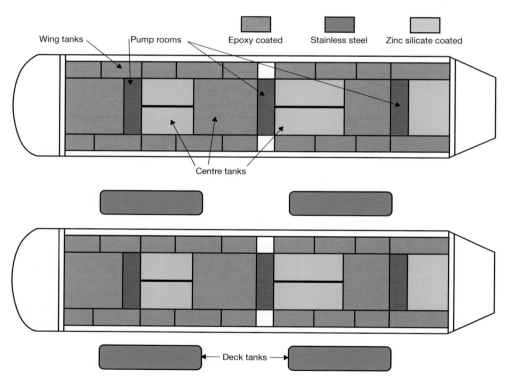

FIGURE 6.31 Tank disposition of example parcel tanker.

Cargo Information – Required Before Loading

1   The Cargo Officer must be informed of the correct chemical name of the cargo to enable the appropriate safety data sheet to be consulted in the *Tanker Safety Guide* (Chemicals).

2 The quantity of cargo and respective weight.

3 Clearance on quality control must be confirmed. Contamination is usually measured in parts per million (ppm) so tanks and pipelines must be assured to be clean.

4 The specific gravity value of the commodity must be advised to allow an estimate of the volume to be occupied for the intended weight of cargo.

5 Incompatibility with other cargoes or specifically other chemicals must be notified. Correct stowage must be achieved so that incompatible cargoes are not stowed in adjacent compartments.

6 Temperature of the cargo:

(i) at the loading stage;

(ii) during the carriage stage.

These criteria are required because temperature of the commodity will affect the volume of the total cargo loaded, while the expected carriage temperature will indicate whether heating of the cargo will be required.

7 The tank coating compatibility must be suitable for the respective cargo.

8 Any corrosive properties of the chemical. This information would also be relevant to the tank coating aspect and provide possible concerns for incurring damage to shipboard fittings.

9 Electrostatic properties can be acquired by some chemicals. With this in mind the principles applicable to hydrocarbons should be applied.

10 Data on the possibility of fire or explosion; 50% of chemicals carried are derived from hydrocarbons and the risk of fire or explosion is similar to the carriage for hydrocarbons.

11 The level of toxicity of the chemical. If high toxic vapours are a characteristic of the cargo, then enclosed ventilation may be a requirement.

12 Health hazards of any particular parcel of cargo.

13 Reactivity with water, air or other commodities.

14 What emergency procedures must be applicable in the event of contact or spillage?

The chemical data sheets of respective cargoes usually provide all the above and additional essential information for the safe handling and carriage of the commodity.

The Protection of Personnel

The hazards of the chemical trade have long been recognised and the need for personal protection of individuals engaged on such ships must be considered as the highest of priorities. A chemical cargo can be corrosive and destroy human tissue on contact. It can also be poisonous and can enter the body by several methods. It may be toxic and, if inhaled, damage the brain, the nervous system or the body's vital organs. Additionally the chemical may give off a flammable gas, giving a high risk of fire and explosion.

The IMO (IBC) Code requires personnel involved in cargo operations aboard chemical carriers to be provided with suitable protection by way of clothing and equipment which will provide total coverage of the skin in a manner that no part of the body is left exposed – i.e. chemical suits.

Protective equipment to include:

1   Full protective suit manufactured in a resistant material with tight-fitting cuff and ankle design.

2   Protective helmet.

3   Suitable boots.

4   Suitable gloves.

5   A face shield or goggle protection.

6   A large apron.

Where the product has inhalation problems for individuals then the above equipment would be supplemented by breathing apparatus.

Where toxic cargoes are carried, SOLAS requires that the ship should carry a minimum of three (3) *additional* complete sets of safety equipment, over and above the SOLAS 74 requirements.

Safety equipment set shall comprise:

1   A self-contained breathing apparatus (SCBA).

2   Protective clothing (as described above).

3   Steel core rescue line and harness.

4   Explosive-proof safety lamp.

An air compressor together with spare cylinders must also be carried and all compressed air equipment must be inspected on a monthly basis and tested annually.

Where toxic chemical products are carried, all personnel on board the vessel must have respiratory equipment available. This equipment must have adequate endurance to permit personnel to escape from the ship in the event of a major accident.

It is normal practice that chemical carriers are built with a designated B/A room carrying full breathing apparatus (SCBA's) for all personnel on board, plus air compressors for recharging and maintaining B/A units.

Associated Operations

Heating of Cargoes

Certain cargoes are required to be carried and/or discharged at high temperatures and to this end, heating while inside the ship's tanks must take place. Heating is usually provided by either heating coils inside the tanks themselves or, in the case of double hull vessels, by heating channels on the outside of the tanks. The medium used is either steam, hot water or oil, but care must be taken that the medium is compatible with the cargo.

Tanks which contain chemicals which could react with each other must not be on the same heating circuit. Another safety factor is that a heat exchanger is used between the boiler and the cargo system. This would prevent the possibility of the cargo product finding its way into the ship's boilers, in the event of a leak occurring in the system.

Inert Gas Systems with Chemical Cargoes

Inert gas, usually nitrogen, is used to blanket some cargoes. These are usually ones that react with air or water vapour in the atmosphere. They are loaded into tanks after they have been purged with inert gas and the tank must remain inerted until cleaning has been completed.

Other cargoes have the ullage space inerted, either as a fire precaution or to prevent reactions which, while not necessarily dangerous, may put the cargo off specification. The nitrogen is supplied by a shipboard generator or from ashore, or from storage cylinders.

Precautions during Loading, Discharging and Tank Cleaning

In addition to the usual safety precautions for tanker practice, if handling toxic cargoes full protective clothing including breathing apparatus should be worn by all persons on deck. Goggles should be worn when handling cargoes which may cause irritation to the eyes. Such vessels are generally equipped with decontamination deck showers together with escape sets for each crew member.

Tank Cleaning

After discharge of the majority of cargoes, the tanks can be washed out with salt water as a first wash, then finished with a fresh water wash. Stainless steel tanks are usually washed only with fresh water because of damage which may be incurred to the steel work by use of sea water.

Washing is often assisted by one of a range of cleansing compounds which can be sprayed onto the tank sides and then washed off. One of the advantages of double hull construction is that all the stiffening members of the tanks are on the outside of the tank and cleaning and drainage is therefore much easier.

Some special chemicals may require special cleaning procedures and solvent use and extreme care should be taken that mixtures created are not of a dangerous nature. Similarly, if washing into a slop tank, a dangerous mixture of unknown chemical properties should not be generated.

Fire Fighting

Fire-fighting arrangements are similar to those aboard petroleum tankers, with the exception that nitrogen is commonly employed as a smothering agent because some cargoes would be incompatible with CO_2. Ships are therefore generally supplied with an adequate supply of nitrogen. Ordinary foam breaks down when used on water-soluble chemicals so a special alcohol-foam is required – so named as being suitable for fires involving alcohol. In addition, large fixed dry powder plants may be provided for use on the tank deck.

Some specialised cargoes require specific fire-fighting techniques and relevant details can be obtained from the shore authorities, prior to loading.

Note: Many cargoes give off harmful vapours when burning and fire parties are advised to ensure they wear protective clothing and breathing apparatus when fighting chemical fires.

Compatibility

Great care must be taken during the cargo planning stage to ensure that chemicals which react with one another do not come into contact. Such planning is often a shore-based operation which is checked by the ship's Master or the Chief Officer prior to the commencement of loading.

Chemicals must be located in an appropriate tank according to the IMO Code and at the same time be compatible with the tank coating as specified in the tables provided by the tank coating manufacturers. Incompatible cargoes must have positive segregation and failure to observe such requirements could give rise to a most hazardous situation involving toxics or flammable gas being given off as a by-product.

Additionally some mixtures of chemicals may react together, but equally some are potentially dangerous on their own. Those that react with air can be contained by inert gas, or provided with vapour return lines as previously described. Some, however, react with water – for example, sulphuric acid – and must be loaded in double skin tanks.

A number of chemicals are self-reactive, in the sense that they may polymerise with explosive violence or cause a generation of considerable heat. Examples of these are vinyl acetate or styrene monomers. If shipped, these have an inhibitor added, but care must be taken with all monomers to ensure that no impurities are introduced which may act as a catalyst and cause polymerisation. Accidental heating with such cargoes should also be avoided.

Volatile Cargoes

Such cargoes of a volatile nature must not be stowed adjacent to heated cargoes. The possibility of flammable or toxic vapour release could be an after-effect which could have disastrous effects should the vapour reach the deck area.

Cargo Handling Reference

Most shipping companies have prepared their own operational and safety manuals, but most are based on the ICS Tanker Safety Guide for Oil Tankers and Terminals (ISGOTT). This includes an index of chemical names, including synonyms. Cargo information from data sheets for the most common chemicals is also included. Checklists are now also commonly employed to ensure correct procedures are observed throughout all cargo operations.

Merchant Shipping Notices

Merchant Shipping Notices stress the danger from asphyxiation and/or effects of toxic or other harmful vapours. They also strongly advise on the entry procedures into tanks and enclosed spaces alongside the Code of Safe Working Practice. Notices emphasise the need for continuous monitoring of the vapour with gas detectors and the necessity of providing adequate ventilation when personnel enter enclosed spaces. Full procedures must include the use of a standby man at the entrance of an enclosed space while personnel are inside that space.

Compatibility Tables

There are various compatibility tables available, but perhaps the most widely applied are the United States Coast Guard – Bulk Liquid Cargoes Guide to the Compatibility of Chemicals. A hazardous reaction is defined as a binary mixture which produces a temperature rise greater than 25°C or causes a gas to evolve.

The cargo groups for the two chemicals under consideration are first established from an alphabetical listing, then cross-referenced in the compatibility table – unsafe combinations being indicated by an 'X' and reactivity deviations within the chemical groups by the letters 'A' to 'I'.

IMO/IBC Code

The International Code for the Construction and Equipment of Ships Carrying Dangerous Chemicals in Bulk and Index of Dangerous Chemicals Carried in Bulk is clearly the main recognised authority regarding the bulk chemical trade. It is recognised as the definitive source of names for products subject to Appendices II and III of Annex II of MARPOL 73/78.

IMO/IGC Code

The International Code for the Construction and Equipment of Ships Carrying Liquefied Gases in Bulk. Applicable to all ships regardless of size, inclusive of those vessels under 500 tons gross, which are engaged in the carriage of liquefied gases having a vapour pressure exceeding 2.8 bar absolute temperature of 37.8°C, and other products as appropriate under Chapter 19 (of the code) when carried in bulk.

(Exception: vessels constructed before October 1994 to comply with Resolution MSC.5(48) adopted on 17 June 1983.)

BULK LIQUEFIED GAS CARGOES

The liquefied gases which are normally carried in bulk are hydrocarbon gases used as fuels or as feed stocks for chemical processing and chemical gases used as intermediates in the production of fertilisers, explosives, plastics or synthetics. The more common gases are liquefied petroleum gases (LPGs) such as propane, butane, propylene, butylene, anhydrous ammonia, ethylene, vinyl chloride monomer and butadiene. Liquefied natural gas (LNG) is also transported extensively in dedicated ships. LNG is a mixture of methane, ethane, propane and butane with methane as the main component.

FIGURE 6.32 Two liquid natural gas carriers lay alongside each other, outside the Dubai dry dock complex. These dedicated ships are prominent for the conspicuous cargo domes covering the gas tanks.

Gas Properties

Liquefied gases are vapours at normal ambient temperatures and pressures. The atmospheric boiling points of the common gases are given as follows:

TABLE 6.2

	Propane	−42.3°C
	Butane	−0.5°C
LPG	Propylene	−47.7°C
	Butylene	−6.1°C
	Ammonia	−33.4°C
	Ethylene	−103.9°C
	VCM	−13.8°C
	LNG	−161.5°C
	Butadiene	−5.0°C

The carriage of gases in the liquid phase can only be achieved by lowering the temperature or increasing the pressure or a combination of both low temperatures and increased pressures.

The carriage condition is classified as either:

'fully refrigerated' (at approximately atmospheric pressure)
or 'semi-refrigerated' (at approximately 0°C–10°C and medium pressure)
or 'fully pressurised' (at ambient temperature and high pressures)

LNG and ethylene are normally always carried in the fully refrigerated condition – they cannot be liquefied by increasing the pressure alone. LPGs, ammonia, vinyl chloride monomer (VCM) and butadiene can be liquefied by lowering the temperature or increasing the pressure; this permits them to be carried in the fully refrigerated, semi-refrigerated or fully pressurised condition. The IMO/IGC Code provides standards for gas tankers and identifies the types of tank which must be employed for the carriage of liquefied gases.

Integral – Tanks which form part of the ship's hull.
Membrane – Non–self-supporting, completely supported by insulation.
Semi-membrane – Non–self-supporting, partly supported by insulation.
Independent – Self-supporting tanks not forming part of the ship's hull.
Independent tanks are subdivided into types A, B and C.

Integral membrane and semi-membrane tanks are designed primarily with plane surfaces. Of the independent tanks, both A and B can either be constructed of plane surfaces or of bodies of revolution, type C is always constructed of bodies of revolution.

Hazards of Gas Cargoes

Hazards associated with gas cargoes are from fire, toxicity, corrosivity, reactivity, low temperatures and pressure.

GAS CARRIER TYPES

Gas Carrier Profile

The more recent builds of LPG carriers include double hull structure with varied capacity. Up to 90,000 m³ cargo capacity is no longer unusual, while LNG construction of 260,000 m³ using self-supporting, prismatic shaped tanks requiring less surface space than the normal construction of spherical 'Moss' tanks are under construction with IHI Marine United Shipbuilders. (LNG carriage is at −162°C and essentially at atmospheric pressure.) Cargo boil-off with LNG is used as fuel for the ship's propulsion system or vented to atmosphere.

FIGURE 6.33 The LPG tanker *Cadiz Khutsen* (IMO No. 9246578) lies at anchor off Gibraltar. The vessel operates with membrane tanks and has a capacity of 138,000 m³.

Fully Pressurised Carriers

These tankers are normally constructed to the maximum gauge pressure at the top of the tank. In all cases, the design vapour pressure should not be less than the maximum allowable relief settings (MARVs) of the tank. This corresponds to the vapour pressure of propane at +45°C, the maximum ambient temperature the vessel is likely to operate in. Relief valves blow cargo vapour to atmosphere above this pressure.

Cargo tanks are usually cylindrical pressure vessels. Tanks below deck are constructed with a dome penetrating the deck on which all connections for the loading, discharging, sampling and gauging for monitoring pressure and temperature are placed. Pumps are not normally installed on this type of ship, the cargo being discharged by vapour pressure above the liquid. No vapour reliquefaction facilities are provided.

Semi-Refrigerated

This type of vessel is normally designed to carry the full range of LPG and chemical gases in tanks designed for a minimum service temperature of –48°C and working under design pressure. Simultaneous carriage of different cargoes is usually possible.

The ships are generally installed with deepwell pumps to facilitate discharge. If delivery is required into pressurised shore storage units, these deepwell pumps operate in series with booster pumps mounted on deck. Cargo heating using sea water is the usual practice. Vapours produced by heat are drawn off into a reliquefaction unit and the resultant liquid is returned to the tank. This action maintains the tank pressures within limits.

Fully Refrigerated Carriers

Cargo tanks are usually designed for a minimum service temperature of –50°C and a maximum design pressure.

Discharge of the cargo is achieved by using deepwell pumps or submerged pumps. Unlike the deepwell pumps, the submerged pump assembly, including the motor, is installed in the base of the tank. As a result it is completely immersed in cargo liquid. Booster pumps and cargo heating may also be installed for discharge into pressurised storage.

Reliquefaction plant is also installed on board for handling boil-off vapours. Fully refrigerated carriers now have capacities up to 250,000 m^3.

Fully Refrigerated Ethylene Tankers

The majority of liquid ethylene tankers can carry the basic LPG cargoes as well. Ethylene cargoes are normally carried at essentially atmospheric pressure. Product purity is very important in carriage and care must be taken during cargo operations to avoid impurities such as oil, oxygen, etc. Reliquefaction plant is also provided on these ships.

GAS TANK CONSTRUCTION (SPHERICAL TANKS)

Gas Tank Construction (based on Moss design)

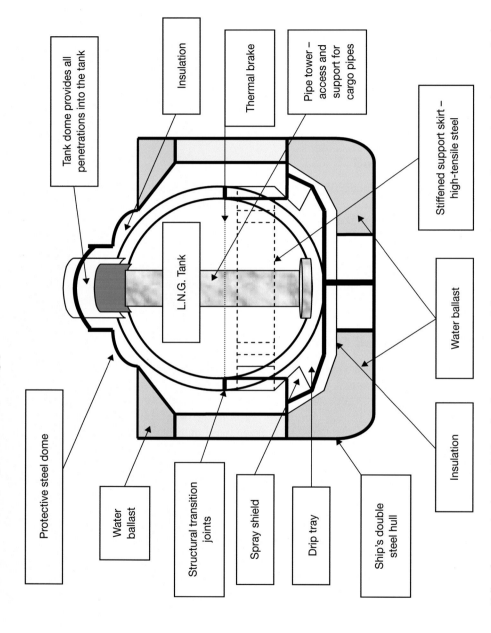

FIGURE 6.34 Fully refrigerated spherical LNG tank – the protective steel dome protects the primary barrier above the upper deck. (No secondary barrier.)

Note: Double hull construction is required in way of all cargo tank spaces.

LNG Spherical Tanks

FIGURE 6.35 Large gas carrier seen manoeuvring with tugs in attendance. The four prominent spherical tank 'domes' are clearly identified above the uppermost continuous deck. Fully refrigerated spherical LNG tank, the protective steel dome protects the primary barrier above the upper deck. (No secondary barrier.) Double hull construction required in way of all cargo tank spaces.

Membrane Tanks

Membrane tanks are now probably the most popular type of build of LNG carrier construction today.

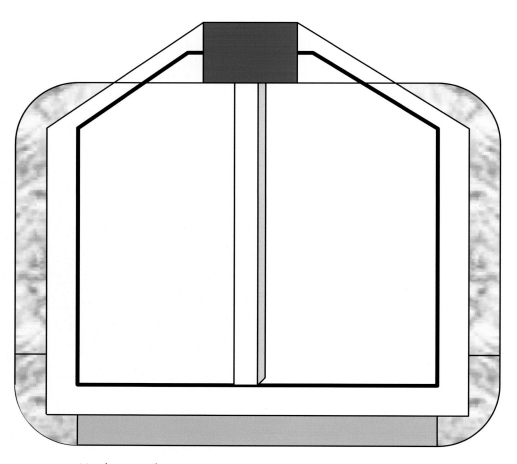

FIGURE 6.36 Membrane tank construction.

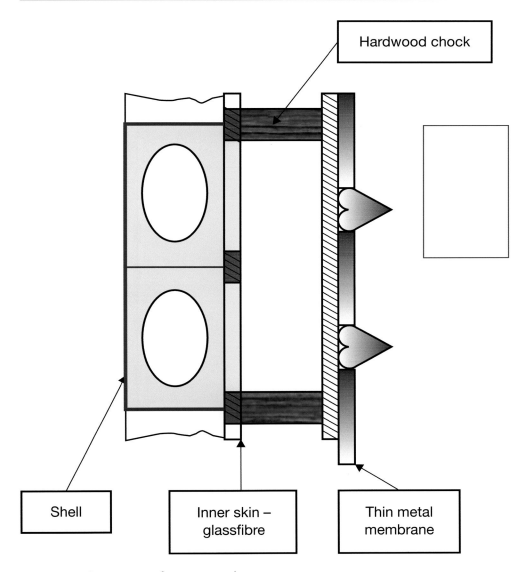

FIGURE 6.37 Construction of prismatic tanks.

Note: Prismatic tanks fully refrigerated, carry cargo at atmospheric pressure, and require a primary and secondary barrier to resist undetermined design stresses. The space between the primary and the secondary barrier is known as hold space and is filled with inert gas to prevent a flammable atmosphere being generated in the event of cargo leakage.

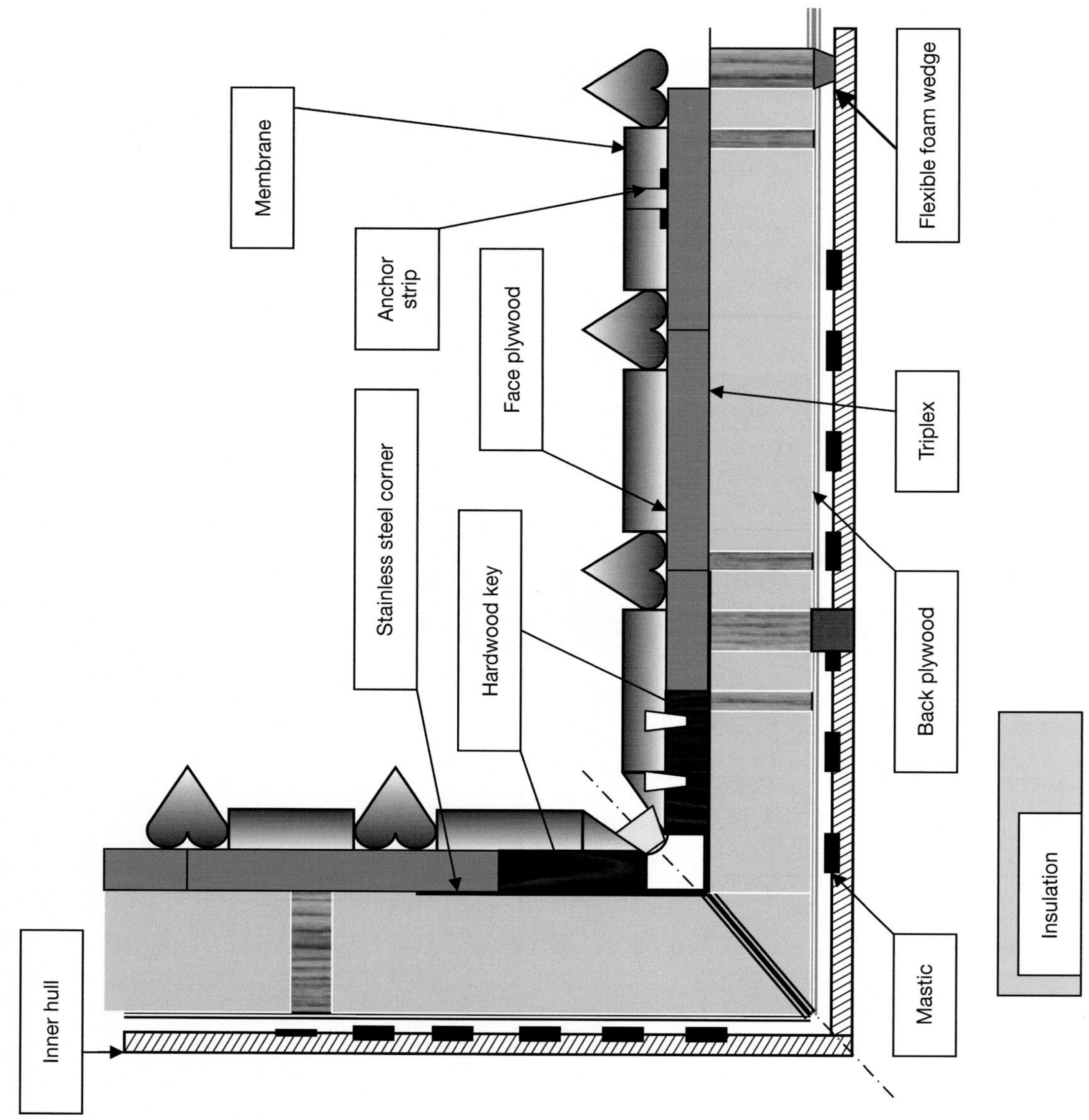

Membrane

Anchor strip

Face plywood

Stainless steel corner

Hardwood key

Inner hull

Flexible foam wedge

Triplex

Back plywood

Mastic

Insulation

FIGURE 6.38 Corner section of membrane tank structure (based on Technigaz system)

Gas Operational Knowledge

One of the main operational features of working on gas carriers is the awareness of personnel to what is and what is not a Gas-Dangerous Space. This is given by the following definition:

A Gas-Dangerous Space, or zone, is a space in the cargo area which is not arranged or equipped in an approved manner to ensure that its atmosphere is at all times maintained in a gas-safe condition.

Further: An enclosed space outside the cargo area through which any piping containing liquid or gaseous products passes, or within which such piping terminates, unless approved arrangements are installed to prevent any escape of product vapour into the atmosphere of that space.

Also: A cargo containment system and cargo piping.

And: A hold space where cargo is carried in a cargo containment system requiring a secondary barrier.

A space separated from a hold space described above by a single gas-tight steel boundary.

A cargo pump-room and cargo compressor room.

Or: A zone on the open deck, or semi-enclosed space on the open deck, within 3 m of any cargo tank outlet, gas or vapour outlet, cargo pipe flange or cargo valve or of entrances and ventilation openings to cargo pump-rooms and cargo compressor rooms.

The open deck over the cargo area and 3 m forward and aft of the cargo area on the open deck up to a height of 2.4 m above the weather deck.

A zone within 2.4 m of the outer surface of a cargo containment system where such surface is exposed to the weather.

An enclosed or semi-enclosed space in which pipes containing products are located. A space which contains gas detection equipment complying with Regulation 13.6.5 and space utilising boil-off gas as fuel and complying with Chapter 16 are not considered gas-dangerous spaces in this context.

A compartment for cargo hoses; or an enclosed or semi-enclosed space having a direct opening into any gas-dangerous space or zone.

A Gas-Safe Space is defined by a space other than a gas-dangerous space.

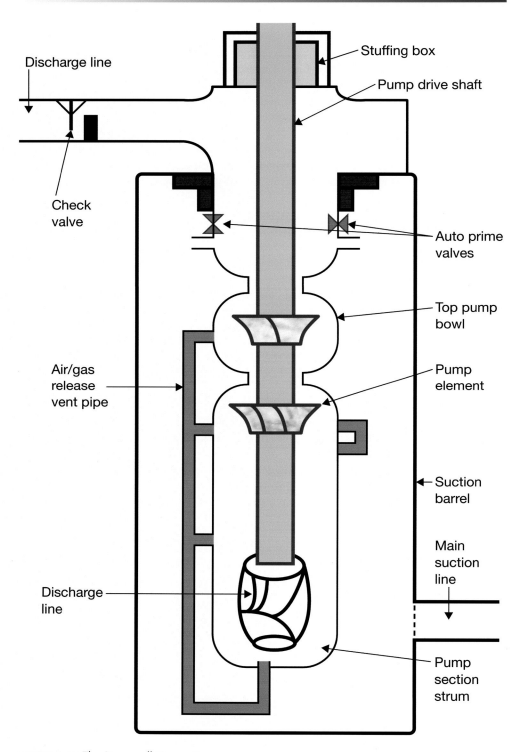

Discharge line

Check valve

Air/gas release vent pipe

Discharge line

Stuffing box

Pump drive shaft

Auto prime valves

Top pump bowl

Pump element

Suction barrel

Main suction line

Pump section strum

FIGURE 6.39 The Deepwell Cargo Pump.

Advantages of the deepwell cargo pumps are:

1 High-speed, high-efficiency and high-capacity pumps.
2 Compact in construction when installed in either the vertical or horizontal position.
3 Choice of power/drive – electric, steam, hydraulic or pneumatic.
4 Self-flooding.
5 Automatic self-priming and eliminates stripping problems.
6 Easy vertical withdrawal for maintenance purposes.
7 Easy drainage, essential on hazardous cargoes.
8 Tolerance of contaminates in fluid (no filters).
9 Improved duty regulation and performance.
10 Air and vapour locks.

Disadvantages include:

Being suspended, the pump can create construction problems during installation and requires essential rigid bracing supports within the tank, in order to prevent swaying. It also has a long drive shaft which is subject to vibration and torsional stresses.

Deepwell pumps must always be operated and handled in accordance with the recommended operating procedures. The net positive suction head (NPSH) requirements of the pump must always be maintained to prevent cavitation and subsequent pump damage.

FIGURE 6.40 The LPG vessel *Gaz Century* (IMO No. 9249685) lies at anchor off Gibraltar.

LNG CARRIERS

The LNG vessels are normally custom-built for the trade and carriage of the cargo at −162°C and essentially at atmospheric pressure. It is usual for LNG boil-off to be used as fuel for the ship's adopted main propulsion system and they subsequently are not always equipped with reliquefaction plant.

Cargo Operations – Safety

Three main safety aspects should be borne in mind when handling liquefied gas:

a   Flammability of the cargo and the need to avoid the formation of explosive mixtures at all times.

b   Toxicity of the cargo.

c   Low temperature of the cargo which could cause serious damage to the ship's hull.

Drying

Once a vessel is ordered to receive a cargo of LNG following overhaul or delivery trials, all traces of water must be removed from the tanks. If this is not done, operating problems due to freezing may result. The dew point of inert gas or air in equipment must be low enough to prevent condensation of water vapour when in contact with the cold surfaces encountered. Purging with dry gas refrigerated driers and dosing with methanol are not uncommon techniques for removing moisture.

Inerting

Once cargo tanks and associated equipment are suitably dried, air must be removed from the cargo system before loading to prevent the formation of explosive mixtures and also to prevent product contamination. Either inert gas from the ship's inert gas generator or a nitrogen supply from shore may be used. Inert gas from a shipboard inert gas generator is of a relatively low purity content in comparison with 'pure' nitrogen from a shoreside supply and usually will contain up to 15% CO_2 and 0.5% O_2. This can lead to contamination problems with cargoes such as ammonia, butadiene, etc. To prevent explosive mixture formation, the oxygen content of the tank must be reduced to 6% for hydrocarbon gases and 12% for ammonia using inert gas or nitrogen.

Purging

When the cargo tanks are suitably inerted, cargo vapours may be introduced to purge the tank of inerts. If the inerts are not completely purged from the tank then operating problems will be encountered in the reliquefaction plant operations. Inert gas is incondensable and can therefore lead to high pressure in the plant condenser, with associated difficulties.

The cargo vapours are introduced either at the top or bottom of the tank depending on the density of the gas and the vapour inert gas mixture is either vented through the vapour return to the shore flare stack or, where local port regulations allow, to the ship's vent stack.

COOLING OF CARGO TANKS

When about to load liquefied gases into tanks which are essentially at ambient temperature, it is important to avoid thermal stresses being generated in the ship's structure by incurring high temperature differences. A correct pre-cooling procedure should be adopted to make sure that the tank is brought down in temperature at a rate not exceeding 10°C per hour. The most common method of achieving this is to spray cargo liquid from ashore through the tank spray line situated at the top of the tank. This procedure is continued until liquid begins to form on the tank base. Cargo vapours are formed during this cool-down and are either returned through the vapour return line to the shore facility or, more commonly, handled by the ship's reliquefaction plant on board.

Loading

When tanks have been cooled down, loading of cargo can commence. The liquid is taken on board via a liquid crossover and fed to each tank through a liquid loading line. This line delivers to the base of each tank to avoid static electricity build-up. The loading rate is determined by the rate at which the vapours can be handled. Vapours are generated by: a) flashing of warm liquid; b) displacement; c) heat in leak through the tank insulation.

The vapour may be either taken ashore for shoreside reliquefaction or handled by the ship's own plant facility.

During the loading operation cargo tanks must be loaded with regard to trim and stability of the vessel at all times. Cargo tanks are fitted with high-level alarms to prevent overfilling. Loading rates should be reduced as the cargo levels approach desired values.

Discharging

Discharging can be accomplished by several different methods depending on the equipment which is available aboard the ship.

a   By use of a compressor alone:
     This is usually only associated with small pressurised carriers. The cargo is pressurised from the tank using a compressor taking suction from another cargo tank or with a vapour supply from ashore.

b   By compressor with booster pump on deck:
     The liquid over-pressured from the tank to the suction of the booster pump.

c   By means of deepwell pumps or submersible pumps installed in the tank.

d   By deepwell pumps operating in series with booster pumps mounted on deck:

This is required when discharging into pressurised or semi-pressurised facilities on shore and is carried out in conjunction with a cargo heater for heating the cargo.

An important feature when discharging cargoes is to remember that the cargo is a boiling liquid and will vaporise very easily under normal conditions. When the cargo has been discharged the vapour remaining in the cargo tank is pumped ashore using the compressor which would be subject to the design vacuum of the tank.

WORKING GAS CARGOES

FIGURE 6.41 The Monrovian LPG gas tanker *Annabella* (IMO No. 9250684) seen starboard side to the gas terminal in Barcelona while engaged in gas cargo operations.

Certificate of Fitness

An international Certificate of Fitness is required to be carried by any vessel engaged in the carriage of gases in bulk. The certificate is valid for a period not exceeding five (5) years or as specified by the Certifying Authority from the date of the initial survey or the periodical survey.

This certificate should be taken to mean that the vessel complies with the provisions of the Section of the Code and is designed and constructed under the International provisions of 1.1.5, and with the requirements of section 1.5 of the International Bulk Chemical Code.

The certificate can be issued or endorsed by another government on request.

No extension of the five-year period of validity will be permitted. It will cease to be valid if the ship is transferred to another flag state. A new certificate of fitness is issued only when the government issuing the new certificate is fully satisfied that the ship is in compliance with the requirements of 1.5.3.1 and 1.5.3.2. of the IGC Code.

Cargo Conditioning While at Sea

During the loaded passage, heat inleak through the tank insulation will cause the cargo pressure and the temperature to rise. The ship's reliquefaction plant should be used to maintain the cargo within the specified limits. In the event that malfunction occurs with the reliquefaction plant, then relief valves will blow vapour off to atmosphere via the ship's vent stack.

Changing Cargoes

Depending on the nature of the cargoes, it is often necessary to gas-free cargo tanks before changing grades.

On completion of the discharge procedure there will always be a little cargo liquid left in the tanks. It is important that this is removed before any gas-freeing is attempted, 1 m^3 of liquid forms 300 m^3 of vapour (depending on the substance) and this vapour formation could greatly extend the gas-freeing operation. The residual liquid is blown from the tank by means of an over-pressure created by inert gas and the liquid is generally blown overside while the ship is on a sea passage.

When the tanks and pipework are known to be liquid-free, the cargo vapours are swept from the tank using inert gas or nitrogen.

Note: Inert gas cannot be used when purging because of its high CO_2 content either before loading or after discharging ammonia cargoes. Ammonia would react with CO_2 to form sticky white carbonates.

The charter's requirements regarding product purity determine the procedure to be adopted on changing cargoes. Some cargoes such as vinyl chloride monomer (VCM) may require visual tank inspection before they can be loaded.

Tank Entry

The same general safety requirements relating to tank entry in oil and chemical carriers apply equally to gas carriers.

Reliquefaction Plant

The function of a reliquefaction plant is to handle vapours produced by heat inflow to the cargo. It is basically a refrigeration plant and may be direct, where the vapours are taken through a vapour compression cycle, or indirect, where the vapours are condensed on refrigerated surfaces such as cooling coils within or external to the tank.

The indirect cycle must be used for gases which cannot be compressed for chemical reasons, e.g. propylene.

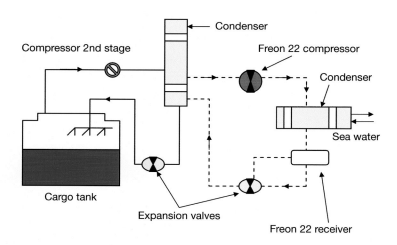

FIGURE 6.42 Reliquefaction direct cycle – cascade system.

The direct cycle can be either single stage or two stage where the cargo condenser is sea water cooled. Cascade cycle, is where the cargo condenser is refrigerated using a suitable refrigerant like Freon 22, within a separate direct expansion cycle.

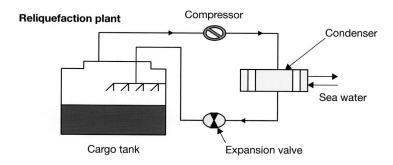

FIGURE 6.43 Direct cycle – single stage.

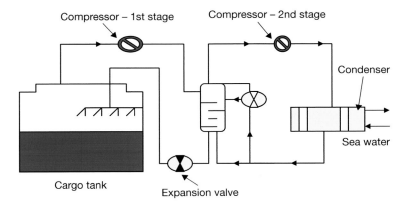

FIGURE 6.44 Reliquefaction plant direct cycle – two stage.

Pump Rooms

To reduce the risk of explosion, cargo compressors and booster pumps are sited in pump rooms divided into at least two compartments with gas-tight bulkheads. The motor's driving compressors are positioned on opposite side of the bulkheads with the connecting drive shafts fitted with bulkhead seals. Integrity of seals must be monitored and maintained at all times.

Note: Pump rooms are considered as 'Enclosed Spaces' and as such the full procedure for safe entry into an enclosed space must be adopted by personnel, as per the Code of Safe Working Practice. They are also equipped with Emergency Escape Breathing Apparatus (EEBAs) and full emergency fire-fighting apparatus is readily available.

Valves

Cargo tanks are protected from over-pressure by relief valves which have sufficient capacity to vent vapours produced under the conditions. Where liquid can be trapped between closed valves on pipework sections, liquid relief valves are fitted to protect against hydraulic pressure developing on expansion.

Liquid and vapour connections on tank domes and crossover are fitted with valves having quick closing actuators for remote operations. These actuators are in addition all interlocked with an emergency shutdown system with emergency operational buttons sited throughout the ship.

Instrumentation

Gas carriers are fitted with a gas detector system which continually monitors for cargo leakage. Sampling points are located, for example, in void spaces, pump rooms, motor rooms and control rooms, etc. The analyser will alarm on any sampling point reaching 30% of the lower explosive limit. In addition, portable gas detection equipment is provided as described under oil cargoes.

FIRE FIGHTING

Under the IMO Gas Code, gas carriers must be fitted with a water spray system capable of covering such areas as tank domes, manifolds, etc. Gas carriers must also be fitted with a fixed dry chemical powder system, actuated by inert gas under pressure, having at least two hand-held nozzles connected to the system.

ENTRY INTO ENCLOSED SPACES

The reader should make additional reference to the Code of Safe Working Practice for Merchant Seamen (MCA publication) regarding the topic of making entry into an enclosed space. A permit to work should also be obtained and a risk assessment completed prior to any person entering an enclosed space.

By definition, an enclosed space is one that has been closed or unventilated for some time; any space which may, because of cargo carried, containing harmful gases; any space which may be contaminated by cargo or gases leaking through a bulkhead or pipeline; any store room containing harmful materials; or any space which may be deficient of oxygen.

Examples of the above include: chain lockers, pump rooms, double bottoms, void spaces, CO_2 rooms, cofferdams, cargo stowage compartments.

Any person intending to enter such an enclosed space must seek correct authorisation from the ship's Master or Officer in Charge. Entry would be permitted in accordance with the conditions stipulated by a 'Permit to Work' for entry into enclosed spaces. The senior officer would also complete a risk assessment prior to entry taking place and all safety procedures being monitored by an appropriate safety check list.

A suggested line of action for permitted entry into enclosed compartments is as follows:

1   Obtain correct authorisation from the ship's Chief Officer.
2   Ensure that the space to be entered has been well ventilated and tested for oxygen content and/or toxic gases.
3   Check that ventilation arrangements are continued while persons are engaged inside the tank space.
4   Ensure that a rescue system and resuscitation equipment is available and ready for immediate use at the entrance to the space.

5 Ensure that persons entering have adequate communication equipment established and tested for contact to a standby man outside the enclosed space.

6 A responsible person is designated to stand by outside the space to be in constant attendance while person(s) are engaged inside the space. (Function of the standby individual is to raise the alarm in the event that difficulties are experienced by those persons entering the space.)

7 Ensure that the space to be entered is adequately illuminated prior to entry and that any portable lights are intrinsically safe and of an appropriate type.

8 Regular arrangements for the testing of the atmosphere inside the space should be in place.

9 Is a copy of the 'Permit to Work' displayed at the entrance of the space to be entered?

10 Have all operational personnel been briefed on withdrawal procedures from the space, in the event that such action is deemed necessary?

When the atmosphere inside an enclosed space is known to be unsafe, entry should not be made into that space.

Where the atmosphere in the compartment is suspect, the following additional safety precautions should be adopted with the use of breathing apparatus (B/A):

11 Ensure that the wearer of the breathing apparatus is fully trained in the use of the B/A.

12 Thorough checks on the B/A equipment must be made and the mask seal on the face of the wearer must be a proper fit.

13 The standby man should monitor the times of entry and exit of all personnel to allow adequate time for leaving the enclosed space.

14 Rescue harness and lifeline must be worn.

15 If the low pressure whistle alarm is activated the wearer must leave the space immediately.

16 In the event of communication or ventilation system breakdown, persons should leave the space immediately.

17 Operational personnel should never take the mask of the B/A off when inside the space.

18 The function of the standby man is only to raise the alarm if necessary. He should not attempt to effect a single-handed rescue with possible consequences of escalating the incident.

19 Emergency signals and communications should be clarified and understood by all affected parties.

20 A risk assessment must be completed by the Officer in Charge to take account of the items covered by the safety check list, the age and experience of the personnel involved, the prevailing weather conditions, the reliability of equipment in use, the possibility of related overlap of additional working practices ongoing, the technical expertise required to complete the task, and the time factor of how long the task is expected to take.

In all cases of enclosed space entry, the use of protective clothing, suitable footwear and the need for eye protection must be considered as an essential element of any risk assessment.

FIGURE 6.45 The *Cap Jean* typical crude oil tanker (IMO No. 9158147) seen exposed in the Lisnave dry dock complex, Portugal after hull and shipboard maintenance.

CHAPTER 7

Specialist Cargoes: Timber, Refrigerated and Livestock Cargoes

INTRODUCTION

The shipping world is actively engaged in trading in virtually every commodity. Many such cargoes fall into specific categories, such as the Container or the Roll On, Roll Off trades, and are easy to collate together under a single title or group. However, when attempting to gather all cargoes under one roof, so to speak, there is bound to be the odd product that falls outside the norm.

Such cargoes as timber, refrigerated ('reefer') and livestock could be discussed to fill a book in their own right. However, the outline of such products falls within the scope of this text which is meant to provide the Cargo Officer with the means to make an educated judgement as to the rights and wrongs of the stowage of such cargo types.

It should be appreciated that cargo handling methods have changed considerably over the years and the container and Ro-Ro trades have greatly affected quantities of raw products that were previously carried in open stow. This is no more so than in the refrigerated trades in foodstuffs and perishable goods, many of which are now shipped in refrigerated container units or Freezer Roll On, Roll Off trucks.

Timber products are shipped in the form of sawn timber in pre-slung bundles, or logs above or below decks. Wood flooring, packaged or on pallets may be shipped alongside wood pulp. Such is the variety of timber cargoes. The securing of timber deck cargoes and the concern for ship security against water absorption is always a concern to a ship's Master. It should be borne in mind that timber, absorbing great quantities of water at a high deck level, while burning off tonnes of fuel from low situated tanks, could dramatically affect the ship's GM and destroy the positive stability of the vessel.

Concern with the specialist cargoes must be exercised at all times. It is the duty of the Deck Officer to ensure that not only is the interest of the shipper to be taken account of, but also that of the ship owner, and the well-being of the crew/passengers must be of a high consideration.

DOI: 10.4324/9781003407706-7

Definitions and Terminology of Specialised Cargoes

Absorption – As associated with timber deck cargoes, an allowance made for weight of water absorbed by timber on deck which could have a detrimental effect on the ship's positive stability.

Cant – A log which is slab-cut, i.e. ripped lengthwise so that the resulting thick pieces have two opposing, parallel flat sides and in some cases a third side sawn flat.

CSWP for ships carrying timber deck cargoes (IMO 1991) – The Code of Safe Working Practice for the Carriage of Timber Deck Cargoes Aboard Ship.

Freon 12 – A chlorofluorocarbon (CFC) used as a refrigerant in reefer ships. It is due to be phased out by the Montreal Protocol and is expected to be replaced by a gas (R134a) which has less ozone depletion potential and less greenhouse potential. (Freon 22 has already been used in place of Freon 12.)

Livestock – A term which describes all types of domestic, farm and wild animals.

Pit props – Straight, short lengths of timber of a cross section suitable for shoring up the roof in a coal mine.

Reefer – A term for a refrigerated carrier.

Timber – Any sawn wood or lumber, cants, logs, poles, pulpwood and all other types of timber in loose or packaged forms. The term does not include wood pulp or similar cargo.

Timber deck cargo – A cargo of timber carried on an uncovered part of a freeboard or superstructure deck. The term does not include wood pulp or similar cargo.

Timber lashings – All lashings and securing components should possess a breaking strength of not less than 133 kN.

Timber loadline – A special loadline assigned to ships complying with certain conditions relating to their construction set out by the International Convention on Load Lines and used when the cargo complies with the stowage and securing conditions of this code.

Wood pulp and similar substances are not included in the timber terminology as far as deck cargo regulations are concerned. The air-dried chemical variety must be kept dry, as once it is allowed to get wet it will swell. This action could cause serious damage to the ship's structure and the compartment in which it is carried. To this end all ventilators and air pipes should be closed off to restrict any possibility of water entering the compartment.
   (S/F 3.06 / 3.34)

EXAMPLES OF TIMBER CARGOES

Timber is loaded in various forms with differing weights and methods being employed.

Package timber is generally handled with rope slings while the heavier logs, depending on size, are slung with wire snotters or chain slings.

Battens – Sawn timber more than 10 cm thick and approximately 15 to 18 cm wide. Usually shipped in standardised bundles and may be pre-slung for ease of handling.

Boards – Sawn timber boards of less than 5 cm thick but may be of any width.

Cord – A volume of 128 cu ft.

Deals – Sawn timber of not less than 5 cm thick and up to about 25 cm in width. A 'standard deal' is a single piece of timber measuring 1.83 m x 0.08 m x 0.28 m.

Fathom – (As a timber measure) equals 216 cu ft (6 ft x 6 ft x 6 ft).

Logs – Large and heavy pieces of timber, hewn or sawn. May also be referred to as 'baulks'. They are stowed above and below decks and individual logs may need to be considered as 'heavy lifts' for the SWL of the cargo handling gear being used.

Pit props – Short, straight lengths of timber stripped of bark and used for shoring up the ceilings of mines. They are shipped in a variety of sizes.

Stack – A measure of timber equal to half a 'fathom' and equating to 108 cu ft.

Note: The metric unit of Timber Measure is known as a 'Stere' and is 1 m^3 or 35.314 cu ft.

TIMBER CARGOES

Generally shipped as logs, pit props, or sawn packaged timber. The high stowage factor of timber (1.39 m/tonnes) generally indicates that a ship whose holds are full with forestry products will often not be down to her marks. For this reason an additional heavy cargo like ore is often booked alongside the timber cargo. Alternatively the more common method is to split the timber cargo to positions both below and above decks.

Where timber forms part of the deck stow, some thought should be made to route planning in order to provide a good-weather route. Prudent selection of a correct route could avoid prevailing bad weather and unnecessary concerns with the cargo absorbing high sea water quantities. The ship being loaded from the onset with adequate GM and sufficient positive stability could be directly affected in the event of shipping heavy seas in conjunction with timber deck cargoes. (Additional reference: Code of Safe Practice for Ships Carrying Timber Deck Cargoes, IMO. (2011) Inclusive of resolution A715 (17)).

Stowage and Lashing of Timber Deck Cargoes

Regulations for the stowage of timber emphasises that timber deck cargoes should be compactly stowed and secured by a series of overall lashings of adequate strength. Where uprights are envisaged as part of the securing, these uprights should be not more than three (3) metres apart. The maximum height of the timber stow above the uppermost deck must not exceed one-third of the ship's breadth when the vessel is navigating inside a seasonal winter zone.

Additional regulations apply if and when timber loadlines are being used, i.e. when the vessel is being loaded beyond the appropriate normal marks. These regulations take account of timber being stowed solidly in wells at least to the height of the forecastle. If there is no superstructure at the after end of the vessel, the timber must be stowed to at least the height of the forecastle. This stow must extend to at least the after end of the aftermost hatchway.

A further consideration is that the securing lashings should not be less than 19 mm close link chain (or flexible wire rope of equivalent strength). These lashings shall be independent of each other and spaced not more than three (3) metres apart. Such lashings will be fitted with slip hooks and stretching screws that must be accessible at all times

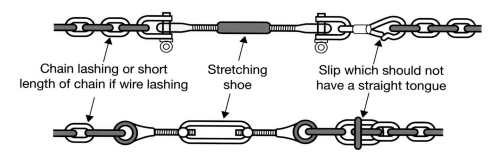

Chain lashing or short length of chain if wire lashing   Stretching shoe   Slip which should not have a straight tongue

FIGURE 7.1 Example of timber securing lashing.

Note: Wire rope lashings must be fitted with a short length of long link chain to permit the length to be adjusted and regulated.

LASHING POINTS

The lashings over timber cargoes are secured to eye plates attached to the sheer strake or deck stringer plate at intervals not exceeding more than 3 metres apart. The end securing point should be not more than 2 metres from a superstructure bulkhead, but if there is no bulkhead, then eye plates and lashings are to be provided at 0.6 m and 1.5 m from the ends of the timber deck stowage position. If the timber is in lengths of less than 3.6 m, the spacing's of the lashings are to be reduced. The following diagrams indicate some of these points.

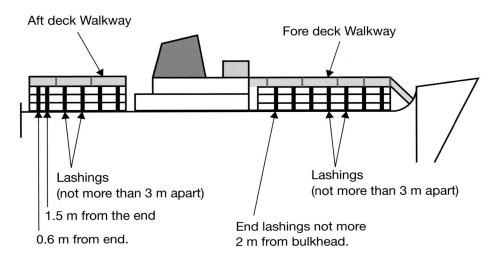

FIGURE 7.2 Lashing over timber deck stow.

Access to parts of the vessel fore and aft must be possible and when a capacity deck cargo is carried a walkway over the cargo is generally constructed.

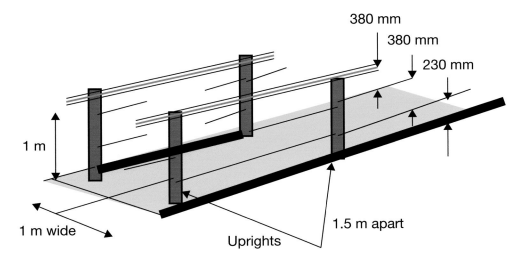

FIGURE 7.3 Example walkway construction.

STOWAGE AND THE WORKING OF TIMBER CARGOES

FIGURE 7.4 A cargo of heavy logs stowed under decks in twin hatches being discharged by multi-fold lifting purchase. Wire snotters are used to manoeuvre the logs to allow slings to be passed under, prior to discharging. Slings are often left in situ after loading to ease discharge, but not always. Sawn bundles of lumber are seen stowed at the hatch side as deck cargo.

Stowage of Logs

The Code of Safe Practice for Ships Carrying Timber Deck Cargoes (2011 TDC Code) (Appendix B) provides general guidelines for the underdeck stowage of logs.

Prior to loading logs below decks the compartment should be clean and hold bilges and lighting tested. A pre-stow loading plan should be prepared considering the length of the compartment and the various lengths of the logs to be loaded.

Recommendations are that logs should be stowed in the fore-and-aft direction in a compact manner. When loading, they should not be in a swinging motion and any swing should be stopped prior to lowering into the hatch. The heaviest logs should be loaded first and extreme pyramiding should be avoided as much as possible.

If void spaces exist at the fore and aft ends of log stows these may be filled with athwartships stowed logs. Logs loaded in between hatch coaming areas should be stowed as compact as possible to maximum capacity of the coaming space.

Logs are heavy and oscillations can expect to cause ship damage. Personnel are advised to maintain a careful watch during the loading/discharging periods.

FIGURE 7.5 A vessel loads short timber logs from quayside trailers to cargo holds. The uprights are identified for the purpose of securing lashings for any timber deck cargo.

Packaged Timber

This will usually be banded and may be pre-slung. Packages may not have standard dimensions and may have different lengths within the package, making compact stowage difficult. Uneven packages should not be loaded on deck and are preferred to be loaded below decks. Where deck stowage is made, the packages should be stowed in the fore-and-aft lengthwise position.

FIGURE 7.6 Tracked gantry crane loading packaged sawn timber, below decks in Napier, New Zealand. The outreaches of the gantry crane are seen extended to allow the hoist carriage to load stacked quayside loads to the cargo holds.

Wood Pulp

One of the most essential commodities used today is 'wood pulp' although the main usage is in the production of paper.* But nothing from the original hard wood and soft wood trees is wasted. The process of manufacture involves cooking wood chips with chemicals under pressure, this separates the wood from the cellulose fibres. The fibres are washed, screened, bleached and dried, prior to shipping as clean white banded bales. The product is then used in the making of newsprint and general paper products including tissue and packaging materials.

The softwoods like conifers, pine, and spruce tend to be made into baby diapers, paper towels, etc., because they have an increased absorbency over the hard woods like birch or eucalyptus.

> Note: Residuals from the manufacture of wood pulp go into edible products, clothing fabrics, bio-plastics and bio-energy products. In the interests of reducing global warming, forestry commissions plant four new trees for each mature tree that is harvested. The larger trees go to sawn timber and plywood products, while the small diameter trees from forest thinning go for pulp production.

From the cargo officer's point of view wood pulp bales must be kept dry during the loading and shipping period. If it is allowed to get wet it is known to expand considerably and could cause on board damage to the steelwork structure of the vessel. Prior to loading wood pulp, the bales should be observed ashore. The ship's interest would be to establish the bales are not wet or damp and have been kept under cover and not exposed to the weather elements.

(*See Chapter 3 for paper cargoes.)

REFRIGERATED CARGOES

The increase in container and Roll on, Roll off trades has to some extent brought about the demise of the conventional 'reefer' ship – one that was dedicated to carrying refrigerated and chilled cargoes in its main cargo carrying compartments, these being constructed with insulation to act as giant refrigerators. Some of these vessels still operate, particularly in the banana trade, but generally the cost of handling cargoes into reefer ships became uneconomic.

Refrigerated cargoes mainly fell into the category of foodstuffs by way of meat, dairy products, fruit, poultry, etc. As such a high degree of cleanliness was expected throughout the cargo compartments. Prior to loading such products the spaces were often surveyed and in virtually every case pre-cooling of dunnage and handling gear had to be carried out.

Bilge bays must be cleaned out and sweetened and the suctions tested to satisfaction. Brine traps should also be cleaned and refilled. Brine traps serve a dual purpose by preventing cold air reaching the bilge areas and so freezing any residual water while at the same time prevent odours from the bilges reaching into cargo compartments.

Compartment Insulation

All compartments are insulated for the purpose of reducing the load on the refrigeration plant and reducing heat loss from the compartment. It also provides time for engineers to instigate repairs in the event that machinery fails.

Qualities of a good insulation material are:

It should not absorb moisture.
It should not harbour vermin.
It should be fire resistant.
It must be odourless.
It should be low cost and available worldwide.
It should be light for draught considerations.
It should not have excessive settling levels as this would require re-packing.
It should have strength and durability.

Examples of insulation materials in use include polyurethane, plastics (PVC), aluminium foil, cork granules, glass wool.

With many frozen/chilled cargoes now being carried by refrigerated containers, designated reefer ships are becoming fewer in number. However, where one is still in operation, it is a requirement to maintain the cargo compartments at reduced temperatures and to this end they are well insulated with a variety of insulation materials. Glass fibre is a popular insulation material because it is lightweight, vermin free, fire resistant and does not readily absorb moisture. It is also readily available and not too expensive.

Cold compartments are usually pre-cooled prior to commencement of loading. Once loaded, locker access points are sealed to prevent loss of internal atmosphere and temperature change.

Where deck openings are present, as for access ladders or doors, these are very often closed or plugged with insulated pontoon covers, usually of deep timber construction, filled with cork and strapped with metal sheathing. Such constructions of inner compartments are prone to damage, especially where forklift trucks are employed to stack palletised cargoes. Essential maintenance becomes a requirement to ensure that insulation protection is maintained and compartment temperatures can be retained at desired low temperatures. Where the insulation is not effective, refrigeration plant needs to work at increased levels to retain the compartments at the required low temperatures.

Whenever frozen and chilled products are carried, efficient ship-keeping becomes essential. Temperature logs need to be maintained with regular spot checks on refrigeration compartments and associated machinery. Deck and engineering officers can expect to be directly involved with monitoring chilled and frozen cargoes. Logs would expect to show any temperature fluctuations in compartments and periods when fans are operated. Loading and discharge temperatures would always be noted.

Fully Refrigerated Cargo Hold (Insulation) Construction

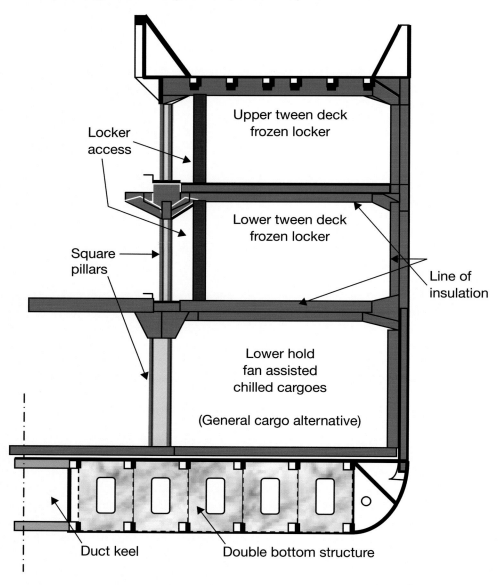

FIGURE 7.7 Athwartships section through fully refrigerated cargo hold by way of insulation.

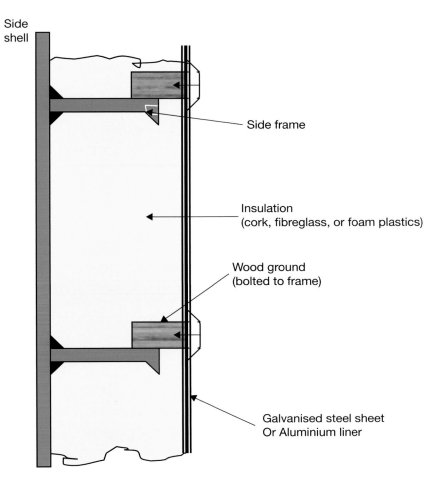

FIGURE 7.8 Side shell insulation.

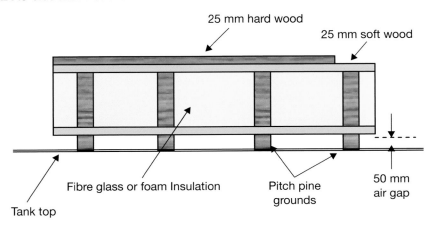

FIGURE 7.9 Tank top insulation.

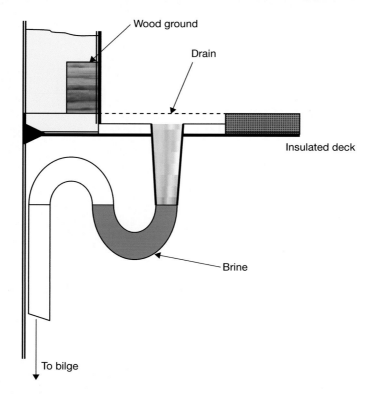

FIGURE 7.10 Example of air (brine) trap.

REFRIGERATION PLANT

Refrigerated cargoes other than those specifically carried in container or Ro-Ro units will be carried under the operation of the ship's own refrigeration plant. Cargo Officers are expected to have a working knowledge of the hardware involved with this cooling plant, and the ramifications in the event of machinery failure.

The majority of refrigeration plants in the marine environment operate on the 'vapour compression' system. (Absorption refrigeration systems are generally not used in the marine environment because they need a horizontal platform.)

The above plant is a direct expansion, grid cooling system. A refrigerant like Freon 12 $(C\ CL_2\ F_2)$ in its gaseous form is compressed, then liquefied in the condenser. It is then passed through into the grid pipeline of the compartment via the regulator valve. As it passes through the pipes it expands, extracting the heat from the compartment and producing the cooling effect.

Its operation is based on the principle that the boiling and condensation points of a liquid depend upon the pressure exerted on it – e.g. the boiling point of CO_2 at atmospheric pressure is about $-78°$; by increasing the pressure the temperature at which liquid CO_2 will vaporise is raised accordingly.

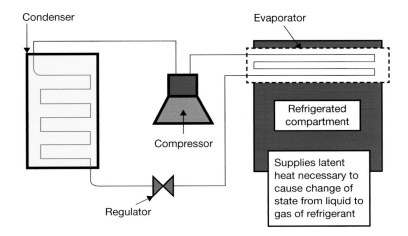

FIGURE 7.11 Operation of a vapour compression refrigeration system.

In the past, many refrigerants have been employed in marine refrigeration plants including carbon dioxide, ammonia and more recently the Freon's, but due to depletion of the ozone layer, more refined products are taking over from Freon 12.

Each refrigerant has specific qualities but the popular ones are those having least ozone depletion potential (ODP) and less greenhouse potential (GP). It should be non-poisonous, non-corrosive and require only a low working pressure to vaporise. Natural refrigerants like CO_2 and ammonia or blends of these are probably the main ones used in any remaining dedicated reefer vessels.

Qualities of a good refrigerant:

A high thermal dynamic efficiency.
Low cost.
Low working pressure and low volume.
Non-toxic, non-inflammable and not explosive.
Easily available worldwide.
High critical temperature.
High value of latent heat.
Non-corrosive.

Refrigeration Plant – Monitoring System

In order to protect cargoes, continual monitoring of the refrigeration machinery is considered a necessity. This can be achieved by the introduction of a Data Logging System to the relevant machinery and to the adjoining compartments. With such a system in operation there is less likelihood of damage because an early warning system would be activated, giving more time to provide corrective action before valuable cargoes are affected by loss of the cooling element.

Sensors and transducers monitor the following points:

Temperature of the cargo compartment.
Temperature of the fan outlet, discharge air.
Brine temperatures entering and leaving the evaporator.
Compressor suction and compression discharge.
Sea water temperature.
External air temperature.

Feedback of the sensed parameters are transmitted to either the cargo control room, the engine control station, or the navigation bridge. Alarm circuits are established to 24-hour manned stations.

Principal Refrigerated Cargoes and Respective Carriage Temperatures

Note: Chilling meat only slows the decomposition process down and it remains in prime condition for about 30 days. This period could be extended by about 15 days if a 10% concentration of CO_2 is introduced into the compartment, assuming the compartment can be sealed and the environment is safe to permit such action.

Product	Carriage temperature
Meats	
Frozen beef	About −10°C (15°F).
Frozen lamb/mutton	From about −8°C to −10°C (15°F to 18°F).
Frozen pork	About −10°C (15°F).
Offal and sundries (includes hearts, kidneys, livers, sweetbreads, tails and tongues)	Carried at as low a temperature as possible and not more than −10°C (15°F). Usually carried in bags or cases. Any which are bloodstained should be rejected.
Chilled beef	Loaded at about 0°C to 2°C, and carried at about −1.5°C (29°F to 29.5°F), unless instructed otherwise by the shipper.
Poultry	Packed in cases and carried at −10°C to −12°C (10°F to 15°F).
Dairy products	
Butter	Liable to taint and should not be stowed alongside other strong-smelling cargoes in the same compartment, e.g. fruit. Generally packed in cartons. Carriage temperature about −10°C (15°F).

Product	Carriage temperature
Cheese	Carriage temperature varies but generally carried at 5°C to 7°C average. Usually stowed on double dunnage.
Shell eggs	Stored in cases and liable to taint. Normally not stowed above 10 cases high with air circulation channels on top of 50 mm dunnage. Carriage temperature 1°C (33°F).
Liquid eggs	Carried in tins at temperatures not over −10°C (15°F).
Bacon	Stow on double 50 mm dunnage, do not over stow. Carrying temperature −10°C (15°F).
Fish	Shipped in boxes or crates and should be stowed on 50 mm dunnage. Fish has a tendency to rapid deterioration, it should be carried at as low a temperature as possible, which should not exceed −12°C (10°F).
Fruit	
	Fresh fruits are generally carried in cardboard cartons or wood boxes, with ventilation holes. They can often be carried in non-refrigerated spaces on short haul runs. Good ventilation must generally be given to prevent a concentration of CO_2 build-up. Carbon dioxide must not be allowed to build up over 3% concentration as this would cause deterioration of the cargo. Frequent air changes are recommended to avoid this.
Apples	Carriage temperature will vary with the variety of apple but is usually in the range of −1°C to 2°C.
Pears	Should not be stowed in the same compartment as apples. Carriage temperature −1°C to 0°C (30°F to 32°F).
Grapes, peaches, plums	Carriage temperature −1°C to 2°C (31°F to 35°F).
Oranges	Oranges must have adequate ventilation as they are very strong-smelling and the compartment must be deodorised after carriage. Carriage temperature 2°C to 5°C (36°F to 41°F).
Lemons	Similar to oranges. Carriage temperature 5°C to 7°C (41°F to 45°F).
Grapefruit	Similar stow to oranges, carriage at about 6°C (44°F).

Product	Carriage temperature
Bananas	The banana trade is specialised and special ships are built for the purpose, many of which use containers. The carriage temperature is critical as too low a temperature can permanently arrest the ripening process. Daily inspection of a compartment would be carried out and any fruit found to be ripe is removed. One ripe banana in a compartment can cause an acceleration of the ripening process throughout the compartment. Carriage temperature usually about 12°C (52°F to 54°F).

The 'Reefer' Trade

It should be realised that many of the said cargoes are now shipped by refrigerated containers or Ro-Ro cold units. Some companies still operate designated refrigeration vessels engaged on the New Zealand to the US West Coast and Europe to South America East Coast meat service.

Other specialised parcels, such as some drugs, often require refrigerated stowage and the instructions as to the carriage temperature would be issued by the shipper.

Prior to loading any refrigerated cargoes it is normal practice for a surveyor to inspect the compartment for cleanliness and to ensure that the compartment temperatures are correct.

Dunnage and any cargo fitments would be pre-cooled and machinery would be tested to satisfaction.

Cooled gas (LNG) and chemical cargoes are referred to in Chapter 5.

Refrigerated Container Units

Lloyds Register has developed 'Rules for the Carriage of Refrigerated Containers in Holds'. These standards take account of the problem of heat emanating from an online refrigeration plant operating below decks in the cargo hold. The heat energy rejected by each unit is from the evaporator fans, the motor and the condenser. Concern for this rejected heat energy into the surrounding air of the hold is currently considered a problem that may or may not be resolved by improved ventilation methods.

The container sector of the industry is exploring ways to carry increased numbers of reefer units below decks. However, such increase would generate increased temperatures into the cargo space areas. An effective ventilation system would probably aim to retain the hold temperature as close as possible to the outside air temperature or a predetermined temperature to suit the internal hold environment.

The majority of refrigerated containers employ insulation, usually polyurethane, within the prefabricated construction of the container. This directly affects the heat transmitted through the insulated unit between the carriage temperature and the external ambient air temperature. Although the insulation will reduce the actual payload capacity of the unit, it is seen as a necessary trade-off.

Example Carriage Temperatures for a 40 ft Container:	
Bananas	13.0°C
Chilled apples	2.0°C
Frozen	−18.0°C
Deep frozen	−29.0°C

Refrigerated Container – Hold Ventilation System

Various ventilation systems operate throughout the industry. The one illustrated is a semi-sealed louvred exhaust duct system. A vertical duct fitted with an air supply fan delivers supply air to each stack of containers, specifically to each container condenser.

The exhaust system operates in a similar manner, with the exception that the fan is an extraction fan as opposed to a supply fan. Isolation valves or flaps could be fitted to isolate 'cells' when not in use, each cell having its own inlet and outlet ducting.

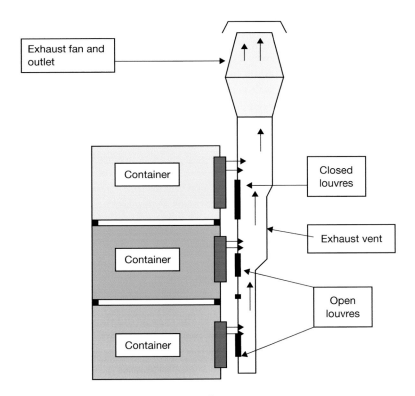

FIGURE 7.12 Refrigerator Container – hold ventilating system.

THE CARRIAGE OF LIVESTOCK

Reference: The Carriage of Livestock by Sea Regulations 2016.

The carriage of animals, either domestic, farm or from the wild, is not an uncommon practice. The carriage in the UK is governed by the regulations laid down by the Department for Environment, Food and Rural Affairs. Further advice can also be obtained from various animal protection societies like the Animal Welfare Foundation, who give advice on cage size, crates, etc. for use with animals.

Where large numbers of animals are to be carried, such as sheep or cattle, designated livestock carriers are available. The ships tend to discharge the beasts directly into quarantine, penned areas. While in transit the animals are kept in pens or stalls which are protected from adverse weather and the sun.

Adequate straw and fodder would also be carried. The feeding and watering of animals would be to the shipper's instructions. It is not unusual for a shipper to send a supervisory attendant, Stock-man where large numbers of animals are carried or where specialist animals like valuable racehorses are carried. If no attendant is carried, members of the crew would be designated to take care of the animals during the passage, cleaning stalls and feeding, etc.

Where one or two animals are carried by a non-designated vessel, they are usually carried in horsebox-type stalls or secured in caged kennels. These are generally kept on a sheltered area of the upper deck away from the prevailing weather. Each animal would be tallied and allotted a carriage number. In the event of the animal dying on passage,* this number must be recorded. All vessels carrying livestock must carry a 'humane killer' with enough ammunition to be considered adequate.

Where a regular livestock trade is featured, for example Australia/Middle East regions, shore facilities for loading and discharging are regularly inspected by the countries' authorities. Ministry officials also inspect the cleanliness and the facilities aboard approved livestock carriers.

Documentation inclusive of veterinary certificates are usually shipped with the animal(s) together with routine welfare instructions. When landed, documentation is usually landed at the same time, being handed to the shipper's representative or quarantine officials.

Animal Transportation by Road

Extensive planning and preparation is generally needed prior to moving any livestock. The logistical preparation is extensive depending on the type of animals and their number, and where they are travelling to and from. In some cases the preparations may take 95% of the effort with about 5% of the effort allocated to the actual movement. It would be expected that an accredited veterinarian can help with such procedures.

It must be anticipated at the outset that road transports may be needed at the loading and discharge stages. Such vehicles should have a 'Certificate of Vehicle Approval' and would expect to have a Transport Authorisation (UK).* The driver should also have a Certificate of Competence (for horses and poultry). Where moving with imports, exports, or transits of livestock of unregistered horses on journeys over eight hours, a journey log must be maintained.

Where a journey log is not kept an animal transport certificate must be carried.

> Note: *Transport Authorisation UK/EU:
>
> Type 1. For journeys over 65 km and up to 8 hours.
> Type 2. For journeys over 8 hours.

*Information regarding the slaughter of livestock is covered by welfare regulations and details may be found at https://www.gov.uk/government/collections/welfare-of-animals at the time of killing.

Welfare of the Animals – Transportation by Sea

Most countries are concerned with the welfare and hygiene conditions that will accompany any livestock movement. To this end Animal Welfare Inspectors may visit the vessel intending to carry out the transport. They would be expected to check on the cages, pens or stable areas that the animals would be accommodated in and the feeding arrangements for the period of the voyage. Adequate water supplies would also be a consideration.

An Animal Welfare Export Certificate (AWEC) will be required for import or export of animals and this certificate would usually be subject to a vessel inspection, before issue.

Specific guides on animal welfare are available for: poultry, laying hens, breeder chickens, gamebirds, pigs, sheep, goats, beef cattle, dairy cows, deer.

Where large numbers of animals are expected to be shipped, they are generally held in pens at the loading port, prior to being loaded.

Ship inspectors will want to ensure that adequate ventilation is available on board the carrier and that suitable sanitary arrangements are in use for the welfare of the livestock. The stores for animal feed/water are clean and contain ample fodder to sustain livestock to cover reasonable unexpected delays.

Pre-departure paperwork and relevant certificates would be inspected. Accommodations for stock-men and/or grooms would need to be allocated, where supernumeraries are travelling with livestock. Animals awaiting boarding would be inspected for their condition and health. Veterinary staff are not normally carried on the voyage except under special circumstances, such as where there are a lot of animals or where they have a particular interest.

> Note: Where animals are shipped in quantity under a Bill of Lading, the B/L may be endorsed stating that the ship is not responsible for the mortality rate of the animals carried during the period of passage.

FIGURE 7.13 Livestock transporter. Typical road transporter for 'Pigs' allows three tier carriage with roof and side ventilation also electrically driven fans and drinking wells to all three levels.

FIGURE 7.14 Pigs being loaded into a road transport through rear doors, direct from a farmyard lead in.

ABBREVIATIONS TO ORGANISATIONS CONCERNED WITH LIVESTOCK MOVEMENT

APHA (UK)	Animal & Plant Health Agency (https://www.gov.uk)
ASEL	Australian Standards for the Export of Livestock
ATA (USA)	Animal Transportation Association (animaltransportation.org)
BCP (EU)	Border Control Posts
DEFRA	Department of Environment, Food & Rural Affairs
DAERA	(N Ireland) Department of Agriculture, Environment and Rural Affairs
DAFM (GB)	Department of Agriculture, Food and Marine
FAWC	Farm Animal Welfare Council

Note: The reader so involved in the carriage of livestock should take further reference from: **'The Animal Welfare on Sea Vessels and Criteria for Approval of Livestock Authorisation'**.
http://www.europarl.europa.eu/thinktank/en/document.html

Domestic Pets

Most countries importing domestic animals would normally expect them to be microchipped and it is obviously in the owner's interest to ensure that registration takes place and that all quarantine regulations at the port of entry are adhered to. On occasion owners may travel with their animals and can expect to take a direct interest in their transport conditions.

Available legislation for pet protection can be found at https://www.gov.uk/animal-welfare-legislation-protecting-pets.

Racehorses

When horses of any kind are transported they are normally accommodated in horse boxes, which are well secured against ship movement. The specific movement of racehorses is particular because of the value and potential earnings of the animal. With this in mind, extra concern for the well-being of the animal is paramount and they are usually accompanied by a groom or grooms.

Cattle

Where cattle are being shipped in large numbers it is subjected to pre-conditioning. This includes getting the beasts ready nutritionally at least 30 days before actual travel takes place, and providing proper vaccinations to meet levels of immunity from disease, can be a lengthy period.

The animals should be accustomed to the people around them to avoid overly high stress and acclimatised to the feed and watering process. This is all to do with ensuring good healthy beasts that are capable of sustaining long haul transportation. Usually stock-tenders start and travel with cattle from ranch/ship/to ranch.

Some cattle can now also be transported in customised 'Cow-tainers'. These are specialised double level 40 ft containers, with good ventilation and non-slip flooring. Built with feed and water facilities to last throughout the period of transport, with a built-in daily clean system. This innovation can hold several animals and can be landed direct to a truck/ship.

Exceptional Transports

Occasionally, exceptional animals are transported, such as elephants or lions. Such beasts have their own precursory requirements. Elephants, for instance, being heavy, are usually lifted on board by means of canvas slings. Also the feeding and hygiene requirements for them will require stowage space and secure ventilated accommodations. Caged animals for lengthy voyages usually have an interlocking cage to permit safe cleaning and feeding arrangements.

Special animals usually travel with designated keepers in attendance to cater for specific animal needs and such persons would subsequently have to be accommodated for the duration of the voyage.

Note: Whenever animals are shipped it should be borne in mind that extensive regulations govern their movement and everything must be geared towards the welfare and continued well-being and health of the animals being carried.

Governments keenly regulate transports in order to prevent the spread of animal diseases crossing over their borders. Animals have compulsory health tests prior to loading and when being discharged.

The majority of entry ports have established quarantine facilities and imposed quarantine regulations apply immediately, when appropriate.

CHAPTER 8

Roll On, Roll Off Operations

INTRODUCTION

Some time after the start of containerisation, came a cargo revolution in the door-to-door service of Roll on, Roll off handling procedures. The Ro-Ro traffic provided a shuttle service for containers as well as cutting delivery times to hours rather than weeks previously experienced with conventional shipping. The Ro-Ro explosion was so great that ports changed their operations and ship design started to incorporate new concepts to handle large vehicles. The coastal traffic saw a new lease in life which opened up numerous avenues in employment, cargo handling methods, service industries and manufacturing. Ferry companies increased their tonnage maximums in a comparative blink of an eye. Port exports climbed beyond previous records, with Ro-Ro activity being the main cause.

Roll on, Roll off was an efficient and cheap method of shipping merchandise which was quickly realised and expanded rapidly beyond anyone's wildest expectations.

The ships' new design included the stern door/ramp, open vehicle deck spaces, drive-through capability with the bow visor. Vehicle lifts became a feature with open and enclosed deck cargo spaces. Units could be carrying liquid or dry cargoes, they could be refrigerated or not, as their load required. However, the most important fact was that they could be delivered in the shortest period of time.

The time factor was critical to ensure that goods reached markets in pristine condition, especially relevant to fresh produce such as flowers, fruits, dairy foods, meats, etc. The ships were enhanced to ensure that deadlines were achieved. Ships docking in and out carrying such cargoes could not be delayed by the need for tugs. Bow/stern thrusters became essential features of ship design. Thruster units came alongside twin controllable pitch propellers, while Masters were given Pilotage Exemption Certificates. Not only were the vessels fast, but the procedures and concepts of ship handling had been changed to meet the needs of the trade.

The Roll on, Roll off trade has now become an essential segment of the shipping industry. Although it might have been seen as the new boy on the block, not now, it is well established as a major commercial shipping sector. Working in close association with the container and passenger sectors.

DOI: 10.4324/9781003407706-8

Roll On, Roll Off Shipping

FIGURE 8.1 The *Super Seacat Two* (IMO No. 9141845) when operating with the Isle of Man Steam Packet Company, on the Irish Sea trades.

FIGURE 8.2 The Ro-Ro vessel *Clipper Pace* (IMO No. 9350678) operating on the Irish Sea between Liverpool and Dublin.

Roll On, Roll Off Definitions and Terminology

Freight only Ro-Ro ship – A Roll on, Roll off vessel with accommodation for not more than 12 (driver) passengers.

High-speed craft – A craft capable of a maximum speed, in metres per second (m/s) equal to or exceeding $3.7V^{0.1667}$ metres per second where V = Displacement corresponding to the design waterline (m^3).

Open Ro-Ro space – Those Ro-Ro spaces which are either open at both ends or have an opening at one end, and are provided with adequate natural ventilation effective over their entire length through permeant openings distributed in the side plating or deck head or from above, having a total area of at least 10% of the total area of the space sides.

Passenger car ferry – A passenger or ferry ship which has Ro-Ro access of sufficient dimensions to allow the carriage of Ro-Ro trailers and/or passengers/cars.

Reefer unit – A mobile/vehicle Ro-Ro unit, designed for and capable of carrying refrigerated cargoes.

Right of ferry – An exclusive right to convey persons or goods (or both) across a river or arm of the sea and to charge reasonable tolls for the service.

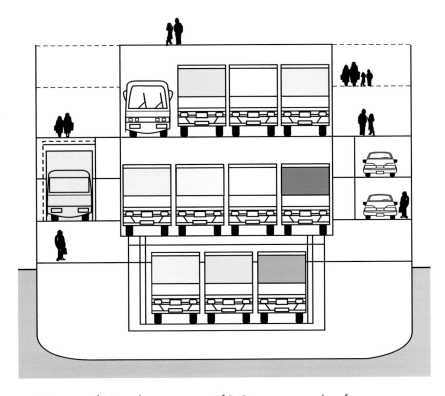

FIGURE 8.3 Diagram of internal arrangement of RoPax passenger/car ferry.

Ro-Ro cargo space – A space not normally subdivided in any way and extending to either a substantial length or the entire length of the vessel in which goods (packaged or in bulk) in or on rail or road cars, vehicles (including road or rail tankers), trailers, containers, pallets, demountable tanks or in or on similar stowage units or other receptacles can be loaded and unloaded normally in a horizontal direction.

Ro-Ro vessel – A vessel which is provided with horizontal means of access and discharge for wheeled, tracked or mobile cargo.

FIGURE 8.4 The modern face of Ro-Ro/Passenger (RoPax) type vessels. High-speed catamaran or trimaran hulls with vehicle access from a stern ramp. Generally engaged on the short sea trade. The *Seacat Isle of Man* (IMO No. 9176072) passenger/vehicle ferry engaged on the Irish Sea trade between Liverpool and the Isle of Man carrying vehicles and passengers.

Short international voyage – An international voyage in the course of which a ship is not more than 200 nautical miles from the port or place in which passengers and crew could be placed in safety. Neither the distance between the last port of call in the country in which the voyage begins and the final port of destination, nor the return voyage shall exceed 600 nautical miles. The final port of destination is the last port of call in the scheduled voyage at which the ship commences its return voyage to the country at which the voyage began.

Special category space – Any enclosed space, above or below the bulkhead deck intended for the carriage of motor vehicles with fuel in their tanks for their own propulsion, into and from which such vehicles can be driven and to which passengers have access.

EXAMPLE: MODERN ROLL ON, ROLL OFF (FREIGHT ONLY) VESSEL

FIGURE 8.5 the *MYKONOΣ* modern Greek operated Roll on, Roll off ferry. Design features include all accommodation forward. A stern door/combined vehicle ramp right aft in the upright closed position when the vessel is at sea.

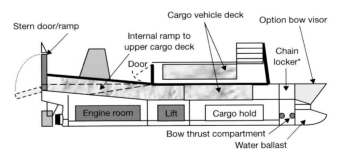

- Open deck stowage of Ro-Ro cargo either side of the engine room smoke stack.
- Chain locker at the ship's sides either side of the fore and aft line to facilitate the operation of separate port and starboard windlass operations.
- Bow visor option to permit drive-through capability – not always featured.

- Lift to lower cargo hold may be mechanical or hydraulic operation.
- All cargo ramps are fitted with wheel tread, anti-skid, steel grips.
- All cargo decks are fitted with insert star lashing points and/or star domes.
- Accommodation for twelve (12) driver/passengers.

FIGURE 8.6 General arrangement – modern Ro-Ro ferry (freight only) 1,900 metres lane length.

THE ROLL ON, ROLL OFF (ROPAX) FERRY

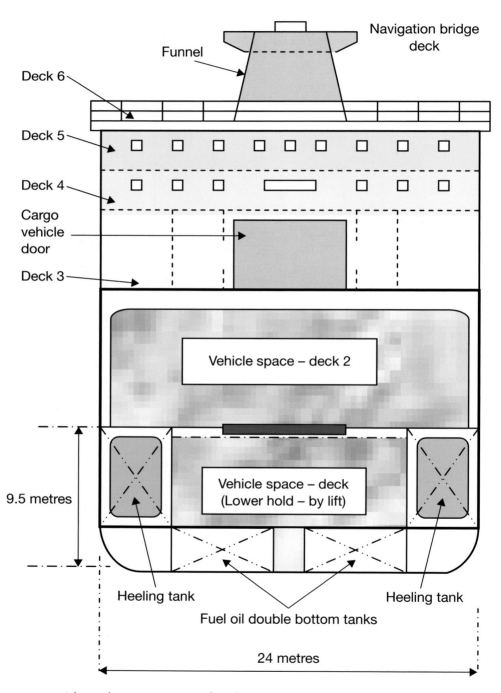

FIGURE 8.7 Athwartships cross section of modern RoPax vessel.

FIGURE 8.8 Example stern door of a RoPax ferry *Al Masour* (IMO No. 7360629) seen in the stowed, closed position as she turns off the berth in the harbour of Tangier.

FIGURE 8.9 Bow visor arrangement on the RoPax vessel *Jupiter* (IMO No. 7360186) (since renamed) seen in the open position against the skyline. The bow-visor fitted with a stern ramp access permits a drive through capability.

VEHICLE RAMPS

The design of Ro-Ro vessels is influenced from the onset of the design stage by the nature of the payload it is intended to transport. Generally the cargo flow, securing and handling equipment can amount to about 5% of the lightweight tonnage. However, to avoid operational problems in the future such fittings need to take account of the types of rolling cargo which is anticipated. Commercial vehicles are limited to about six types (unlike military vehicles) and these need to be accommodated by respective access widths, ramp slopes, clearing heights, lane lengths, turning areas or drive-through facilities.

Similarly, shoreside receptions must be compatible with ship's facilities. Ramp slopes and break angles for commercial traffic will generally fall at about 1:8 or 1:10 in order to avoid the vehicle grounding while in transit from the ship to the shore. Where tidal waters are present and average rise or fall is expected, floating shore links or adjustable link spans tend to overcome excessive tidal movement while at the same time keeping the break angle with the ship's ramp manageable.

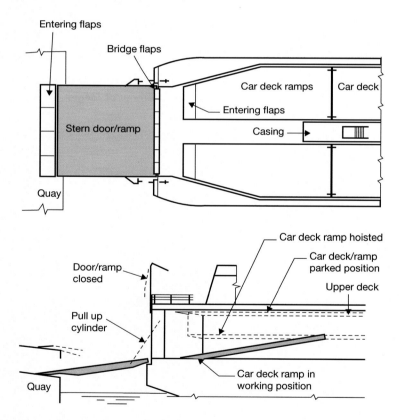

FIGURE 8.10 Example stern door and vehicle ramp arrangement.

Stern Door/Ramp Qualities

The design of equipment will be to the requirements of Lloyds Register or similar classification society, but would include specific features to satisfy operational needs. In order to match these needs and provide a suitable end product, a designer would include the following features:

1 Length of ramp (overall)

2 Width of ramp (overall)

3 Total load on ramp (anticipated maximum)

4 Maximum axle loads

5 Hinging arrangement (top, bottom or guillotine)

6 Number of ramp sections and hinges within the structure

7 Maximum/minimum operating angles

8 Watertight sealing/securing arrangements

9 Cleating/locking arrangements

10 Power requirements (electric, hydraulic) with limitations

11 Operational lifting/lowering times

12 Supporting and preventer arrangements

13 Roadway landing area.

Many stern ramp arrangements open up all the transom to provide maximum width and height clearance. This effectively gives wide access to a variety of vehicles of differing lengths with comparative short load/discharge times involved. Other designs have employed stern quarter ramps (with or without bow quarter ramps). Such ramps are still required to meet the design criteria of the classification society but must also satisfy design features to meet specific vehicle traffic, such as 'car carriers'.

Ramps tend to be manufactured in steel with 'chevron' pattern anti-skid bars on the working surface. They are usually operated by twin hydraulic cylinder actions or winch arrangement. Watertight integrity is achieved with hydraulic pressure cleating in conjunction with a hard rubber seal, with the hinge arrangement being positioned above the waterline.

Example Axle/Vehicle Loads

Stern ramps and vehicle doors are designed to accept most types of vehicle. The following ramp examples are for a 16 m length (inclusive of the flaps) and a total width of 18.6 metres:

Forklift truck: Axle load 15.8 tonnes, over four wheels.

Mafi-trailer: Axle load 27.5 tonnes, over four wheels Bogie load 55 tonnes, two axles.

Road trailer: Three axles (20 t – 20 t – 10 t).

Axle load on pneumatic tyres 20 tonnes.

Tug master: Axle load 35 tonnes.

FIGURE 8.11 Example stern door/ramp access into the enclosed vehicle deck of a Ro-Ro ferry operating on the Irish Sea trade. Fixed ramp aspect seen on the port side of the vessel leads to the upper cargo deck where drop units have already been stowed.

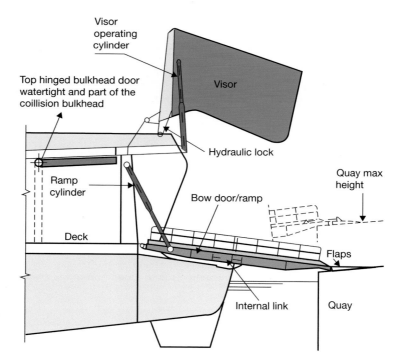

FIGURE 8.12 Bow visor with combination inner bow door and vehicle ramp.

FIGURE 8.13 The RoPax Ferry *Ben-My-Cree* (IMO No. 9170705) seen stern to the vehicle link in Heysham harbour. The stern door/ramp is open and the vessel is loading vehicle cargo. Upper deck is already part loaded with drop trailer units.

FIGURE 8.14 Example upper vehicle deck of the *Ben-My-Cree* RoPax ferry operating in the Irish Sea region. Drop trailer units are seen stowed on trestles in vehicle lanes. Load binder chain lashings are seen securing the units to star dome lashing points.

Ferry Features

The current generation of Ro-Ro vessels have moved into multi-deck construction with a totally enclosed main vehicle deck with access from either a stern ramp or bow door arrangement. This deck may be fitted with an elevator access to a lower hold while an internal ramp to a partially covered upper vehicle deck permits access to the higher, uppermost continuous deck.

Some ferry transports are built with an internal tilting ramp from the main deck to an upper deck level. Such a tilting ramp could be deployed to facilitate a bow/forward discharge to avoid freight wagons having to turn around when leaving the park position and allows a faster discharge.

Modern builds tend to incorporate as many labour-saving devices as practicality permits. Such designs tend to reduce crew complement and effectively reduce operational costs.

Where a vessel is fitted with a lower hold and hydraulic lift, it allows increased units to be loaded and improves the freight. However, operating the lift requires trained, manual labour and depending on the number of units, can be time consuming for cargo officers.

FIGURE 8.15 A runner (Tractor Unit) moves along the internal fixed ramp to the upper stowage vehicle deck.

Internal Cargo Operations – Ro-Ro Vessel

Vehicles require wide open deck space to be able to manoeuvre. Such deck areas are lane marked to ease vehicle stowage and alignment of mixed types of vehicle, e.g. private cars and commercial trucks. The deck areas are always well illuminated by overhead lighting

FIGURE 8.16 The upper vehicle main deck of the ferry *Clipper Point*. Vehicle lanes are clearly identified and trestles are positioned at the sides for use with drop trailer units.

FIGURE 8.17 The same vehicle deck of the ferry *Clipper Point* (IMO No. 9350666) fully loaded with drop trailers and runner units on the ramp on the port side.

and fitted with extraction fans to change the air volume ten times every hour. Such atmosphere replenishment prevents the build-up of exhaust gases from drive on, drive off operations.

Cargo vehicle decks are protected by sprinkler and/or water drenching systems and well provided for with fire extinguishers at every 40-metre length. Additional extinguishers are also positioned at access points. Such protection dictates that the decks must also be fitted with an adequate drainage system to clear residual waters quickly.

Vessels without 'bow visor' facilities are generally denied drive-through capabilities and usually must provide sufficient deck space to permit the turning of wagons ready for stern discharge at the arrival port. Vehicle decks have always been considered as a hazardous environment for both shore and shipboard personnel, especially where vehicles are turning. To this end speed of vehicles is strictly controlled by stowage marshals who usher units into designated lane spaces. With this in mind deck spaces are clearly sign painted to reflect basic instructions to driver personnel and car/passenger travellers.

FIGURE 8.18 View inside the lower hold of the ferry *Clipper Point*. A lower ramp from the main vehicle deck is seen on the starboard side and the vehicle stowage lanes are indicated on the deck of the hold. Some Ro-Ro vessels do not have a fixed ramp into lower hold spaces and employ a built in hydraulic lift operation to load units below.

VENTILATION SYSTEM

It is a requirement that RoPax vessels carrying more than 36 passengers must be provided with a powered ventilation system (fans) sufficient to give ten (10) air changes per hour

FIGURE 8.19 Typical bulkhead markings prominently displayed around vehicle decks to ensure safe and efficient loading of vehicle lanes.

in spaces designated to carry vehicles (with fuel in their tanks for their own propulsion). If the vessel carries less than 36 passengers then the venting system need only provide six (6) air changes per hour. When carrying more than 36 passengers, then ten changes per hour are required.

Ventilation ducting serving such spaces should be constructed in steel and the system should be completely separate from other ventilation systems aboard the vessel. It must be capable of being controlled from outside the vehicle spaces and be operable at all times when vehicles are occupying specific areas.

> Note: Where special category spaces are employed the administration may require an additional number of air changes when vehicles are being loaded or discharged.

Ventilation systems must be fitted with rapid means of shut-down in the event of fire occurring. They must also have a means of monitoring any loss or reduction in the venting capacity, with such data being indicated on the navigation bridge.

In cargo ships ventilation systems are expected to run continuously when vehicles are on board. If this is impractical they should be operated for a limited period before discharge and afterwards the space should be proved to be gas-free. Portable gas-detecting instruments should be carried for this purpose.

Drainage Systems

The hazards of slack water on large vehicle decks and the subsequent loss of stability which could occur are well known. The fixed pressure water spraying system installed for fire prevention, if operated, could cause an accumulation of water on vehicle deck or decks. To ensure adequate stability at all times a suitable drainage system must be installed to effect rapid discharge of slack water.

Scuppers should be fitted to ensure discharge directly overboard. Special category spaces situated above the bulkhead deck, and in all RoPax vessels which have positive means of closing scuppers by valve action, must keep such valves open while the vessel is at sea in accord with the Load Line Convention.

In the case of special category spaces, the administration may require additional bilge pumping and drainage facilities over and above the specifications of SOLAS, Regulation II-1/21.

Note: A special category space is defined as those enclosed vehicle spaces above and below the bulkhead deck into and from which vehicles can be driven and to which passengers have access. Special category spaces may be accommodated on more than one deck provided that the total overall clear height for vehicles does not exceed 10 m.

BILGE PUMPING ARRANGEMENTS

Cargo ships and passenger vessels are required to have in place an efficient bilge pumping system, capable of pumping from and draining any watertight compartment. Passenger ships are required to have at least three power pumps connected to the bilge main.

A feature of modern Ro-Ro construction, where the need is to provide a flat uniform surface is that bilge cover plates are strong and counter sunk to provide a flush driving surface so as not to impede traffic movement or cause an obstructed deck.

FIRE EXTINCTION

Vehicle spaces and Ro-Ro spaces which are not special category spaces and are capable of being sealed from a location outside of the cargo spaces shall be fitted with a fixed gas fire-extinguishing system which shall comply with the provisions of the Fire Safety Systems Code.

Ro-Ro and vehicle spaces not capable of being sealed and special category spaces shall be fitted with an approved fixed pressure water spray system for manual operation which shall protect all parts of any deck and vehicle platform in such spaces.

Fire extinguishers must also be provided at each deck level in compartments where vehicles are carried, spaced not more than 20 m apart on both sides of the space. Also, at least one placed at each access to each space(SOLAS Reg, 20, 6.2.1.)

CARGO (VEHICLE) DEFINITIONS

A **vehicle** is defined as a vehicle with wheels or a track-laying vehicle.

A **flat-bed trailer** is defined as a flat-topped open-sided trailer or semi-trailer and includes a roll trailer and a draw-bar trailer.

A **freight vehicle** is defined as a vehicle which is a goods vehicle (flat-bed trailer) (road train) (articulated road train) combination of freight vehicles or a tank vehicle.

A **semi-trailer** is defined as a trailer which is designed to be coupled to a semi-trailer towing vehicle and to impose a substantial part of its total weight on the towing vehicle.

A **tank vehicle** is defined as a vehicle fitted with a tank which is rigidly and permanently attached to the vehicle during all normal operations of loading, discharging and transport and is neither filled nor discharged on board and driven on board by its own wheels.

Reefer unit: Container box unit fitted with refrigeration plant. Employed to transport frozen or chilled produce by road and sea. Power for the freezer unit is generated by the drive motor of the unit when on the road and supplied from the ship's supply while the vessel is at sea. (Special stowage space is required for reefer units to ensure that they are positioned aboard the ferry close to a power supply connection.)

FIGURE 8.20 A long concrete load seen chained to a sixteen-wheel extending trailer after leaving the undercover hold of the ferry *Clipper Point* (IMO No. 9350666).

FIGURE 8.21 Long load seen on the main vehicle deck after the vehicle lashings have been removed and the unit is ready to discharge from the ferry *Clipper Point*.

RO-RO VEHICLE TYPES

The majority of freight vehicles engaged in Ro-Ro vessels vary in size and type, for the shipment of not just dry goods but liquid cargoes as well. Probably the most widely used is the drop trailer vans (40 ft container box/van stowed on a horse or trestle and the rear wheels of the unit). Other varieties include:

Curtain-sided trailers
Semi-trailer without sideboards (drop sides)
Semi-trailer with sideboards
Semi-trailer with sideboards and hood cover
Fully enclosed goods vehicle
Open flat top truck
Flat top truck with canvas covered load
Articulated trailer
Road tanker
Framed container/tank
Freight container (20' x 8' x 8')
Freight container (40' x 8' x 8')
Draw-bar combination (two units)
Draw-bar combination (three units)

Refrigerated (reefer) vans
Low-loaders (for heavy machinery/plant and tracked vehicles)
Adjustable (stretch) low-loader (for exceptional long loads)

Additional private vehicles such as coaches, furniture removal vans, buses, caravans, boats on trailers, military transports, etc. are also regularly shipped, known as runners.

Freight units of one kind or another, once discharged, may be reloaded but not in every case. Many units are often returned by the same ferry or a sister vessel in an empty state.

ROLL ON, ROLL OFF UNIT TYPES

FIGURE 8.22 Possible heavy duty vehicles seen stowed on the upper weather deck of Ro-Ro vessel, prior to being fitted with chain lashings as per the Cargo Securing Manual (CSM). Farm tractors, digger vehicles etc, moving on tyres, are treated as runners. Tracked vehicles like military tanks or bulldozes are usually secured to a low-loader by chains and stowed as an integral unit.

Example of Vehicle Ferry Cargo Plan

FIGURE 8.23 Roll on, Roll off vehicles being stowed on the internal cargo deck of a large vehicle ferry. Cargo officer monitors tight stowage in the 'vehicle lanes' seen walking past the inner ramp to the upper cargo deck.

VEHICLE STOWAGE AND SECURING

It is essential with vessels carrying vehicles that a stable deck is maintained and this is why virtually all Roll on, Roll off ferries are now built with stabilisers of one form or another. However, cargo movement can still expect to occur in very rough sea conditions even when stabilisation systems are operational. To this end, individual vehicles are secured by various means to prevent movement at sea.

The stowage/securing arrangements of units should be supervised by a responsible ship's officer assisted by at least one other competent person. Vehicles should as far as possible be aligned fore and aft, with sufficient distance between vehicles so as to allow access through the vehicle deck. The parking brake on each vehicle/unit should be applied and where possible the unit should be placed in gear. Where drop loads or uncoupled units are being carried, these should be landed on trestles or equivalent support, prior to being secured by chain or other suitable securing constraint.

All vehicle/cargo units should be secured prior to the vessel leaving the berth and such securings should be at the Master's discretion to be most effective. While on route these lashings should be regularly inspected to ensure they remain effective during the time at sea. It should also be realised that personnel so engaged on vehicle deck inspections should take extreme caution against injury from swaying vehicles. As such, Masters may feel it appropriate to alter the ship's course while such inspections are ongoing to reduce the motion on the vehicle deck.

Vehicles stowed on slanting decks should have the wheels 'chocked' and the hand-brakes observed to be on and working. Suitable lashings against the incline should be secured and the unit left in an opposing gear. Any vehicle which is lashed should be secured at the correct securing points so designed on the vehicle and at the deck position.

All lashings applied, whether of a 'hook' type or other variety, should be secured in such a manner that in the event of them becoming slack, they are prevented from becoming detached. They should also be of a type which will permit tensioning in the event of them becoming slack during the voyage.

> Note: Lashings are considered to be most effective at between 30° and 60° to the deck line. Alternatively additional lashings may be required. Crossed lashing should where practical not be used, as limited restraint against tipping is experienced with this style of securing.

Lashings should only be released once the ship is secured at the berth and personnel so engaged should take care when clearing securing's. These may be under high tension following transit and cause injury if released without forethought.

Unit Securing – Chain Lashings

Ro-Ro units are secured in accordance with the Cargo Securing Manual of the vessel. In some short sea voyages, during the summer season and with a predominantly good weather forecast units may not even be secured other than by the handbrakes and left in gear. However, at the Master's discretion chain lashings could be applied by the crew if and when circumstances dictate that securing becomes necessary.

In virtually all cases, hazardous units would automatically be chained down. Chain lashings vary but tend to have a common theme of being able to be applied between a deck 'star' lashing point and the unit itself, then tensioned by a load-binding lever.

Such lashings can be secured and tensioned quickly and lend to labour saving. The number of lashings per unit will be variable, depending on the weight and size of the vehicle. However a standard 40 ft unit would usually be fitted with a minimum of six lashings.

Vehicle decks are built with star lashing points or 'elephant's feet' type anchor points. Lashings will have a club foot fitting into these points, with a hook at the opposite end. Alternatively, as shown in the diagram, hooks at each end.

Drop Unit Stowage

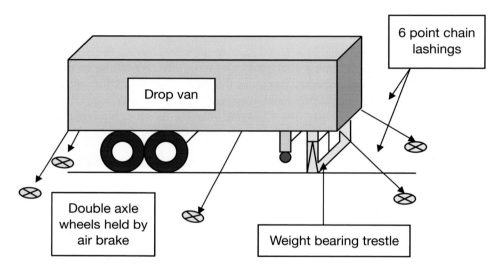

FIGURE 8.24 Drop unit stowage and securing.

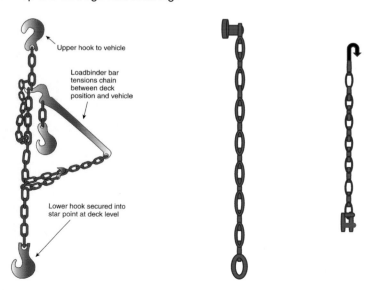

FIGURE 8.25 Chain lashing options.

FIGURE 8.26 Deck star dome, raised or flush for the anchoring of chain lashings for Ro-Ro cargo units by hook or club foot attachment to the chain lashing end.

FIGURE 8.27 Wheeled trestle is positioned under the unit before being detached from the motor tug. Trestles are fitted with spring-loaded wheels to permit easy manoeuvring under the cargo unit.

Note: Cargo units must be stowed and secured in accord with the ship's Cargo Securing Manual (CSM), as approved by the Marine Authority. The CSM is required to be carried aboard all types of ships engaged in the carriage of unitised cargoes with the exception of bulk cargoes, like oil or grain.

RO-RO SHIP STABILITY

Modern Roll on, Roll off shipping has experienced some painful losses over the years, the most notable being the *Herald of Free Enterprise* (1987), the *Estonia* (1994) and, more recently, the *Tricolour* (2003) with 2,800 cars and the *Hyundai No.105* (2004) with more than 4,000 vehicles on board. Clearly, the losses and subsequent salvage operations have rocked the marine insurance markets, generating tighter legislation to cause improved conditions on Ro-Ro vessels.

Improvement features now include the following:

1   The stability of the vessel must be assured as adequate with the main deck flooded to a depth of 50 cm of water.

2   Cargo loading computers must have a direct link to the shoreside administration.

3   The vessel will be fitted with automatic draught gauges.

4   All access points to inner compartments will be monitored by CCTV and have light open/shut indicators displayed to the navigation bridge.

5   Increased drainage facilities must be fitted to vehicle decks.

6   Individual units will be weighed and respective kg measured ashore for transmission to the vessel's Cargo Officer.

Inherent Dangers Associated with Ro-Ro Vessels

The ships themselves generally have high freeboards and expect to experience high windage over and above the waterline. Cargo units are by necessity loaded with a high kg value, which can be detrimental to the overall ships GM. In the event of bad weather conditions these features tend to lend to the vessel rolling heavily, which may generate units shifting.

To improve these conditions most Ro-Ro vessels are equipped with stabiliser units of either the fin varieties (fixed or deployable) or tank sluice systems or a combination of both tanks and fins. Tank systems can be extremely useful when loading/discharging as they tend to keep the vessel upright throughout cargo operations. Over-reliance on mechanical systems must not allow the vessel to list over.

If the vessel is allowed to list the vehicle ramp(s) are likely to become twisted. This may cause damage to the ramps themselves but will inevitably stop all cargo units passing over the ramps.

Ship to Shore Access

A RoPax vessel is a passenger ship with Ro-Ro cargo spaces or special category spaces as defined by SOLAS Regulation II-2/3. Passenger traffic usually tend to board through a terminal arrangement which separates mobile units from foot passengers for obvious safety reasons.

FIGURE 8.28 The P&O passenger/ferry terminal at the Port of Hull in the UK.

Passenger terminals usually have ticket reception, passenger lounge, security and passport control, basic refreshments with occasional shops. Covered companionways allow and maintain a passenger flow between terminal buildings to shipboard access.

Road traffic is usually directed initially to a customs holding compound where documentation and customs clearance is obtained. Movement from compounds tends to be via a roadway and link span direct to the stern or bow door arrangement of the vessel.

Link Spans

Access to Roll on, Roll off vessels must be capable of landing vehicles at all states of tide and in order to operate successfully, the shipboard end of the link must be able to adjust for the rise and fall of the tidal conditions prevailing. A hoist structure with associated lifting machinery is built at the shipboard end of the link to allow movement of the span to suit the rise of tide and the freeboard of respective vessels.

FIGURE 8.29 Canopied passenger access from terminal assembly to ship.

FIGURE 8.30 Ro-Ro mobile units entering the freight compound prior to customs clearance and redirection to the link span for loading on to the drive on ferry.

FIGURE 8.31 A typical road companionway adjustable link to a concrete structured dolphin at the port of Heysham, Cumbria. The height of the dolphin platform is built to accommodate the range of tide to allow acceptance of the vessels stern ramp. The ships stern door/ramp is then lowered for the angled flaps to lay flush to the companionway.

FIGURE 8.32 A typical mechanical link span machinery housing for hoisting and lowering the link span down to a position above the waterline. The stern ramp of the Ro-Ro ferry then lowers her stern ramp onto the links driveway to permit discharge of vehicles. The rise and fall of tide of 10.5 metres is countered by the adjustment of height to the link span to allow continuous operations no matter what state of tide.

High Speed Craft

The image of the ferry world has changed considerably over the last decade. The sleek lines of mono- and multi-hull craft now operate as RoPax vessels all over the world. They provide a fast and regular service mostly on the short sea trades together with some more long haul ventures.

FIGURE 8.33 The High-speed RoPax Catamaran vessel *Millennium Dos* (IMO No. 9237644) seen loading vehicles via the stern access, lying port side to the terminal in Barcelona, Spain.

Such high-speed craft operating commercially must operate under a Permit to Operate High Speed Craft and have a High Speed Craft Safety Certificate. They are constructed to carry cargo only or passengers only, or a combination of both cargo and passengers.

The HSC code applies to vessels on international voyages, but by the very nature of the business they are generally found on short defined, regular service runs from a base port.

PURE CAR CARRIERS (PCCS) AND PURE CAR AND TRUCK CARRIERS (PCTCS)

These vessels are designated for the carriage of cars. It was estimated that between 74 and 78 million new cars were sold in 2020. The main trade countries for such cargoes being mainly Japan, China and South Korea. The Car Carrier ships are employed with multi-deck, side-loading facilities and internal ramps to facilitate high-speed loading/discharging rates.

The ships are designed with exceptionally high freeboards and as such are susceptible to wind pressure causing considerable leeway, slowing service speed and detrimentally affecting fuel burn. More recent designs have taken this into account and the new generation car carriers are fitted with an aerodynamically rounded bow and bevelled along the bow line with a view to reducing wind pressure from head winds. Eukor Car Carriers Inc. are one of the world's largest vehicle-carrying companies operating large vessels (PCCs and PCTCs) carrying about 10,000 units at any one time.

The modern carrier is frequently fitted with hoistable decks which permit flexible loading of buses, trailer loaded cargoes and industrial mobile units of a varying height. These vessels tend to operate with all the major car manufacturing companies including Hyundai, Jaguar, Chrysler, Ford, Peugeot, Volkswagen, Kia, BMW, etc.

FIGURE 8.34 The high sided *Hual Trotter* (IMO No. 8116910) car carrier manoeuvres with tug assistance fore and aft in the port of Barcelona, Spain.

Features of the Car Carrier

The multi-deck configuration of the car carrier is in itself a striking constructional feature. The decks are interlinked by a fixed internal ramp system and elevator to lower holds. Rates of movement of car units vary directly with design but 1,000 CEUs per eight-hour shift would not be unusual. The vessel turns round from empty in a 48-hour period.

Some decks are set at different heights to allow different head vehicles to be carried, particularly relevant where high-sided trucks may become an optional cargo. Other features of the same deck might also include higher and heavier structure to cater for the heavyweight wheeled load. Some designs incorporate hoistable car decks offering alternative head room as an added feature, providing additional flexibility to maximise cargo load.

A vehicle cargo mix tends to offer more options to shippers as well as being convenient in permitting direct dealing with a single carrier, the speed of cargo operations being a direct influence for shippers and on the ship's running costs. Loading and discharge are generally achieved by a minimum of two vehicle ramps, one about the amidships area, while a side-loading quarter ramp has become a popular feature of many car carriers and PCTCs.

Fixed deck loading is usually about 2 tonnes/m^2 throughout, though this may vary where hoistable decks are engaged. Decks are fitted with forced ventilation fan systems to clear exhaust fumes during loading and discharge periods.

The main disadvantage of these ships is in their construction, producing very high-sided vessels which are subject to massive wind effect when in open aspect sea conditions. As such they experience considerable leeway which can generate increased fuel burn over a passage. Some efforts in design features like the rounding of the bow area and bevelled bow lines have been incorporated in some of the latest builds in an effort to increase fuel efficiency.

The ships tend to be fitted with a high ballast capacity because of the designated trade not lending to full return cargoes per voyage, although some mutual exchange cargoes which are suitable for the design decks, like palletised cargo/forklift or tractor loading, can sometimes be arranged.

Note: Car carriers do not conform to conventional Ro-Ro regulations.

Large car carriers are shipping up to 10,000 car units at any one time, usually on a one-way trip, with limited prospects for return cargoes. With this in mind a high ballast capacity is generally a main feature of their operation. Where return cargoes are booked the Pure Car and Truck Carriers have greater flexibility for the carriage of general commodities.

The new car trade is predominantly from South Korea, China, Japan, Europe and Scandinavia, routing to Canadian, United States, European and Australian markets.

The largest companies operating PCCs and PCTCs, are: MOL, Grimaldi, Hual and Wallenius Wilhelmsen.

CAR CARRIER CONSTRUCTION

Typical build features:

Gross tonnage 60,587 gt.
Draught 9.82 m.
Air Draught 52.0 m
Length O/A 121.08 m
Breadth 32.23 m
Service speed 21 kts.

Panamax sized vessel. Serviced by ship to shore ramps, one at the stern (Stbd Quarter) the other amidships (beam on). Also has an option to carry refrigerated cargo on decks 5, 6, and 7 instead of doing the return voyage in ballast.

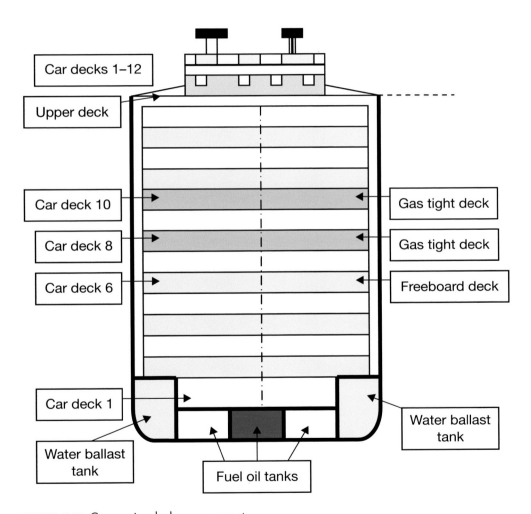

FIGURE 8.35 Car carrier deck arrangement.

Car Carriers

This type of ship is noticeable because of its exceptionally high sides and is often fitted with an angled quarter vehicle ramp in addition to an amidships cargo ramp. Their main feature is the high number of decks, usually at least twelve (12) to accommodate the numerous vehicle cargo. Return voyages are often empty as return cargoes are difficult to come by, but some palletised cargoes may be suitable.

Slewing ramps are popular with car carriers, providing versatility in load/discharge procedures. They are often incorporated with an amidships ramp to allow a double method of loading and/or discharge. As with other types of vehicle ramp, they tend to have a herringbone, non-skid surface and are fitted with removable stanchions and fencing. Angled flaps at each end provide flush joins to deck and quay landings, provided the vessel is upright, on an even keel.

FIGURE 8.36 Angled stern quarter vehical ramp, extensively fitted to multi-deck car carriers.

Hinged Stern Slewing Ramp

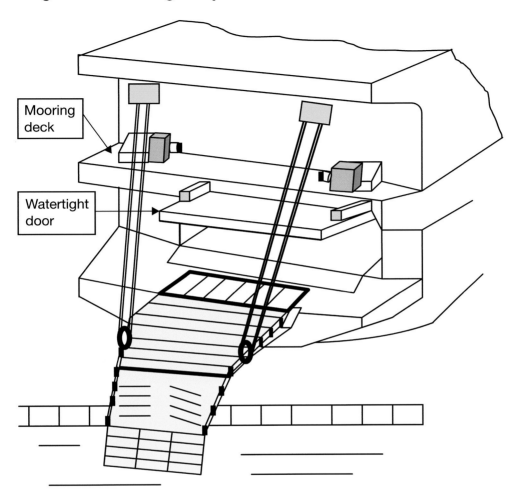

FIGURE 8.37 Hinged stern slewing ramp. Slewing ramps are popular with car carriers, providing versatility in load/discharge procedures. They are often incorporated with an amidships ramp to allow a double method of loading and/or discharge. As with other types of vehicle ramps they tend to have a herring bone, non-skid surface and are fitted with removable stanchions and fencing. Angled flaps at each end provide flush joins to deck and quay landings, provided the vessel is upright, on an even keel.

Quarter Vehicle Ramps

FIGURE 8.38 The quarter vehicle ramp of the vessel *Asian Breeze* (IMO No. 8202381) seen in the stowed and secured position while the vessel is at sea. A side ramp also stowed upright in a position just forward of the amidships position on the starboard side. The vessel is also fitted with a prominent freefall lifeboat in a launching position directly astern, a regular feature now of high-sided car carriers.

THE BACAT (BARGE CATAMARAN)

A double hull catamaran-shaped vessel which accommodates barges of up to about 140 tonnes. Barges are floated in from the stern and lifted from the water tunnel between the hulls by an elevator system. Additionally the 'LASH' barges (375 tonnes) can be transported by the water tunnel with a stern door being closed up after the completion of loading.

THE LASH (LIGHTER ABOARD SHIP) SYSTEM

A lift on, lift off system where lighters are raised to the upper deck by means of a moveable gantry crane. They are often loaded into holds or on deck in a similar manner to containers.

Alternatively they are operated on a similar principle as the floating dock, where the parent vessel is ballasted down and the lighters are floated in via the stern, between the high-sided bulkheads. As the vessel de-ballasts, the barges are lifted into the transport.

THE SEABEE (SEA BARGE)

This system uses barge units of about 800 tonnes deadweight which are floated towards a stern elevator. An automatic transporter rolls under the barge, when at the required deck level it is carried forward to the desired stowage position.

Note: LASH and SEABEE systems can also accommodate the carriage of containers.

CHAPTER 9

Containers and Containerisation

INTRODUCTION

The first recognised container vessel was a converted World War II tanker named the *Ideal X*, owned by Pan Atlantic. Her first container voyage shipped 58 containers on specially rigged decks from Port Newark, New Jersey in April 1956.

Malcom P. McLean (1914–2001), a liner shipping pioneer, was probably the accepted founder of containerised traffic. He received the 'Admiral of the Ocean, Sea Award' in 1984 from President Reagan and *Lloyd's List* nominated him as one of the three most influential men of the twentieth century, alongside Aristotle Onassis and Ted Arison.

The first fully 'cellular container ship' was a converted cargo vessel, the *Gateway City*, altered to carry 225 container units of 35 ft size. Her maiden voyage was between the Mexican Gulf and Puerto Rico but dock labour refused to work the vessel and the ship returned to the United States with her cargo.

Then the first transatlantic container line was started in 1966 and, as they say, the rest is history.

The container sector of the industry is now by far the largest of all the shipping sectors within the maritime trading world. It has dominated over virtually all general cargo practice and become deeply involved with the Ferry/Roll on, Roll off sector. Very few dry goods, other than heavy lift parcels, are outside the limits of containers. Even liquid loads have long been carried in container framed tanks.

The largest container ships are operated by the Mediterranean shipping company, the largest being the 'MSC Irina' carrying up to 24,300 TEUs, currently the largest container ship in the world. Container transport has now become the most dominant sector of commercial transport, handling thousands of cargo tons daily and moving approximately 65% of all commercial cargoes through many expanding terminals of the world.

This door-to-door service has met a huge customer demand and revolutionised the shipping industry and continues to expand. Terminals around the globe have increased the numbers of gantry cranes, created deep water berths to accommodate for an increased number of container ships, operating with deeper draughts.

DOI: 10.4324/9781003407706-9

LIST OF RELEVANT CONTAINER DEFINITIONS AND TERMS

Administration – The government of a contracting party, under whose authority containers are approved.

Approved – Indicates approval by the Administration.

Approval – The decision by the Administration that a design type or a container is safe within the terms of the present convention.

Bay – That space of 20' or 40' allocated to accommodate containers, set athwartships across the vessel.

Bay Plan – The ships 'Bay Plan' ranges the length of the ship and illustrates the position of each container loaded Bay. Bays are positioned athwartships and defined by that area set between the cell guides.

Cargo – Any goods, wares, merchandise and articles of every kind whatsoever carried in the containers.

Cell – The space which could be occupied by a single vertical stack of containers aboard a container vessel. Each stowage/hatch space would contain multiple cells, each serviced during loading/discharging by 'cell guides'.

Cell guide – A vertical guidance track which permits loading and discharge of containers in and out of the ship's holds in a stable manner.

Container – Defined as an article of transport equipment:

  a) Of a permanent character and accordingly strong enough to be suitable for repeated use.
  b) Specially designed to facilitate the transport of goods, by one or more modes of transport, without intermediate reloading.
  c) Designed to be secured and/or readily handled, having corner fittings for these purposes.
  d) Of a size such that the area enclosed by the four outer bottom corners is either:

   i at least 14 m^2 (150 sq ft) or
   ii at least 7 m^2 (75 sq ft) if it is fitted with top corner fittings.

The term 'container' includes neither vehicles nor packaging; however, containers when carried on chassis are included.

Container spreader beam – The engaging and lifting device used by gantry cranes to lock on, lift and load containers.

FIGURE 9.1 Empty 'Cell guides', numbered from the centre line, odd to starboard and even to port, found at the ends of the bays.

Corner fitting – An arrangement of apertures and faces at the top and/or bottom of a container for the purposes of handling, stacking and/or securing.

Existing container – A container which is not a new container.

Flexible boxship – A term which describes a container vessel designed with flexible length deck cell guides, capable of handling different lengths of containers, e.g. 20, 30 and 40 ft units.

Gantry crane – A large heavy lifting structure found at container terminals employed to load/discharge containers to and from container vessels. Some container vessels carry their own travelling gantry crane system on board.

Hatchless holds – A container ship design with cell guides to the full height of the stowage without separate or intermediate hatch tops interrupting the stowage.

International transport – Transport between points of departure and destination situated in territory of two countries to at least one of which the present International Convention

FIGURE 9.2 Gantry cranes engage in container cargo operations over the *Zim California* (IMO No. 7043556) berthed in Barcelona, Spain.

FIGURE 9.3 The container stack on the deck of the *Zim Jamaica* (IMO No. 9113680) being discharged by terminal gantry cranes in the port of Barcelona, Spain.

for Safe Containers (CSC) applies. The CSC will also apply when part of a transport operation between two countries takes place in the territory to which the present convention applies.

Karrilift – Trade name for a mobile ground-handling container transporter. There are many variations of these container transporters found in and around terminals worldwide. Generally referred to as 'Elephant Trucks' or 'Straddle Trucks'.

Lashing frame/Lashing platform – A mobile or partly mobile personnel carrier by which lashing personnel can work on twist locks at the top of the container stack without having to climb on the container tops.

Maximum operating gross weight (Rating) – The maximum allowable combined weight of the container and its cargo.

Maximum permissible payload (P) – The difference between the maximum operating gross weight or rating and the tare weight.

New container – A container the construction of which was commenced on or after the date of entry into force of the present convention.

Owner – The owner as provided for under the national law of the contracting party or the lessee or bailee, if an agreement between the parties provides for the exercise of the owner's responsibility for maintenance and examination of the container by such lessee or bailee.

Prototype – A container representative of those manufactured or to be manufactured in a design type series.

Rating (R) – see Maximum operating gross weight.

Rows – Are defined by being the lines of the container stack fore and aft and are numbered from the centre line (00), outboard. The number helps the identification in any container tracking system. Odd numbered rows are set to starboard and even number rows set to port. Numbering the row numbers helps to identify the position of the container in the bay.

Safety approval plate – An information plate which is permanently affixed to an approved container. The plate provides general operating information inclusive of country of approval and date of manufacture, identification number, maximum gross weight, allowable stacking weight and racking test load value. The plate also carries 'end wall strength', 'side wall strength' and the maintenance examination date.

Stack – A term when referring to containers, which represents the deck stowage of containers in 'tiers' and in 'bays'.

Tare weight – The weight of the empty container including permanently affixed ancillary equipment.

Terminal representative – The person appointed by the terminal or other facility where the ship is loading or unloading, who is responsible for operations as conducted by the terminal or facility with regard to that particular ship.

TEU – Twenty foot equivalent unit. Used to express the cargo capacity of a container vessel.

Tier – The term tier is defined by the number of layers in the height of the stacked containers.

Type of container – The design type approved by the Administration.

Type series container – Any container manufactured in accordance with the approved design type.

WORKING CONTAINERS

FIGURE 9.4 Extended jib of a ship loading gantry crane seen engaged in container movement, ship to shore.

Source: © Shutterstock/sattahipbeach.

Loading Containers

The order of loading, when the large container vessels are carrying currently up to 24,000 TEUs, must be well planned and considered as a detailed operation. Planners are usually employed ashore to provide a practical order of loading – particularly important when the vessel is scheduled to discharge at two, three or more terminal ports.

Once loading in the cell guides is complete, the pontoon steel hatch covers, common to container vessels, are replaced and secured. Containers are then stowed on deck in 'stacks' often as high as 6 tiers. The overall height of the deck stowage container stack may well be determined by the construction of the vessel. It must allow sufficient vision for bridge watch-keepers to be able to carry out their essential lookout duties. The stability criteria of the vessel, when carrying containers on deck, must also be compatible with the stowage tonnage below decks. (Currently the greatest stowing height on the larger container vessels is 8 tiers.)

Any deck stowage requires effective securing and this is achieved usually by a rigging gang based at the terminal. As the 'stack' is built up, each container is secured by means of specialised fittings, between containers themselves and to the ship's structure.

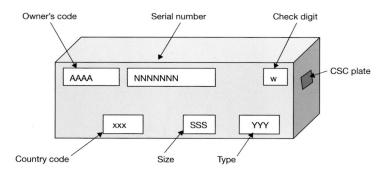

Length (m)	Width (m)	Height (m)	Gross weight (kg)	Tare weight (kg)	Pay load (kg)	Usable capacity (m³)	Imperial size (ft)
6.05	2.43	2.43	20,320.9	1590.30	18,730.6	30.75	(20')
9.12	2.43	2.43	24,401.2	2092.92	23,308.3	46.84	(30')
12.19	2.43	2.43	30,481.4	2593.64	27,887.0	62.92	(40')

FIGURE 9.5 Markings on containers.

CSC Plate

The Plate is a legal safety requirement that all containers to be used in transport of goods, is approved at the time of its manufacture. It is fastened to every shipping container at the time of manufacture and is valid for a period of five years. The container is then monitored under the Periodic Examination Scheme (PES).

The plate usually secured to the left door acts as a Certificate for Safety of Containers (CSC) and will show the following information:

Container Model (Type), Manufacturers name, Date of manufacture, Country of approval reference, Country of approval certificate and Manufactures serial number.

The container Number, Owners name and address, Title of timber treatment, Gross weight, Allowed stacking weight, and the ACEP number (continuous examination when operating under PES).

Container Transport

A fully laden container vessel is unlikely to be loaded down to her loadline marks despite having a container stack on deck of three or four high. Containers may weigh up to about 30 tonnes gross weight each, when fully packed, but may also be empty. Hence a full capacity load may not necessarily equal the maximum permissible deadweight.

If containers are carried on deck they must be well secured by means of the iron rod lashings with associated rigging screws, fixed as part of the ship's structure. Empty or light containers could be affected by buoyancy when seas are shipped and Deck Officers should be especially diligent when checking the upper deck stow and respective securings.

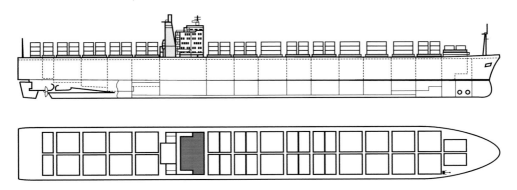

FIGURE 9.6 Container vessel – upper deck stow showing stack of three tier high on the top of pontoon hatches. Below decks, containers secured in cell guides. The navigation bridge is not obscured for the vessel's operational needs.

FIGURE 9.7 The *OOCL Fontune* (IMO No. 9188518) a small container vessel seen in a loaded condition at sea.

FIGURE 9.8 The container vessel *Valentina* (IMO No. 9344722) seen loaded in a seagoing condition. The ship is equipped with its own container cranes.

FIGURE 9.9 The container vessel *Independent Voyager* (IMO No. 9481532) seen loaded, starboard side to the container berth in Liverpool, UK, with gantry cranes operational.

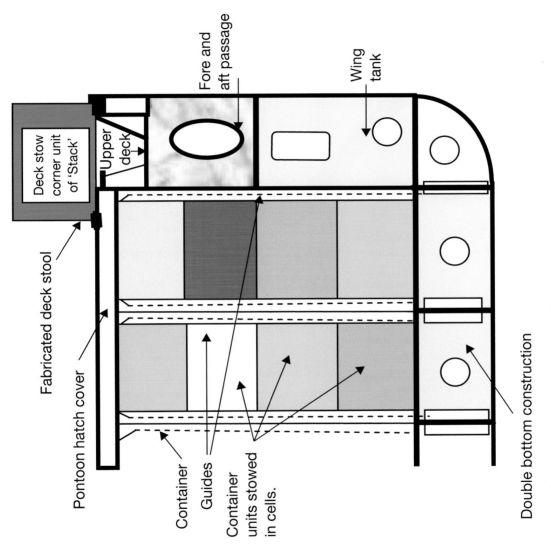

FIGURE 9.10 Container vessel construction and stowage below decks.

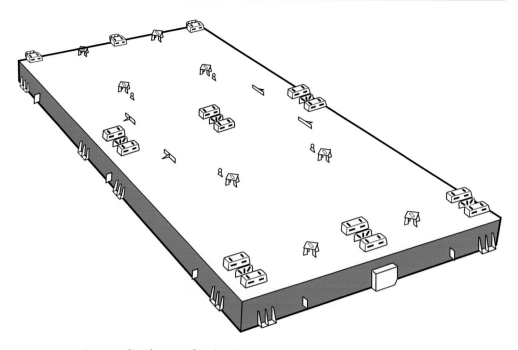

FIGURE 9.11 Pontoon hatch cover fitted with deck stools and hoist lock positions.

FIGURE 9.12 The P&O *Nedlloyd Susana* (IMO No. 9286774) lies port side to the container terminal in Lisbon, Portugal, part-loaded prior to sailing.

FIGURE 9.13 Typical shore side, container gantry cranes silhouetted against the Lisbon sky.

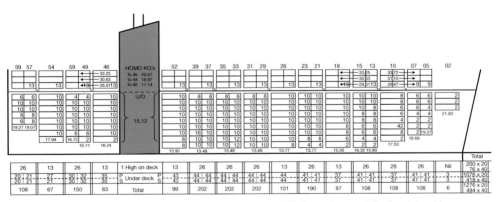

The modern type of container vessel will normally operate a container 'box' tracking system which allows continuous monitoring of any single container at any time during its transit. The plan allows a six-figure number to track and identify its stowage position aboard the vessel. Distinct advantages of such a system tend to satisfy shipper enquiries as well as showing that the shipping company is efficient in its business.

Other aspects of security are also clearly beneficial in a security-conscious age.

An example tracking system could be typically:	The first two numbers of the six-digit number	– The identification of the 'bay' of stowage
	The second two numbers	– The 'cell' of stowage
	The last two numbers	– The level/tier of stowage

FIGURE 9.14 Container ship cargo plan and tracking system.

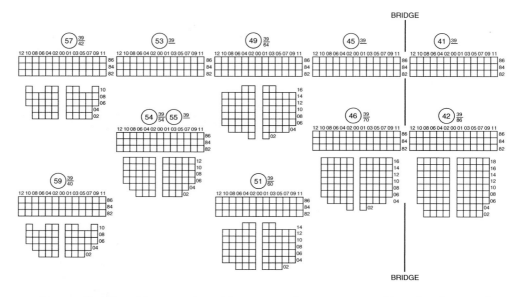

FIGURE 9.15 Bay Plan. Most container ships now employ computer tracking identification of individual containers.

LOADICATOR AND LOADING PLAN COMPUTERS

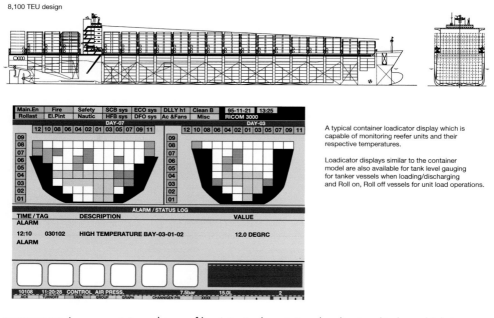

FIGURE 9.16 Large container ship profile. A typical container loadicator display which is capable of monitoring reefer units and their respective temperatures.

Many ships are now equipped with loadicator systems or a loading computer with appropriate software. It is usually a conveniently sited visual display for the Master and the Loading Officers and is gainfully employed on Ro-Ro vessels, container ships, tankers and bulk carriers. The system should be interlinked with the shoreside base to enable data transmissions on unit weights/tonnages or special stow arrangements. The computer would permit the location and respective weights of cargo/units to be entered quickly and provide values of limiting KG and GM together with deadweight's at respective draughts/displacements. It would also have the capability to provide a printed record of the state of loading and show a visual warning in the event of an undesirable stability condition or overload occurring.

Distribution of the ship's tank weights, stores and consumables affecting final calculations and total displacement would also be identifiable within the completed calculations. The primary aim of the loading computer is to ensure that the vessel always departs the berth with adequate stability for the voyage. If this situation can be achieved quickly, costly delays can be eliminated and safety aspects are complied with.

The data required to complete the stability calculations would need to be supplied by the shoreside base with regard to cargo weights. This in turn would be certificated by the driver, for Ro-Ro unit loads, obtaining a load weight certificate authorised from an approved 'weigh bridge' prior to boarding the vessel. Draught information would inevitably come from a 'draught gauge system' for the larger vessel and be digitally processed during the period of loading.

Ship's personnel could expect to become familiar with manipulation of the changing variables very quickly alongside the fixed weight distribution throughout the ship. This would permit, in general, few major changes to the programme, especially so on short sea ferry trade routes where limited amounts of bunkers, water and stores are consumed and values stay reasonably static.

Fixed weights are applicable to a variety of units or vehicles and, as such, where units are pre-booked for the sea passage, an early estimate of the ship's cargo load and subsequent stability can often be achieved even prior to the vessel's arrival.

The loadicator programmes provide output in the form of:

Shear forces and bending moments affecting the vessel at its state of loading.
Cargo, ballast and fuel tonnage distributions.

A statement of loaded GM, sailing draughts and deadweight.

Container Movement

FIGURE 9.17 The container spreader beam operates secured to the gantry crane travelling the length of the gantry jib to lift the containers on and off the vessel. The corners of the spreader beam are fitted with hinged droppable guides to ensure the beam locks can accurately locate the container corner recesses. The beam is also used to lift off pontoon hatch covers but, when doing so, does not deploy the hinged guides.

Source: © Shutterstock/donvictorio.

CONTAINER TYPES

There are many container types in operation to suit a variety of trades and merchandise. Sizes also vary and they can be shipped in the following sizes: 8 feet in width and 8 feet or 8 feet 6 inches in height, with lengths of 10, 20, 40 or 45 feet.

Conventional unit (general purpose) – Also known as a dry container, made from steel and fully enclosed with a timber floor. Cargo securing lashing points are located at floor level at the base of the side panelling. Access for 'stuffing' and 'de-stuffing' is through full height twin locking doors at one end.

Open top container – Covered by tarpaulin and permits top loading/discharging for awkward-sized loads which cannot be easily handled through the doorways of general purpose containers. These may be fitted with a removable top rail over and above the door aperture.

Half-height container – An open top container which is 4 feet 3 inches in height, i.e half the standard height of a general purpose container. They were designed for the carriage of dense cargoes such as steel ingots, or heavy steel cargoes or stone, etc. Since these cargoes take up comparatively little space in relation to their weight, two half-height containers occupy the same space as one standard unit.

Flat rack container – This is a flat bed with fixed or collapsible ends and no roof. They are used to accommodate cargoes of non-compatible dimensions or special cargoes that require additional ventilation.

Bulk container – Designed to carry free-flowing cargoes like grain, sugar or cement. Loading and discharging takes place via three circular access hatches situated in the roof of the unit. They also incorporate a small hatch at the base which allows free flow when tipping the unit. Such containers are usually fitted with steel floors to facilitate cleaning.

Tank containers – Framed tank units designed for the carriage of liquids. The cylindrical tank, usually made of stainless steel, is secured in the framework which is of standard dimensions to be accommodated in loading and discharging as a normal general purpose container unit. The tanks can carry hazardous and non-hazardous cargo and are often used for whisky or liquid chemicals.

Ventilated containers – Generally designed as a general purpose container but with added full-length ventilation grills at the top and bottom of the side walls of the unit. They were primarily designed for the coffee trade but are equally suitable for other cargoes which require a high degree of ventilation during shipping.

Open-sided containers – These units are constructed with removable steel grate sides which are covered by PVC sheeting. The side grates allow adequate ventilation when used to carry perishable goods and/or livestock. Such containers permit unrestricted loading and discharging with the grates removed.

Insulated containers – These containers are insulated and often used in association with refrigeration air blower systems to keep perishable cargoes fresh, e.g. meats, fruits, vegetables, etc. The container has two porthole extractors fitted to one end of the unit to allow cool air circulation to operate from the cooling plant. They are generally stowed under deck and close to, or adjacent to, the ship's circulation ports.

Other types of container in this category rely only on the insulation and are not fitted with cooling plant and these can be stowed in any position on the ship.

Refrigerated containers – More generally known as the reefer container, they are totally insulated and fitted with their own refrigeration plant. They must be connected to the ship's mains and require close stowage to a situated power point. They are usually employed for foodstuffs, meat and dairy products being prime examples. These units have become prolific and have caused a major reduction in the number of dedicated 'reefer ships' as was. Although reefer ships still operate, they tend to be limited to specific trades like bananas.

FIGURE 9.18 The *OOCL Shanghai* (IMO No. 9198111) lies port side to the container terminal in Barcelona, Spain, after completing cargo loading with a full container load. The deck stack is at a six tier height. The terminal gantry cranes are seen in the upright and clear position. (The ship is not fitted with its own cranage.)

Container (Internal) Hatch Stowage

FIGURE 9.19 Container 'bay' (hatch) with part load containers lying in the cell guides of the lower hold cargo space.

REEFER CONTAINERS

With many of the chilled and frozen products being transported by sea containers there was bound to be an influence on the reefer trade – so much so that designated 'reefer' ships have been greatly reduced in number, other than possibly in the banana trade. Roll on, Roll off units as well as the specified refrigerated containers have now dominated the reefer commodity shipping markets.

The container units themselves are built with insulation and pre-cooled prior to being loading at the handling station. A shore power supply is used to activate the unit's cooling plant. Once packed and sealed, the temperature of the unit is lowered to the desired level and monitored by a temperature sensor attached to the container. As soon as the unit is packed, the refrigeration machinery is activated either by the continued use of a shore supply or linked directly to the transporter's (mobile) power source.

Terminals and container parks have specialised park areas to enable mobile units to switch to a static shore power supply, once the mobile transport supply is stopped. Disconnection of units takes place just prior to loading on board the ship. The supply is reconnected to the ship's mains once the unit is stowed in its allocated position aboard the vessel.

The modern container vessel can expect to carry numerous units with refrigerated cargoes, all plugged into the ship's power supply fitted to specified loading bays. They would in the main be fitted with a reefer container monitoring system to ensure that temperatures are retained within acceptable limits.

'Reefer' Container Monitoring

Various types of monitoring systems are available for shipping operators, either stand-alone or integrated operations which could include tank gauge systems, ballast control, power management, fire-fighting, etc.

The local control unit indicated could monitor up to 3,000 cargo units or numerous tanks for pressure, temperature, volume, viscosity, etc.

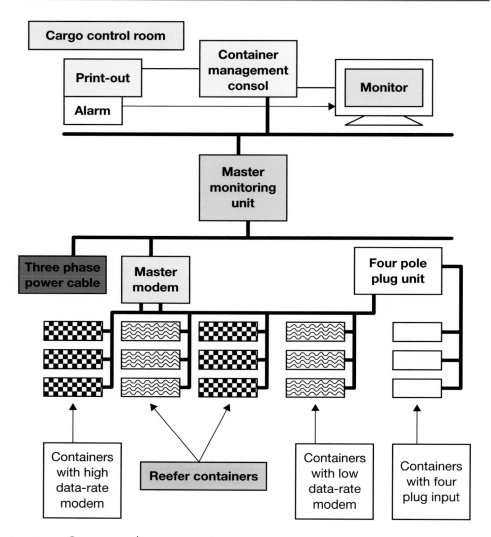

FIGURE 9.20 Cargo control room – container monitoring.

Containers on Deck

It is regular practice to carry containers on deck on both designated container vessels and general cargo/service vessels. Further recommendations on deck stowage are advised by 'M' Notice 1167.

Deck containers should be stowed and secured taking account of the following:

1   Containers should preferably be stowed in a fore and aft direction.

2   They should be stowed in such a position as not to deny safe access to those personnel necessary to the working of the vessel.

3   They should be effectively secured in such a manner that the bottom corners will be prevented from sliding and the top corners will be restrained to prevent tipping.

4   The unit will be stowed in a manner that it does not extend over the ship's side. (Many containers are stowed part on the hatch top and part on extending pedestal supports, but the perimeter of the unit is kept within the fine lines of the vessel.)

5   Deck containers should be carried at a single height (one high). However, this may be increased if twist locks are used to secure the bottom of the container to a fabricated deck stool.

6   Deck loads should not overstress the deck areas of stowage. Where units are on hatch tops, these hatch covers must be secured to the vessel.

7   No restraint system should cause excessive stress on the container.

8   Restraint systems and securing's should have some means of tightening throughout the voyage period.

FIGURE 9.21 The container vessel *Wan Hai 602* (IMO No. 9327798) seen from astern under the gantry cranes in Singapore.

Container Deck Stowage

Container decks and reinforced pontoon hatch tops to take the deck load capacity are generally constructed with increased scantlings to satisfy Classification and Construction Regulations. Both open decks, as seen below, and the pontoon hatch covers, when fitted, are usually equipped with container stools to permit the 'boxes' to be locked into position – the first tier being the foundation for second and subsequent tiers to be stowed on top.

FIGURE 9.22 The exposed container forward cargo deck of the *Baltic Eider* (IMO No. 8801917) seen with the container deck stool securing points in uniform rows to form the basis of an even stow. The vessel is moving forward through 10/10ths light pancake ice in the Baltic Sea.

FIGURE 9.23 The foredeck of the *Baltic Eider* loaded with a container deck cargo. The stow includes single and double tier standard containers inclusive of liquid tank containers seen aft and to starboard. The vessel is seen passing through a pancake ice sea surface in the Baltic Sea.

Container Lashing Fitments

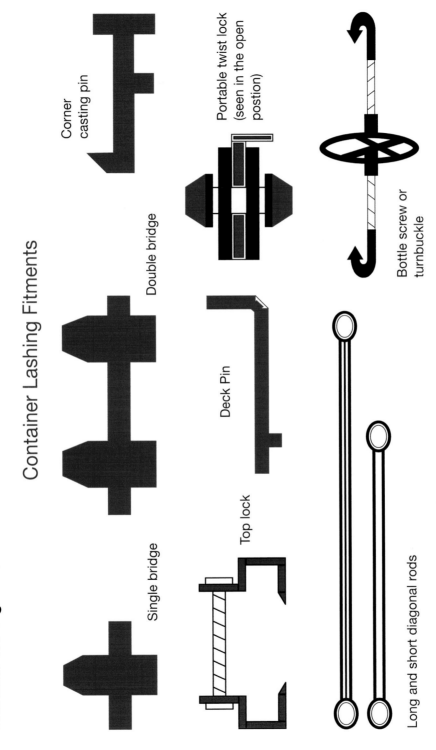

Container Lashing Fitments

Corner casting pin

Portable twist lock (seen in the open postion)

Bottle screw or turnbuckle

Double bridge

Deck Pin

Single bridge

Top lock

Long and short diagonal rods

FIGURE 9.24 Container lashing fitments.

Parametric Rolling – Container Vessels/Deck Securings

The large container vessels are now able to load deck containers up to ten (10) and eleven (11) tiers high. However, in so doing the ship becomes more vulnerable to what is known as Parametric Rolling when at sea. If this condition occurs and is allowed to be sustained there is an increased risk of container tiers toppling and units being lost overside.

The new container vessels are constructed with a wide beam and generally have a large 'flare' off the bow. This construction lends to the carriage of increased cargo units. When in a seaway, the ship will experience natural roll and pitch motions in any event, but when combined with its construction, the buoyancy changes and wave excitation, the vessel can be pushed sideways.

The effects of this movement can lead to a synchronised motion which can then lead to heavy rolling. If the motion is coupled with a pitching motion, the positive stability of the vessel could be compromised with fluctuations in the waterplane area and the GM value. Such a situation can be common to head seas and has also been experienced as hazardous with following seas. Passing waves causing variation in the waterplane area leading to possible instability in any roll motion.

Should such conditions prevail the ship can expect to experience:

- Heavy stresses in the fore and aft parts of the ship.
- Extreme stresses on the container stack and its securing's.
- Variation in load on the ships engines, causing speed fluctuations.
- Unpleasantness for the crew if allowed to continue.

Buoyancy Changes

A change of course to change the period of encounter of the wave pattern, will probably make conditions more comfortable and reduce the risk of container tiers from toppling or units being lost overside.

PARAMETRIC ROLLING: CAN OCCUR WHEN

- The rolling period is twice the wave encounter period
- Wave lengths are in the range of the vessels length

Example Container Deck Stowage and Securing

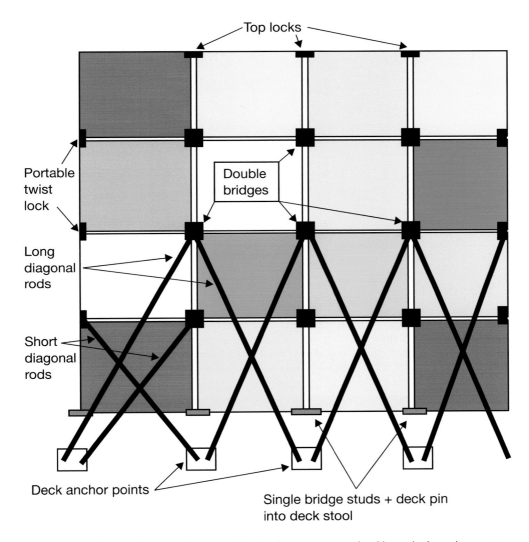

FIGURE 9.25 Short and long rods secured by bottle screw or turnbuckle to deck anchor points.

Deck Stowage and Securing of Container Stack

FIGURE 9.26 The deck stow is seen in way of the edge of the pontoon hatch cover. The lashings are to the base, third tier container level by long bar. The second tier container level being secured by short bar lashings with the base containers locked to the pontoon.

FIGURE 9.27 Stowed containers first and second tier of deck stow secured by lashing rods.

FIGURE 9.28 Container lashing bars, seen secured from the main deck to the first and second tier of deck containers.

FIGURE 9.29 The container hoist engaged in lifting the pontoon hatch covers clear of the cellular holds.

TERMINAL OPERATIONS

The sheer size of container terminals around the world must generate cause, for the tremendous volume of work which is involved in the transport, storage and shipping of the many units. The general public would only visibly see the number of units which a terminal has inside its perimeter at any one time. However, the maintenance of the gantry cranes, the ground handling transports, the documentation concerning a single 'box unit' become the invisible operations that generate a successful terminal. They employ considerable manpower with various skills, from the wharf men to personnel engaged in 'stuffing' container units, security personnel, administration staff, maintenance workers, ships' planners, etc., not to mention the insurance and legal professionals engaged in the background.

FIGURE 9.30 A single gantry crane and a mobile dockside crane work the container cargo of the feeder container vessel *Providence* (IMO No. 9080417). The ship's own container cranes are seen turned outboard to facilitate the shoreside loading systems in the port of Barcelona.

The largest terminals in the world for container handling are:

TABLE 9.1

Terminal Port	Volume 2019 Million TEU's	Volume 2020 Million TEUs
Shanghai	43.0	43.0
Singapore	37.0	37.0
Ningbo-Zhoushan	28.0	27.0
Shenzhen	24.0	25.6
Guangzhou China	24.0	24.0
Quingdao	21.0	22.5
Busan	21.0	21.0
Tianjin	17.6	17.5
Hong Kong	17.0	17.0
Rotterdam	14.5	15.0

Main container ports for Europe are Rotterdam and Hamburg. The largest container port in the UK is Felixstowe. Main container ports for the United States are Los Angeles and New York.

The increased growth of unit movements are based on figures from 2019/2020. World trade is influenced by many factors, not least the strength of national economies, the strength of the US Dollar, the emergence of China and charter rates, to mention but a few of the relevant influences. However, what is clear is that if the location of the ports is noted, the geography would indicate that the United States, Europe, the Far East and in particular China have emerged as the main trading blocks for containerised traffic. Feeder operators to Australasia, India, Baltic and Mediterranean regions continue to be active in support of the major operators.

Note: 2020 All ports and trades suffered a major disruption from the effects of the coronavirus around the globe. The container sector now being the largest of all the shipping sectors of the industry.

Container Terminals

FIGURE 9.31 Automated stacking cranes provide unit movements to the terminal's container stack. The containers are being loaded to driverless ground handling transports.

FIGURE 9.32 Automated driverless ground handling transports deliver container units from the terminal stack to the underside of ship/shore gantry cranes.

FIGURE 9.33 Tracked container gantry crane for the loading of containers. Semi-automated driverless transports deliver container units to its underside prior to loading onto the container vessel.

FIGURE 9.34 Tracked gantry operations stack and deliver container units in and around a container terminal park. The units are stacked in rows and bays inside the Lisbon terminal to ensure orderly location and delivery to the ship loading gantries.

CONTAINER OPERATIONS

Shipping and Booking

In order to ship a container certain procedures and documentation processes are required. The freight office of a shipping company would require the following information:

1 Name and address of company booking the unit for export.
2 Bill of Lading, with name and address of shipper if different from above.
3 The quantity of cargo to be shipped including weight, measurement, marks and number of packages.
4 Name of port of discharge.
5 Commodity details, hazardous, refrigeration required, and/or precise description of goods.
6 The place of delivery and acceptance.
7 Place of packing the container.
8 Earliest date of container availability.
9 Customs assigned number.
10 Customs status of cargo declaration.

The container would then be designated a booking reference number to allow a constant trace to be maintained on the unit while being exported.

EXPORT CONTAINER PACKING DOCUMENT (ECPD)

A detailed packing list of the container is required and this serves as not only a list of container contents but also includes:

a Quarantine declaration (if required).
b Transport document for receipt from shippers for empty container; receipt from shipping company for full container.
c Stated conditions which relate to the use of equipment at shipper's premises.
d Declaration of Customs Status.

Bills of Lading

The shipping company will produce a Bill of Lading once they have been informed of all relevant details regarding the nature of the cargo. It would be normal practice that a freight invoice would also be issued at this time, as the Bill of Lading and the freight invoice are both computer-generated. (See Chapter 11 Documentation for Bills of Lading detail.)

GROUND HANDLING OF CONTAINERS

Equipment and Methods

FIGURE 9.35 Container transporter with lift, transport, and stacking capability. A type of container karri lift based on the forklift truck principle which is widely employed in container terminals worldwide.

Lifting, Stacking and Transporting

FIGURE 9.36 Tractor unit fitted with extending arm capable of carriage and stacking and loading containers to flat top transport vehicles.

FIGURE 9.37 Mobile units working and stacking containers inside a container park. Heavy duty forklift truck engaged in ground handling the transport of container cargoes. Straddle trucks (elephant trucks) are also deployed at major terminals.

CONTAINER SUMMARY

The container sector has clearly dominated the commercial shipping industry around the globe. It continues to develop on many fronts including innovation in the hardware of working actual units. Dual trolley gantry cranes, twin/tandem lifting spreaders and automated/driverless carriage for units from container parks to shipboard lifting points.

Automated computer controlled loading/discharging and remote handling systems have become effective around the world, inclusive of Japan, Europe, Canada and both the East and West coasts of the United States. Many innovations in both large and smaller ports are increasing their facilities and their water depths off container berths, to be able to accept the ever growing size of vessels with their increased draughts.

Computerised tracking systems for units is now well established and meeting the needs of shippers and their customers, improving delivery times for retail outlets and meeting societies needs overall. Modern ships, with increased capacity are seeing much faster turn-rounds in way of more automated terminals.

The geography of the Panama Canal, previously seen as an obstruction limit on trade, is being further developed to allow increased ship size transit in the future. By 2025 it is expected, the updated Panamax tonnage will increase its length to 1200' x 160' wide at 49' depth due to major structural changes to the Canal.

The entrances from both the Atlantic and the Pacific Oceans will be widened and deeper channels established. A new lock system will be introduced to permit access to the larger Panamax container vessels. These changes to the Canal, can expect general tonnage to double and container traffic to triple. With extensive physical changes to such main container ports as New York, Baltimore, Los Angeles and Long Beach.

With such growth, ports have realised that they can expect increased trans-shipment traffic to the smaller ports and such preparations are currently underway in many shipping, ports and harbours, in order to compete for the overspill from an increased Canal size.

Container traffic is not reducing; it is becoming more efficient and work in the sector is demanding a greater input by those persons associated within the sectors perimeter.

CHAPTER 10

Special Cargoes, Hazardous Goods and Deck Cargoes

INTRODUCTION

All cargoes must be considered special in one way or another, particularly so, if it is the first time that an individual has had experience of that specific cargo. It would take a lifetime for a mariner to carry every commodity and even then, certain products would be absent from the list. The interpretation of 'special cargoes' can encompass many types of cargo, but it is generally accepted that those parcels which require special or additional attention for their safe transport and discharge, fall into this 'Special Cargo' category.

Clearly hazardous goods covered by the two volumes and the code supplement of the IMDG Code, are deeply entrenched under this particular umbrella of special cargoes. However, it is not just hazardous materials; valuables such as bullion, bank notes, stamps or personal effects, requiring lock-up stowage conditions, are also considered as specials.

Within the scope of this chapter falls 'deck cargoes'. They are stowed on the weather deck and exposed to the elements and are often the first to suffer from any misadventure which may befall the ship's voyage. Deck cargoes by their very nature may fall into the class of hazardous goods in their own right or they may, like timber deck cargoes, have their own inherent dangers which may threaten the well-being of both ship and cargo.

Whatever goods are shipped, it is essential that correct stowage procedures are taken from the onset. They should be clearly noted on the stowage plan and relevant persons should be made aware of the nature of potential hazards or special precautions that should accompany the transport, i.e. increased securings.

DEFINITIONS AND TERMINOLOGY RELATING TO HAZARDOUS CARGOES

Auto ignition temperature – The lowest temperature at which a substance will start to burn without the aid of an external flame. Spontaneous combustion begins, provided that conditions are right, when auto ignition temperature is attained.

Carrier – Any person, organisation or government undertaking the transport of dangerous goods by any means of transport. This includes both carriers for hire or reward

DOI: 10.4324/9781003407706-10

(known as common or contract carriers) and carriers on own account (known as private carriers).

Control temperature – The maximum temperature at which certain substances (such as organic peroxides and self-reactive and related substances) can be safely transported during a prolonged period of time.

Cylinders – Transportable pressure receptacles of a water capacity not exceeding 150 litres.

Dangerous goods – Substances, materials and articles covered by the IMDG code.

Defined deck area – The area of the weather deck of a ship or of a vehicle deck of a Roll on, Roll off ship which is allocated for the stowage of dangerous goods.

Emergency temperature – The temperature at which emergency procedures shall be implemented.

Flammable liquid – A liquid having a flashpoint lower than 37.8°C. A **combustible liquid** is a liquid having a flashpoint of 37.8°C or above. E.g. gasoline is a flammable liquid, whereas kerosene is a combustible liquid.

Flammable range – The limits of flammable (explosive) range lie between the minimum and the maximum concentrations of vapour in air which forms a flammable (explosive) mixture. Usually abbreviated to Lower Flammable Limit (LFL) and Upper Flammable Limit (UFL). These are synonymous with the lower and upper explosive limits.

Flashpoint – The lowest temperature at which a liquid gives off sufficient vapour to form a flammable mixture with air near the surface of the liquid, or within the apparatus used. Flashpoint represents the change point from safe to risk.

Harmful substances – Those substances which are identified as marine pollutants in the IMDG code.

International Maritime Dangerous Goods Code (IMDG code) – A mandatory code for the carriage of dangerous goods at sea, as adopted by the Maritime Safety Committee of the IMO. Effective from 1 January 2020 and is applicable to all ships to which the SOLAS convention applies. (Resolution MSC. 122 (75))

Medical First Aid Guide – A section of the supplement to the IMDG code which details guidelines for the application of first aid to persons exposed and affected by hazardous goods.

Packaged form – The form of containment specified in the IMDG code.

P & I Club – A protection and indemnity or P & I Club, is a non-governmental , non-profitable mutual co-operative association of marine insurers. Its members include Ship owners, charterers and seafarers.

Settled pressure – The pressure of the contents of a pressure receptacle in thermal and diffusive equilibrium.

Shipper – That person for whom the owners of the ship, or the charterers agree to carry cargo or freight to a specified port at an agreed price as written in the Bill of Lading.

Sift-proof – Packaging which is impermeable to dry contents including fine solid material produced during transport.

Tank – A portable tank (including a tank container), a road tank vehicle, a rail tank wagon or a receptacle with a capacity of not less than 450 litres to contain solids, liquids or liquefied gases.

Water reactive – Any substance which in contact with water emits flammable gas.

Working pressure – The settled pressure of a compressed gas at a reference temperature of 15°C in a full pressure receptacle.

THE IMDG CODE

The International Maritime Dangerous Goods (IMDG) code is the recognised code of practice for the carriage of hazardous cargoes and is covered by two volumes, plus a supplement. The 2018 edition (mandatory from 1 January 2020) is based on the safety considerations set down in Part A and A-1 of Chapter VII of the 1974 SOLAS convention. It has recommendations for individual substances, materials and articles for good operational practice regarding: terminology, packing, labelling, stowage, segregation, and handling and also emergency response action.

IMDG Code Volume 1

This contains a general introduction and covers standards on:

1   General provisions definitions and training.
2   The classification of goods.
3   The packaging and tank provisions for goods.
4   Consignment procedures, documentation required when shipping.
5   Construction and testing of packaging's, IBC's large packaging's, portable tanks MEGC's and road tank vehicles.
6   Provisions of Transport operations.

Various sections cover the above standards for Class 1 to Class 9 hazardous goods in a more detailed format. Following the introduction the code provides details on modes of packaging to UN standards. An alphabetical general index of all the substances inclusive of its UN number, class and packaging group follow. This index should be employed as the first step to retrieve information affecting a particular cargo substance.

IMDG Code Volume 2

Part 3 (Dangerous Goods List, special provisions and exceptions)

Appendix 'A'   A list of generic Shipping Names.

Appendix 'B'   Glossary of terms.

Supplement

- Emergency Response procedures (Ems Guide)
- Medical first aid guide (MFAG)

> IMDG Code for Windows v 14 (2018) IMDG Code (Volume 1 & 2) Amendments 40–20.
> (IMO sales No ZL200E) (ISBN 978-92-801-1684-7)

Features unique to the IMDG code for Windows, include:

Search by substance or UN Number.
Multiple windows (MDI) for viewing multiple pages or substances.
Extensive cross referencing.
On screen colour displays of hazard labels, signs and marks.
Search by names in French and Spanish substances indexes.
Easy generation and saving of a Dangerous Goods Note.
Easy to use menus, on screen user manual, as well as help screens.
Printing facility and downloading.
Single user or network versions.

Author Recommendation: A tutorial on the digital format of the use of the IMDG code is ideal for any person needing to use the code. Once opened the main menu will be displayed, allowing access to the dangerous goods list

> Supplement of IMDG Code 2018 edition.
> IMO Product code 1J210E* ISBN 978-92-801-1683-0

This contains emergency procedures (Ems) and schedules for particular commodities, plus details of specialised equipment required for handling spills and fires. The Medical First Aid Guide (MFAG) provides information on symptoms and the body's reaction to exposure following an accident. Additionally, the supplement covers safe practice for handling of solid bulk cargoes, particularly concentrates, together with reporting procedures for vessels involved in incidents.

SHIPPING PROCEDURE FOR THE LOADING AND TRANSPORT OF HAZARDOUS GOODS

To transport dangerous goods by sea, they must pass through the following procedure:

1   The shipper is responsible for obtaining 'Export Licenses' for the goods in question.
2   The shipper would also be responsible for marking and labelling the goods to be shipped in accord with the IMDG code.
3   Following contact with the shipping company agents, the following should be provided:
    The number of packages, together with their weight.
    The value of the goods.
    Special requirements for carriage of the goods.
4   Customs clearance would be required as for any other cargo.
5   The Bill of Lading would be sighted and seen to be free of endorsements.
6   The goods would be entered on the ship's manifest and marked on the cargo stowage plan.
7   Ship's officers would check the UN number, the details of the commodity, the labelling of the package and the condition of the packaging. Any special stowage arrangements would be noted and observed at this stage.
8   The ship's Master has the right to accept or reject the cargo prior to loading.

Once the goods are stowed on board the vessel, the requirements of the IMDG code would be followed throughout the period of the voyage.

Note: Reference should also be made to Annex III of MARPOL, regarding the regulations for the prevention of pollution by harmful substances carried at sea in packaged form.

If appropriate, a 'Document of Compliance' for the carriage of certain hazardous goods may be required by the ship.

Documentation for Shipping Dangerous Goods

1   Where dangerous goods are to be carried by sea, all documentation relating to the goods must carry the correct technical name where the goods are named. The use of a trade name alone must not be used.
2   Any shipping documents prepared by the shipper must include or be accompanied by a signed certificate or declaration that the shipment offered for carriage is correctly packaged and marked, labelled, etc., and is in proper condition for shipment.
3   The person responsible for the packing of dangerous goods in a freight container or road vehicle must provide a signed container packing certificate or a vehicle packing declaration, which states that the cargo in the unit has been correctly packed and secured and that all applicable transport requirements have been fulfilled.
4   In the event that a freight container or road vehicle containing dangerous goods is not compliant with the above, then such vehicle or container shall not be accepted for shipment.

5   Every ship carrying dangerous cargo shall have a special list or manifest of such dangerous goods on board contained within a detailed stowage plan. Such documents will identify by class and location all such dangerous goods on board the vessel. Copies of these documents will be available prior to departure to a person as designated by the Port State Authority.

6   In the case of marine pollutants, the signed shipping documents must also state that the parcel offered for shipment is a marine pollutant and that as such it is in a proper condition for carriage by sea.

Note: A copy of the stowage plan must be retained ashore up to the time that the harmful substances have been discharged.

Documentation Detail – for Shipping Dangerous Goods

One of the prime functions of any documentation which accompanies dangerous goods for shipping is to provide basic information associated with the hazardous substance. To this end the shipping document for each product, material or article offered for shipment must include the following:

1   The proper shipping name.

2   The class and when assigned and the division of the goods.

3   The United Nations number.

4   The packaging group for the substance carried under a 'Not Otherwise Specified' (NOS) notation or other generic entry which may include the possibility of the assignment of more than one packaging group.

5   For Class 7 (radioactive materials) only, the Class 7 schedule number.

6   Any empty packages or any packages containing residual dangerous goods must be marked by the words 'EMPTY UNCLEANED' or 'RESIDUE–LAST CONTAINED', before or after the proper shipping name of the substance.

7   Where dangerous goods waste (except radioactive waste) is being transported for disposal, the proper shipping name should be preceded by the word 'WASTE'.

8   The number and kind of packages together with the total quantity of dangerous goods covered by the description.

9   The minimum flashpoint, if 61°C (141°F) or below (closed cup test), or other additional hazard which is not communicated in the description of the dangerous goods.

10   The identification that the goods are 'MARINE POLLUTANTS' and, when declared under an NOS or generic entry, the recognised chemical name of the marine pollutant in parentheses.

11   For Class 4.1 (self-reacting substance) or Class 5.2 (organic peroxide), the control and emergency temperatures, if applicable.

Additional information is required where special classes of dangerous goods are carried and this information is applicable for:

All Class 1 goods, gases, infectious substances, radioactive materials, certain substances in Class 4.1 which may be exempt from display of an explosive subsidiary label, and certain organic substances which are also exempt from displaying the explosive subsidiary label.

Slinging Arrangement and Movement of Nuclear Waste Flasks

FIGURE 10.1 Secure stowage of nuclear waste flasks inside the hold of a designated cargo transport vessel, IMDG class 7. The carriage framework for these cargo parcels is bolted to deck beam bearing supports to ensure the integral movement with the ships motions when at sea. Steadying lines are secured to the lifting frame to provide control when loading/discharging by means of a four legged bridle arrangement.

CLASSES OF DANGEROUS GOODS

Dangerous goods are classified as follows:

Class 1: Explosives.

Class 2: Flammable gases, poisonous gases, or compressed, liquefied or dissolved gases which are neither flammable nor poisonous.

Class 3: Flammable liquids, subdivided into three categories:

  3.1 Low flashpoint group of liquids having a flashpoint below $-18°C$ ($0°F$) closed cup test, or having a low flashpoint in combination with some dangerous property other than flammability.

  3.2 Intermediate flashpoint group of liquids having a flashpoint $-18°C$ ($0°F$) up to but not including $23°C$ ($73°F$) closed cup test.

  3.3 High flashpoint group of liquids having a flashpoint of $23°C$ ($73°F$) up to and including $61°C$ ($141°F$) closed cup test.

Class 4:

  4.1 Flammable solids.

  4.2 Flammable solids or substances liable to spontaneous combustion.

  4.3 Flammable solids or substances which in contact with water emit flammable gases.

Class 5:

  5.1 Oxidising substances.

  5.2 Organic peroxides.

Class 6: Poisonous (toxic) substances.
Infectious substances.

Class 7: Radioactive substances.

Class 8: Corrosives.

Class 9: Miscellaneous dangerous substances – that is, any other substance which experience has shown, or may show, to be of such a dangerous character that this class should apply to it.

Stowage of Class 1: Explosives

Explosives are categorised for stowage in one of the following methods:

1   Stowage Category I – Goods not requiring a magazine stowage.

2   Stowage Category II, Type 'A' – Fixed magazine structure. This magazine should be close boarded on the inner sides and floor, although cargo battens are sufficient on the ship's sides and bulkheads if they are not more than 150 mm apart.

3   Stowage Category II, Type 'B' – Fixed magazine structure. Similar to Type 'A' but close boarding of sides and floor is not a requirement.

4 Stowage Category II, Type 'C' – Fixed magazine structure similar to Type 'B', but restrictions are placed on the permitted distance from the ship's side.

5 Stowage Category II – Approved portable units.

6 Stowage Category II – Freight containers.

7 Stowage Category III (Pyrotechnics) – Similar stowage to Category I, except that goods should not be overstored with other cargo.

8 Stowage Category IV – Goods requiring this stowage should be placed as far as possible away from living accommodation and should not be overstored. Deck stowage is preferred.

Package Requirements for Dangerous Goods

All dangerous goods intended for carriage by sea must conform to the specifications and performance tests as recommended by the IMDG code.

Packaging must be:

1 Well-made and in good condition.

2 Sealed to prevent leakage.

3 Package material should not be adversely affected by the substance it has contained within. If necessary it should be provided with an inner coating capable of withstanding ordinary risks of handling and carriage by sea.

Where the use of absorbent material or cushioning material is employed, that material shall be:

a Capable of minimising the dangers to which the liquid may give rise.

b So disposed as to prevent movement and ensure that receptacle remains surrounded.

c Where reasonably possible, of sufficient quantity to absorb the liquid in the event that breakage of the receptacle occurs.

When filling packages/receptacles with liquids, sufficient ullage should be left to make an allowance for expansion which may be caused by rises in temperature.

Gas cylinders for gases under pressure must be adequately constructed and tested, maintained and correctly filled. When pressure may develop in a package by the emission of gas from the contents due to a rise in temperature, such a package may be fitted with a vent, provided that the gas emitted will not cause danger in any form to the surround.

Packages of 'Dangerous Goods' must be transported in accordance with the provisions of the IMDG code. Packages containing a harmful substance shall be durably marked with the correct technical name (trade names alone should not be used). They should be marked to indicate that they are a marine pollutant and identified by additional means, such as by use of the relevant UN number.

Markings on packages containing harmful substances must be of such a durable nature as to withstand three months' immersion in sea water. They must be adequate to minimise the hazard to the marine environment having due regard to their specific contents.

LABELS, MARKS AND SIGNS

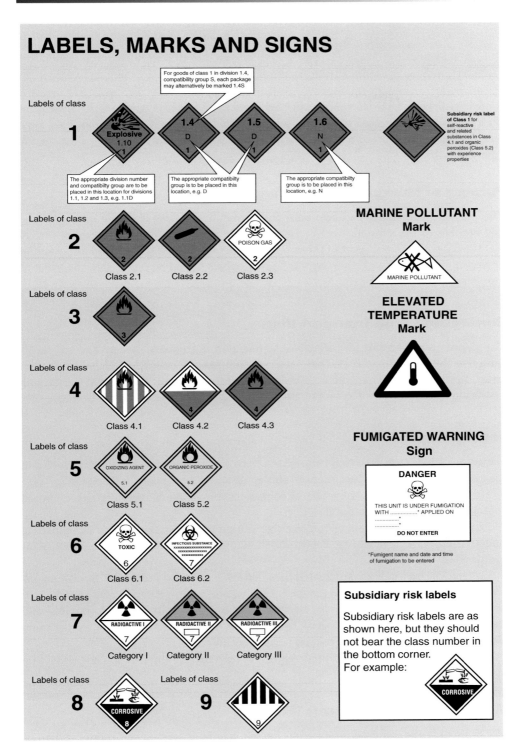

FIGURE 10.2 Marking and labelling of dangerous goods (Ref. IMDG code). Reproduced with kind permission from IMO.

Note: packages which contain small quantities of harmful substances may be exempt from the marking requirements. Exemptions are referenced in the IMDG code.

Empty packages which have previously been used for the transport of harmful substances shall themselves be treated as harmful substances unless adequate precautions have been taken to ensure that they contain no residues that are of a harmful nature to the marine environment.

Purpose of Marking and Labelling

The purpose of marking packages with the correct and proper shipping name and the UN number of the substance is to ensure that the material or substance can be readily identified during the transport of the goods. This identification is particularly important in determining the nature of emergency treatment which would be required in the event of a spillage or accident occurring.

Carriage in Cargo Transport Units

The shipper is responsible for providing the transport documents, namely a signed certificate that the unit offered for carriage is properly packaged, marked and labelled or placarded, as appropriate. If dangerous goods have been packed in such a unit and the packing certificate is not available, the cargo transport unit should not be accepted for carriage.

Segregation

Dangerous goods that have to be segregated from each other must not be transported in the same cargo transport. Exceptions to this apply and are contained in Chapter 7.

Further advice on the segregation of containers containing dangerous goods on board container vessels is given in the IMDG code.

Dangerous goods that may react together during transportation or any incompatible cargoes must be segregated (e.g. Foodstuffs should not be packed or stored with any dangerous goods).

Dangerous chemicals are segregated based on their chemical properties. Similar chemical properties in goods may be grouped. Groups are: acids, ammonia compounds, bromates, chlorates, chlorites, cyanides, heavy metals and their salts, hypochlorite, lead and lead compounds, liquid halogenated hydrocarbon, mercury and mercury compounds, nitrites, perchlorates, powdered metals, peroxides and azides.

Similar conditions for Ro-Ro units apply and reference to the IMDG code should be made.

Material Safety Data Sheets (MSDS)

MSDS provide the required information for safe handling, storage and transportation of dangerous goods. The MSDS card/sheet will provide the necessary information as per the following sections:

- **Chemical Product and Company Identification**: Provides product name, chemical formula and manufacturers name details.
- **Composition/Information on Ingredients**: Provides composition and ingredient details of the goods.
- **Hazards Identification**: Provides necessary information on the associated hazards such as corrosive, oxidising, toxic and flammability, etc.
- **First Aid Measures**: Provides necessary information on first-aid regarding inhalation, skin/eye contact, etc.

DANGEROUS/HAZARDOUS CARGOES

(in dry cargo/container ships or Roll on, Roll off vessels)

In the event of any dangerous goods or harmful substances being carried aboard the vessel, reference to the International Maritime Dangerous Goods (IMDG) code should be consulted. Additionally, the Chemical Data Sheets contained in the Tanker Safety Guide (Gas and Chemical) issued by the International Chamber of Shipping may be appropriate.

Such goods/substances must be classified, packaged and labelled in accord with the Merchant Shipping Regulations. Such trailers or vehicles should be given special consideration when being loaded and inspected for leakage prior to loading on the vessel. Such vehicles/containers should also be provided with adequate stowage that will provide good ventilation in the event of leakage while in transit – e.g. upper deck stowage, exposed to atmosphere is recommended as a general rule.

Deck (Cargo) Officers should pay particular attention to the securing of such transports to prevent negative movement of the unit. Special attention should also be given to the securing of adjacent units to prevent escalation of cargo shifting in a seaway. Tank vehicles may not necessarily be carrying hazardous goods, but any spillage of the contents could act as a lubricant on surrounding units and generate a major cargo shift on Ro-Ro vessels in heavy seas.

In the event that a cargo parcel/unit is found to be 'leaking' or have exposed hazards, the nature of the cargo should be ascertained and personnel kept clear of the immediate area until the degree of hazard is confirmed. In any event the unit should not be accepted for shipment and rejected until satisfactorily contained.

Where a hazardous substance is discovered at sea to be a threat to personnel, full information should be sought as soon as possible. Any action taken would reflect on the nature of the substance and the emergency actions stipulated in carriage instructions. It may become prudent to seek additional instructions from the manufacturer/shipper of the substance and act accordingly.

Note: With reference to Regulation 54 of SOLAS (1996 Amendment) in ships having Ro-Ro cargo spaces, a separation shall be provided between a closed Ro-Ro cargo space and the weather deck. The separation shall be such as to minimise the passage of dangerous vapours and liquids between such spaces. Alternatively separation need not be provided if the arrangement of the closed Ro-Ro space is in accordance with those required for the dangerous goods carried on the adjacent weather deck.

Top copy — Shore
Second copy — Ship
Third copy — Ship

P&O Irish Sea

European Ambassador
Hazardous Cargo Stowage Plan

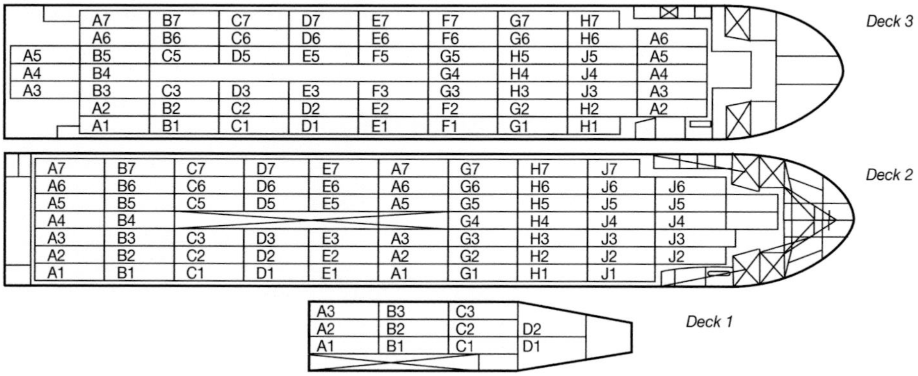

Deck 3

A7	B7	C7	D7	E7	F7	G7	H7			
A6	B6	C6	D6	E6	F6	G6	H6	A6		
A5	B5	C5	D5	E5	F5	G5	H5	J5	A5	
A4	B4					G4	H4	J4	A4	
A3	B3	C3	D3	E3	F3	G3	H3	J3	A3	
A2	B2	C2	D2	E2	F2	G2	H2	A2		
A1	B1	C1	D1	E1	F1	G1	H1			

Deck 2

A7	B7	C7	D7	E7	A7	G7	H7	J7	
A6	B6	C6	D6	E6	A6	G6	H6	J6	J6
A5	B5	C5	D5	E5	A5	G5	H5	J5	J5
A4	B4					G4	H4	J4	J4
A3	B3	C3	D3	E3	A3	G3	H3	J3	J3
A2	B2	C2	D2	E2	A2	G2	H2	J2	J2
A1	B1	C1	D1	E1	A1	G1	H1	J1	

Deck 1

A3	B3	C3	
A2	B2	C2	D2
A1	B1	C1	D1

DK3: CLASS 1 LANE 3,4,5 UP TO FR. 39   Q = DECK 3 AFT OF FR.52 *ONLY*
LANE 1,2,6,7 FWD. FR. 25 TO FR. 39   DECK 1 & 2 O.K.

PROHIBITED: DECK 1 & 2; CLASS 1.1- 1.6, 2.1, 3.1, 3.2, 5.2, 6.1(B), 8(B)
DECK 3: AS ABOVE BETWEEN FR. 35/39

Position	Class	Reg. No.	UN No.	EMS	MFAG	Remarks: Packaging, etc.

Date _____

Time _____

Voyage _____

Signature: Loading Officer

FIGURE 10.3 Stowing and monitoring of hazardous goods in Ro-Ro units.

Precautions when loading/discharging hazardous goods:

1 All documentation regarding the 'Dangerous Goods' should be in order and include the Container Packing Certificate, the Shipper's Declaration and relevant Emergency Information.

2 All cargo operations should be supervised by a responsible officer who will be in possession of operational and emergency information.

3 No unauthorised persons or persons intoxicated or under the influence of drugs should be allowed near to hazardous cargoes.

4 The compartment or deck area should be dry and clear, suitable for the stowage of the cargo.

5 Where cargo handling equipment is to be used, such equipment should be inspected to be seen to be in good order before use.

6 Dangerous goods should not be handled under adverse weather conditions.

7 All packaging, labelling and segregation of the goods is carried out as per the IMDG code.

8 Tanks, where applicable, should not be overfilled.

9 Suitable 'Emergency Equipment' should be kept readily available for any and every potential hazard associated with the goods.

10 Fire wires should be rigged fore and aft of the vessel.

11 Packages should be identified and stowed in an appropriate place to protect against accident.

12 Safe access to packages must be available in order to protect or move away from immediate hazards.

13 Emergency stations with suitable protective clothing should be identified in respect to the location of the cargo.

14 Correct signals, i.e. 'B' Flag should be displayed during the periods of loading and discharging.

15 Stowage positions should be such as to protect the goods from accidental damage due to heating. Combustible materials should be stowed away from all sources of ignition.

16 Cargoes requiring special ventilation should be positioned to benefit from the designated ventilation system.

17 The Port Authority should be informed of all movements of hazardous goods.

18 Suitable security should be given to special cargoes such as explosives.

19 All hazardous parcels should be tallied in and tallied out of the vessel.

20 Some packages may require daylight movement. Some operations may also be affected by rain or strong sunlight and appropriate loading schedules should reflect related hazards.

Note: Emergency information on cargoes should include:

- The correct technical name of the product and its UN number.
- The classification and any physical or chemical properties.
- Quantity to load and the designated space to load.
- The stated action in the event of leakage.
- Firefighting and spillage procedures and any specific equipment required.

Author's Note: It has never been the intention by this author to substitute this text for the necessary and practical use of the IMDG code. The practical application and the importance for mariners is an absolute necessity to become directly involved with the working of the code when carrying example hazardous cargo parcels.

The enclosed text of *Cargo Work* is in no way meant to detract from use of the code.

DECK CARGOES

The term 'weather deck' refers to an open deck which is exposed to the weather on a minimum of two sides. The phrase is synonymous with deck cargo being carried on exposed decks and running a greater risk of loss because of the stowage location.

Below decks and unaffected by the elements of weather provides a level of assumed security completely opposite to the weather risks associated with deck cargoes.

Numerous codes of Safe Practice, conventions and recommendations have been published to advise on the securing and safe transport of specific deck cargoes, including timber, containers, vehicles, livestock, dangerous goods, steelwork, etc.

The losses incurred over the years would indicate that the force and power of the elements can generate extreme forces on exposed cargoes and cause restraints to part and cargo parcels to be lost overside. Such losses, if noted frequently, would probably deter the carriage of any deck cargoes at all. However, certain cargoes must be categorised and classed as deck cargo because of inherent dangers if they were carried below decks – e.g. certain hazardous goods.

General observance of SOLAS Chapters VI and VII, together with the Code of Safe Working Practice for Cargo Stowage and Securing of Cargo Units including containers have become recognised sources of information. Alongside these, full reference to the ship's Cargo Securing Manual should go some way to assisting the Cargo Officer with decisions concerning the number and positioning of securing's and restraints on deck cargo loads.

Note: The carriage of cargo on deck without the agreement of the shipper may result in a breach of contract of carriage.

The contract of carriage must expressly state that the cargo is to be carried on deck and that the cargo is in fact carried on deck.

It is usual practice that a clause excludes the carriers for goods carried as deck cargoes.

A typical clause in the contract could read:

The cargo is carried on deck at the sole risk of the shipper. The carrier shall in no case be responsible for loss or damage to the deck cargo whatsoever and howsoever caused, even if caused by the negligence of the carrier or his servants or agents.

High Profile Deck Cargoes

FIGURE 10.4 Overhead view of the port side of the *Happy Sky* (IMO No. 9457220) seen en route transporting two container gantry cranes as deck cargo. Such a load would have dictated a detailed stability assessment and would operate with an accurately assessed air draught.

FIGURE 10.5 The *Happy Delta* (IMO No. 9551935) seen at sea transporting 13 stainless steel tanks on deck. The largest tanks were 14 metres in diameter at a height of 20.8 metres. The largest tank had a weight of 106 mt. The total volume of 40,000 cubic meters was the main issue as the total weight of the cargo was only 900 mt. Such a large exposed cargo attracts a greater windage which in turn could influence increased leeway when on passage.

The Regulations require that the following criteria are met when carrying deck cargo:

1   That the vessel will have adequate stability at all stages of the voyage for the amount of cargo it is proposed to load. It should be borne in mind that cargoes like coke and timber may absorb up to about one-third of their own weight by water. Also losses of bottom weight like fuel oil and water from double bottom tanks would work against the positive stability of the vessel.

2   Adequate provision must be made for the safe access of the crew when passing from one part of the vessel to another. Deck cargoes which prevent access to crew's quarters either along or underdeck must be provided with a walkway over the cargo, and in any event walkways are required for ships with timber deck cargoes.

3   Steering gear arrangements must be protected against damage and in the event of a breakdown of the gear, enough deck space must be available to operate an emergency system.

4   If cargo is to be stowed on hatches, these hatches must be correctly battened down and of adequate strength to support the carriage of the cargoes.

5   Decks designated for the stowage of deck cargo must be of adequate strength to support the stowage.

6   Deck cargo parcels are to be well secured and if necessary protected from the weather elements, including the heat of the sun. The height of any cargo should not interfere with the navigation of the vessel and obstruct the keeping of an efficient lookout.

7   The stowage of the cargo above the deck or any other part of the ship on which it stands will not interfere or obstruct safe and efficient access of the working of the ship and will not, in particular, obstruct any opening giving access to those positions or impede its ability to being readily secured weather tight.

Example Deck Cargoes

Acids and corrosives – Liquid acids and dangerous corrosive substances are usually carried in glass containers known as carboys. These containers are straw protected by a steel wire frame and are often crated for shipping. They are always allocated deck stowage away from crew quarters in accordance with the IMDG code and would need to be well lashed and secured against movement. In the event of spillage, the accompanying documentation should be consulted and any persons involved in clear-up procedures should be issued with protective clothing inclusive of goggles, gloves and suitable footwear.

Chemicals – The type of chemical substance and its form will dictate its style of packaging. Obviously the numerous chemicals shipped vary considerably and stowage method would be advised by shippers and supplied documentation. Special attention should be paid to instructions in the event spillage occurs, as some chemicals react with water or air and become harmful to personnel if incorrect procedures are adopted.

Containers – Regularly carried on open decks of container vessels in the 'stack'.

However, containers carrying hazardous goods are identified and given appropriate segregated stowage. Where single containers with dangerous goods are carried on open decks on other than dedicated container vessels, suitable stowage and securing are expected to be provided. The main concern for Cargo Officers is that the goods themselves are secured inside the container and packed under correct supervision and delivered for shipment with a Container Packing Certificate, together with relevant documentation regarding the actual goods inside the container.

Gases – Carried in cylinders of various sizes. These must be well secured against unwanted movement. They should not be stowed near any heat source and protected from the sun's rays, usually by a tarpaulin.

Livestock (see Chapter 7) – Most livestock would be carried on a sheltered part of the upper decks. Shipper's instructions for the welfare, feeding and hygiene would normally accompany any loading of animals.

Oil (drums) – Can be carried below decks as well as above decks. Part cargoes are often carried as deck cargo to provide an improved stability condition without having to shut cargo out. Drums are usually of 50 gallon size and should be tightly packed. The most common commodity carried is generally lubricating oil. Once stowed, they should be securely lashed and bowsed into the side bulwarks. If total deck coverage is employed then a walkway, similar to that for timber deck cargoes, would need to be constructed to provide crew access to fore and aft parts of the ship.

Steelwork – May be shipped as heavy lifts in a variety of forms: castings, bulldozers, railway lines, etc. Must be stowed on timber bearers and not steel on steel. The bearers are meant to reduce friction between the deck and load but also to spread the deck load capacity weight. In every case, heavy steel cargoes should be well secured, preferably with chains and bottle screws. A combination of chains and wires is also considered as being suitable, depending on the nature of the load. Some steel loads may lend to being welded to the deck to prevent unwanted movement.

Deck Cargoes – General Principles

Deck cargo should be stowed and distributed in a manner that will avoid undue stress on deck areas and ensure that adequate stability is retained throughout the voyage.

Certain deck cargoes, like timber, have the associated danger of absorbing moisture at a position higher than the ship's centre of gravity. With the combined burning off of fuel and the consumption of fresh water from the lower tanks of the vessel, the danger of generating a loss of GM or even creating a negative GM is readily apparent. Ice accretion of cargoes, particularly container deck stows, could be extremely detrimental to the stability of the vessel.

Other cargoes may be large or heavy and generate their own restrictions on the ship.

Deck cargoes must not impair the working of the vessel, particularly by obstructing the lookout's duties or preventing access to the working spaces of the vessel. Large cargo parcels could increase the windage experienced by the ship and cause excessive leeway effects and such effects would need to be monitored by Navigation Officers.

During loading, Chief Officers are advised to ensure that decks are not over-stressed by 'point loading' and that supporting structures about the loaded area are adequate to cater for the size and volume of load. Reference to the load density plan should be made prior to loading. All loads should be suitably secured to prevent movement in a seaway and in the event of heavy weather, prior to sailing.

All deck cargoes should be loaded in accord with the Merchant Shipping (Load Lines) (Deck Cargo) Regulations and S.I. No. 1089 of 1968.

Heavy Weather and Cargo Procedures

In the event of heavy weather due to affect the ship's passage, certain obvious precautions, depending on the nature of the cargo carried, can be adopted to protect the cargo condition.

1   Investigate an alternative route for the vessel clear of weather-affected areas.
2   Improve the ship's stability and reduce any free surface effects.
3   Tighten up on any cargo lashings, especially deck cargo lashings and heavy lifts.

FIGURE 10.6 The steel fore deck of the *Baltic Eide* (IMO No. 8801917) fitted to receive containers overall, but seen above with a part load of containers, as its deck cargo, during a transit of the Kiel Canal.

4   Reduce speed in ample time to avoid the vessel pounding.

5   Adjust the ship's head to avoid excessive rolling.

6   Close up ventilators to avoid water ingress.

7   Check all hatch and access seals are secure.

Securing Example Deck Cargoes

Efficient and effective securing of deck cargoes tend to follow recommendations from the Code of Safe Practices for stowage and securing of Cargo (CSS) and also references from the Cargo Securing Manual (CSM) as approved by the administration.

The ships deck load capacity plan, should be referred to ensure that 'point loading and uneven distribution does not take place and that no unnecessary damage takes place to decks and hatch covers'.

Spreading the load by use of timber bearers will spread the total weight and reduce friction movement of loads. Coupled with secure chain and wire lashings, cargo movement can be eliminated even when in rough sea conditions.

Where multiple 50 gallon oil drums are being carried on deck, netting and bowsing wires can be ideal to solidify the concentrated load against bulkheads.

FIGURE 10.7 Deck load seen on offshore supply vessel *Highland Courage* (IMO No. 9249491). Alongside an FPSO in the North Sea. Loads are normally left pre-slung to ease discharge to FPSO or to the installation.

Offshore Supply Vessels – Deck Cargoes

A major section of the industry is occupied with oil and gas recovery from offshore waters. Offshore installations, from the colossal production platforms to the smaller drilling rigs, have the need to be re-supplied on a continuous basis. Cargoes vary in this sector of the industry from the unusual in the form of 'mud' and/or cement, carried in underdeck tanks, and to the more mundane pipe sections and general stores packed in small containers.

The offshore supply vessel, once in close proximity to the installation, is discharged by use of the rig's own cranes. The position of the vessel is held precariously close to the structure of the installation by Dynamic Positioning (DP) and by expert ship handling skills of the vessel's Master, weather permitting.

FIGURE 10.8 Steel pipes seen stowed on upper decks. Chain lashings are stretched across at intermediate lengths and tensioned by ratchet gear once in position. The pipes have been left pre-slung with wire snotters for speed of discharge.

FIGURE 10.9 Capped drilling pipes seen loaded on the wide beam cargo deck of an offshore supply vessel. The pipes are capped at each end to prevent water retention in the event of the vessel encountering rough weather. Any fluid in quantity being retained among deck cargoes could seriously affect the positive stability of the vessel.

FIGURE 10.10 A semi-submersible drilling rig under tow by offshore multi-purpose supply vessels out of Invergordon, Cromarty Firth, Scotland. Installation moving and positioning is a regular employment of offshore vessels.

FIGURE 10.11 A jack-up oil drilling rig 'Galaxy III' is moved under tow by the anchor handling vessel (AHV) *Highland Courage*.

Offshore Operations

The Offshore sector of the industry operates many stand-by, supply and various support vessels to sustain its operations in a safe and economical manner.

The fact that all installations must have a stand-by vessel within 2 miles is a standard and statutory requirement for continued operations.

FIGURE 10.12 The loaded aft cargo deck of the *Highland Courage* (IMO No. 9249491) offshore supply vessel, in close proximity to the stern of the offshore storage unit 'Usiage Gorm' in the North Sea.

These vessels are usually constructed to be multi-tasking and frequently double up as supply but also as an ice-breaker, safety boat, towing vessel, anchor handling (AHV) or as a Diving Support vessel (DSV). Many are fitted with helicopter landing pads themselves. They generally supply tank containers of aviation fuel to offshore installations.

Offshore vessels are generally constructed with specialised tank systems in order to supply mud/cement and fresh water to operational drilling platforms. Dry stores for production platforms and similar installations are usually containerised for hoisting aboard the installations by their own cranage.

Timber Deck Cargo

Special regulations apply to the carriage of timber on deck (see Chapter 7). Separate load-lines may apply for full timber loads and specific securing arrangements are recommended as per the Code of Safe Practice for Ships Carrying Timber Deck Cargoes.

The main concern with timber being carried on deck is from the absorption factor from seas breaking over the cargo. Any absorption could dramatically increase the top weight of the exposed cargo and be detrimental to the positive stability of the vessel.

Vehicles (On deck)

It is not unusual to see vehicles carried as deck cargo on board ships other than designated Ro-Ro types of vessel. Tractors and other farm vehicles, heavy lift bulldozers and similar tracked military vehicles are frequently secured on low-loaders on deck. Private cars are generally carried below decks where some protection from the weather elements, where possible, can be achieved.

Securing of vehicles on deck by means of rope, wire or chains for heavy plant vehicles, is expected for ocean-going vessels. Some form of chocking or 'tomming off' may also be desirable. Cargo Officers should pay particular attention to the securing of these cargoes. They are often the last parcels to be loaded and rigging gangs may be tempted to cut corners to be off the ship quickly prior to sailing.

Some short sea trades in the ferry sector can take livestock carriers like horse boxes or domestic pets in private vehicles. These usually require carriage on an open deck to benefit from the ventilation afforded by an above deck stow. Racehorses may also be accompanied by a groom(s), and voyages are generally less than 8 hours.

Once at sea, a prudent Chief Officer would order vehicles to be stowed in gear with the handbrake on. Deck cargo lashings to be tightened, especially in the event of a heavy weather being forecast.

CHAPTER 11

Security and Cargo Documentation

INTRODUCTION

Since the implementation of the ISPS code in July 2004, all ship's officers have been made aware of the need to be security-conscious. This is not to say that before this time personnel were ignorant of the dangers and security risks which have always been associated with the maritime industries. The fact that ports have now installed better security fences, X-ray detection methods, close monitoring of dock transports and tighter controls of crews seems to have provided some degree of improved marine security.

For the Cargo Officer, vigilance is essential and with most ships security, starts with ensuring correct documentation is presented by crew members on joining. Together with close inspection of the cargo manifests, correct shipping papers for specific cargoes, etc. It is from such information that the safety of the ship can be realised with confidence. The Chief Officer is able to take account of the vessel's stability criteria for all stages of the voyage. Hazardous parcels can be secured and monitored for the protection of personnel and cargo alike. It should be borne in mind that the function of the ship's crew is to protect the ship owner's interests and effect the delivery of all cargoes in good condition and in a safe manner.

It has been said that information is power. It is also abundantly clear that cargo information is an essential element of the ship's well-being. To this end ports around the world are moving rapidly to comply with the security measures required by the code. Maritime authorities are continuing to work under the umbrella of the International Safety Management (ISM) system and monitoring the safe operation of vessels on the high seas.

The industry sectors such as safe navigation operate with such external assistance as VTS schemes, communication networks and hydrographic departments around the globe. The safe transport of cargoes now similarly employs equal support, in the way of customs, police and in some cases the military to ensure a secure working environment in the modern world. However, these people cannot be all things to all men and it has become clear that the ship's officer is closer to the front line of safety and security aboard ships than any other individual.

DOI: 10.4324/9781003407706-11

THE ISPS CODE AND CARGO SECURITY

It is difficult to visualise Cargo Officers being directly involved in the cargo security aspects of a 20,000 TEU container vessel other than being vigilant during loading, discharging and while in transit. The practicalities of searching excessive numbers of containers is clearly beyond their scope. Sampling possibly a few containers at random must be considered the maximum that anyone could expect as being practical.

The task of security of cargoes must therefore be considered at the start of the container's journey when it is empty, prior to the packing stage. The 'stuffing' of the unit must be carried out under supervision and receive a packing certificate. The goods would be subject to customs controls and inspection before being sealed.

Units should be provided with secure holding before delivery to the terminal. Once inside the container park, units fall under the security cordon expected by the ISPS code. Full documentation of the unit is listed with the shipping agents and seals would be inspected prior to loading the unit on board the vessel.

The level of terminal security will vary from port to port and the degree of ship/port interface will be established with experience. The ability to detect security threats and take preventative action will be paramount. It will be envisaged that the Port Security Officer (PSO) will liaise with the Ship Security Officer (SSO) regarding all aspects of cargo security. Such liaison is expected to ensure that: a) tampering of cargo is prevented, and b) cargo which is not intended for shipment is prevented from being accepted and stored on board.

In order to retain a safe environment it is anticipated that such measures will be in place to include inventory and control applications, such precautions being supported by the identification of all cargo parcels on board the vessel. To this end, container companies have installed methods which allow the tracking of all 'box' units and Roll on, Roll off units, showing as being approved for loading and shipping by the vessel.

Screening of stores, cargo parcels and unaccompanied baggage tends to rest with the port facility and is meant to be covered by the Port Facility Security Plan (PFSP). Such screening may include the searching of baggage both ashore and on board. Scanning equipment and/or dogs may very well be used, to ensure the security of packages.

Security

The Master of the vessel should not be constrained by the company, the charterer or any person for making a decision, which in his professional judgement as the Master is necessary to maintain the safety and security of the ship. This includes refusing to load cargo, including containers or other enclosed cargo transport units which may have a direct threat to the well-being of the vessel or personnel on board.

The International Ship and Port Facility Security (ISPS) Code – Application

The ISPS code is applicable to the following types of vessel on international voyages:

1   Passenger ships, inclusive of high-speed passenger craft.

2   Cargo ships, including high-speed craft of 500 gross tonnage and upwards.

3   Mobile offshore drilling units.

4   Port facilities serving such ships engaged on international voyages.

Declaration of Security – made by the Master of the ship or the ships Security Officer on behalf of the ship, will complete the declaration of security that may be required for a port visit when specific security requirements exist. Made to the Port Facility Security Officer or to any other body responsible for shore side security on behalf of the port facility.

   The declaration addresses the security requirements that could be shared between the port facility and the ship and states the responsibility of each. The government of the country being visited by the ship will determine when a Declaration of Security is required by assessing the risk the ship/port interface or ship activity poses. The port and ship security levels being at an agreed set level.

The Contents of the Declaration of Security

Includes the details that the ship and port facility need to know, so as to set the security level that is most appropriate for both the port and the ship.

   Ship details: Name of Ship. Port of Registry, IMO Number, & Name of Port Facility.
   Gross Tonnage, Nature of cargo and Hazardous parcels, (if any).
   Passenger number if appropriate.
   Summary of Actives: Validity dates, List of activities covered, Security level(s) for
      the ship.
   Security level(s) for port facility.
   Security measures agreed between ship and port.
   Monitoring restricted areas to ensure that only authorised personnel have access.
   Controlling access to the port facility.
   Controlling access to the ship.
   Monitoring of the port facility includes berthing areas and areas surrounding the ship.
   Monitoring of the ship includes berthing areas and areas surrounding the ship.
   Handling of cargo.
   Delivering of ship's stores.
   Handling unaccompanied baggage.
   Controlling the embarkation of persons and their effects.
   Ensuring that the security communication is readily available between the ship and
      port facility.

Quarantine

Since the outbreak of the Covid pandemic all vessels entering ports have had to give information in their Health declaration regarding request for 'free pratique'. Infectious case numbers have to be declared and the authority will decide on whether the vessel must go to a quarantine anchorage.

Definitions Effective within the ISPS Code

Ship Security Plan – Defined as a plan developed to ensure the application of measures on board the ship designed to protect persons on board, cargo, cargo transport units, ship's stores or the ship from risk of a security incident.

Port Security Plan – Defined as a plan developed to ensure the application of measures designed to protect the port facility and ships, persons, cargo, cargo transport units and ship's stores within the port facility from risk of a security incident.

Ship Security Officer (SSO) – The person on board the ship accountable to the Master, designated by the company as responsible for the security of the ship, including implementation and maintenance of the ship security plan and for liaison with port facility security officers.

Note: The ship's Master can now be the designated Ship Security Officer (SSO).

Company Security Officer – The person designated by the company for ensuring that a ship security assessment is carried out; that a ship security plan is developed, submitted for approval and thereafter implemented and maintained; and for liaison with port facility security officers and the SSO.

Port Facility Security Officer – The person designated as responsible for the development, implementation, revision and maintenance of the port facility security plan and for liaison with the SSO and the Company Security Officer.

Security incident – Any suspicious act or circumstance threatening the security of the ship, or of a port facility or any ship/port interface or ship-to-ship activity.

Security level – The qualification of the degree of risk that a security incident will be attempted or will occur.

> **Security 'Level 1'** – The level for which minimum appropriate protective security measures shall be maintained at all times.

> **Security 'Level 2'** – The level for which appropriate additional protective security measures shall be maintained for a period of time as a result of heightened risk of a security incident.

Security 'Level 3' – The level for which further specific protective security measures shall be maintained for a limited period of time when a security incident is probable or imminent, although it may not be possible to identify the specific target.

Ship – The term 'ship' as used within the context of the code includes mobile offshore drilling units and high-speed craft as defined by Regulation XI-2/l.

Ship/port interface – The interactions that occur when a ship is directly and immediately affected by actions involving the movement of persons, goods or the provisions of port services to or from the ship.

Ship-to-ship activity – Any activity not related to a port facility that involves the transfer of goods or persons from one ship to another.

Security Threats

Clearly in this day and age any threat to the ship or the port facilities could have a direct consequence to personnel working aboard or within the port confines. The Port Facility Security Plan (PFSP) is meant to identify such threats and prioritise protective security actions. Such threats may take a variety of forms, from damage to the ship or port facilities from an explosive device, arson, tampering with cargo or smuggling activities, to the extreme of nuclear, biological or chemical attack.

Controlled Access to the Vessel

Once in port and secured alongside the vessel is exposed to a greater risk from intruders who may have criminal intent. A tight control of the access to the vessel, inclusive of a security watchman and 'gangway log' maintained to identifying visitors should be an established practice. Person's identification should be checked and individual's security tagged to conduct legitimate business aboard.

Time of boarding and time of visitors leaving should be recorded as a standard operation. Prior to sailing, the vessel can be searched for stowaways after all valid visitors have disembarked.

Communications and crew monitoring on/off board, should be maintained and if assistance is required by the security watchman then such assistance should be communicable and readily available.

Cargo Concerns

Anything that generates cause to affect the well-being of cargo parcels is of concern to the ship's officer. Pilferage from open stow cargoes has long been an expensive activity and to some extent containerisation went a long way to curb theft. However, thieves are known to hijack the whole container, often with insider knowledge as to its contents. Other cargoes are of a higher profile and more readily visible, requiring immediate and increased security, such as nuclear waste flasks.

FIGURE 11.1 Nuclear waste flasks discharged with full heavy lift precautions onto special transports in Japan monitored by security mobiles and specialised personnel. Open aspects and security fencing surround the working area of the port facility.

FIGURE 11.2 Security personnel monitor nuclear waste flasks following discharge into the Japanese port facility handling grounds. Customised transports are designated to each flask and movement is checked at every stage of transit.

FIGURE 11.3 Nuclear waste flasks being lifted from carriage cradles from a stowed position inside a lower hold. The units are bolted onto the athwartships beams of the ship and treated as heavy lift, supervised loads. Once landed to the transports on the quay, the bedplates of the units are bolted onto the ground transports to become an integral unit, preventing any unwanted movement of the flask.

Shipboard Security Activity

Many aspects of shipboard activity are exposed to abuse and threats to security. Some of these are listed and officers should be mindful of the security elements associated with:

1   The handling of cargo which may contain harmful substances or terrorist personnel.
2   The handling of unaccompanied baggage.
3   The handling and loading of ship's stores.
4   Controlling access of persons who may have criminal intent.
5   Monitoring berthing areas in close proximity to the ship's hull.
6   Monitoring offshore areas to prevent incursion by waterborne craft.
7   Controlling the embarkation of persons and their effects (especially so with high-profile vessels carrying increased numbers of passengers).

FIGURE 11.4 High-profile passenger ship *Queen Mary 2* (IMO No. 9241061) lies port side to alongside the berth in Southampton. Such vessels require maximum security within the port facility and on all access points to the vessel. The vessel is seen undergoing a lifeboat drill inside harbour limits.

Security Progress

In order to be compliant with the ISPS code, SOLAS has been amended to include relevant requirements:

1   All ships on international voyages will be equipped with AIS.
2   Companies will be expected to install Ship Security Alert Systems on their vessels.
3   Ships will run a continuous synopsis record while in service from the time of launching for new builds.
4   Companies must report to a flag state-appointed, recognised security organisation and create internal positions for dedicated company security officers (CSOs) as well as on-board ship security officers (SSOs).

5   Companies will develop their own security plans to suit each individual vessel, which must be implemented on board.

6   Vessels which are compliant will be issued with an International Ship Security Certificate (ISSC) by the company's Recognised Safety Organisation. This organisation will also be responsible for conducting internal audits.

Application

The ramifications of the ISPS code will mean that practical activities to safeguard the ship and cargoes will begin to operate – for example: Masters being informed beforehand of the presence of self-igniting chemicals; containers being scanned inside the terminal, prior to being loaded aboard the vessel – while tighter access controls into terminals and onto ships will expect to create a more security-conscious environment.

BILLS OF LADING

A Bill of Lading (B/L) is a list of cargo which is signed by the ship's Master or his agent. It forms the basis of contract for the transport of the goods aboard the designated vessel. The bill contains terms and conditions for which the goods are to be received on board, shipped and delivered. It acts as a receipt for goods shipped as well as being a negotiable document of title.

The B/L is normally made up by the carrier or by the shipping agent. Where a B/L is issued under a Charter Party, it remains in the hands of the shipper or the charterer. When the B/L is passed on, it acts as a receipt for the goods while in the hands of the charterer or the shipper. Once the B/L is passed to the consignee or an endorsee, it acts as evidence of the terms of carriage.

The B/Ls are usually issued in sets of three (3) parts, which are actual copies. The number of negotiable copies must bear a declared entry on each of the B/Ls and all must be signed. They may be accompanied by additional, non-negotiable copies.

There are several types of Bills of Lading, each with a relevance directly affecting the reception, loading, transport and delivery of the goods being shipped:

A Received for Shipment B/L is issued once goods have been delivered to the ship owner or his agent, probably before the ship has arrived at the loading port. It denotes goods retained in custody, prior to the issue of a B/L which includes the numbers, weight and labelling of such goods, together with the condition of those goods.

A Shipped B/L describes goods which have actually been loaded aboard the vessel. In the event that a Received for Shipment B/L has been issued, then this must be surrendered or converted to a Shipped B/L by endorsement.

A **Direct B/L** is intended to cover goods carried in one ship from one port to another.

A **Through B/L** is issued for goods which are conveyed to their destination partly by sea and partly by overland transport, or where sea transport is by two or more vessels. A through bill is evidence of contract between the person that signs and delivers the bill. In theory that person is responsible for any loss or damage incurred during the entire movement of the goods. However, in practice it is usual for the ship owner to only take responsibility for that period of time that the goods are under his care. Goods being carried are subject to the conditions imposed by the carrier. Normally freight would be pre-paid on the whole journey, and the ship owner's agents would then pay for any future transhipment onwards to the final destination.

An **Open B/L** is one which gives no indication of to whom the goods are consigned. Any person holding the B/L would be able to take delivery of the goods, although the receiver would not have a legal hold on the goods against the rightful owner of those goods.

A **Bearer B/L** is one that makes the goods delivered to 'the bearer' and the goods are handed over at delivery. A Bearer B/L does not require any endorsement.

A **Straight B/L** is one which is issued in the name of the consignee only. Unless it has been endorsed by the consignee, no other person could take possession of the goods. It is an American term and such bills are rarely issued except for high-value commodities.

An **Order B/L** – A popular B/L which is widely used. One part of it can be exchanged for the goods or alternatively a delivery note. It can be endorsed in two ways: a) by that person to whom the goods are to be delivered, thus making the B/L into a bearer bill; or b) by special endorsement to a specific person or persons, turning it effectively into a Straight B/L, as stated above. In the case of (b), if the person designated wishes to pass the delivery on to a third person, the bill would require an additional endorsement.

A **Liner B/L** is one which is unrelated to the charter party and issued by the owner or charterer of a vessel engaged in general trade. The B/L so issued may have a variety of clauses affecting its use and content.

An **Outward B/L** – A B/L reflecting exported goods from that country, under the Carriage of Goods by Sea Act, previously (1924) and more recently the 1971 Act, providing rights under the Hague Rules.

A **Homeward B/L** – A B/L which is issued abroad to cover the goods being imported into the United Kingdom.

BILL OF LADING

SHIP FROM	Bill of Lading Number:_____
Name: Address: City/State/Zip: SID#: FOB: ☐	BAR CODE SPACE

SHIP TO	CARRIER NAME: _____
Name: Location #:____ Address: City/State/Zip: FOB: ☐	Trailer number: _____ Seal number(s): **SCAC:** **Pro number:**

THIRD PARTY FREIGHT CHARGES BILL TO:	
Name: Address: City/State/Zip:	BAR CODE SPACE

SPECIAL INSTRUCTIONS:

Freight Charge Terms (Freight charges are pre-paid unless marked otherwise)

Prepaid _____   Collect _____   3rd Party _____

CUSTOMER ORDER INFORMATION

CUSTOMER ORDER NUMBER	# PKGS	WEIGHT	PALLET/SLIP Y or N	ADDITIONAL SHIPPER INFO
GRAND TOTAL				

CARRIER INFORMATION

HANDLING UNIT		PACKAGE		WEIGHT	H.M. (X)	COMMODITY DESCRIPTION Commodities requiring special or additional care or attention in handling or stowing must be so marked and packaged as to ensure safe transportation with ordinary care.	LTL ONLY	
QTY	TYPE	QTY	TYPE				NMFC #	CLASS
							RECEIVING	
							STAMP SPACE	
					GRAND TOTAL		

Where the rate is dependent on value, shippers are required to state specifically in writing the agreed or declared value of the property as follows:

"The agreed or declared value of the property is specifically stated by the shipper to be not exceeding

_____ per _____ ."

COD Amount: $_____

Fee Terms: Collect: ☐   Prepaid: ☐
Customer check acceptable: ☐

NOTE Liability Limitation for loss or damage in this shipment may be applicable. See 49 U.S.C. - 14706(c)(1)(A) and (B).

RECEIVED, subject to individually determined rates or contracts that have been agreed upon in writing between the carrier and shipper, if applicable, otherwise to the rates, classifications and rules that have been established by the carrier and are available to the shipper, on request, and to all applicable state and federal regulations.

The carrier shall not make delivery of this shipment without payment of freight and all other lawful charges.

Shipper Signature

SHIPPER SIGNATURE / DATE This is to certify that the above named materials are properly classified, packaged, marked and labeled, and are in proper condition for transportation according to the applicable regulations of the DOT.	Trailer Loaded: ☐ By Shipper ☐ By Driver	Freight Counted: ☐ By Shipper ☐ By Driver/pallets said to contain ☐ By Driver/Pieces	**CARRIER SIGNATURE / PICKUP DATE** Carrier acknowledges receipt of packages and required placards. Carrier certifies emergency response information was made available and/or carrier has the DOT emergency response guidebook or equivalent documentation in the vehicle.

FIGURE 11.5 Example of a Bill of Lading.

Each Bill of Lading will usually include the following information:

- The name of the vessel in which the goods are to be shipped.
- The name and address of the shipper.
- The ports of loading and unloading.
- The name and address of the consignee.
- The description of the goods, including packaging and labelling details.
- The order and condition of the goods.
- The number of parcels/packages, their weight or quantity.
- The number of B/Ls within the set.
- The place and date of issue.
- The place where freight is paid.
- The date of shipment.
- Designated signatures.

A B/L is usually issued after the Mate's Receipts have been surrendered, as a 'clean Bill of Lading'. Where damage or fault is found with the goods or their packaging, the bill would be endorsed and considered as a 'foul bill' or more commonly described as a 'dirty Bill of Lading'.

Bills of Lading (BL or BoL) Summary

A Bill of Lading (B/L) is a legal document issued by the transporting company to a shipper and details the type, quantity and destination of goods being carried. It also serves as a receipt of shipment when the carrier delivers the goods at a predetermined destination.

This document must accompany the shipped goods and must be signed by an authorised representative from the carrier, the shipper and the receiver. It is a document of title as well as being a receipt for shipped goods and forms a contract between a carrier and a shipper.

Bills of Lading contain the terms and conditions for which the goods are to be received on board, shipped and delivered. It has three main functions:

1   It is a document of title to the goods described on the Bill of Lading.

2   It is a receipt for the shipped products.

3   It represents the agreed terms and conditions for the transportation of the goods.

THE CARGO SECURING MANUAL

Shipping Procedures

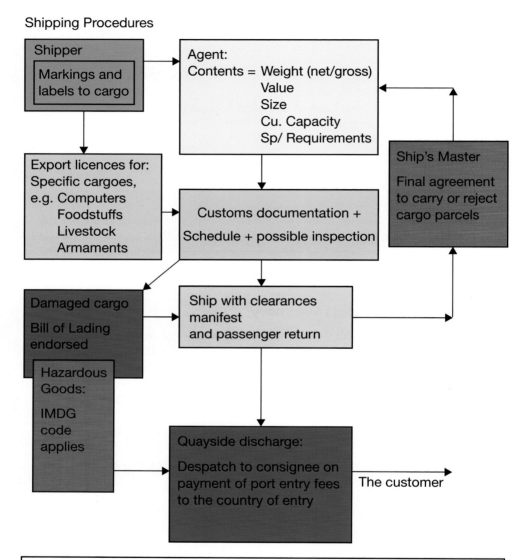

NB. It is normal practice for the shipper to have a financial bond in place prior to shipment. The function of that bond being to guarantee payment at the country of entry of the goods.

FIGURE 11.6 Cargo flow chart procedure.

Reference should be made to the IMO publications:

1   Cargo Securing Manual and Regulations V1/5 and V11/6 of the 1974 SOLAS Convention.
2   Code of Safe Practice for Cargo Stowage and Securing (CSS Code) regarding the compilation of Cargo Securing Manuals.

Application: A Cargo Securing Manual is required to be carried by all cargo units and cargo transports, loaded, stowed and secured with dangerous goods, solid cargoes and containers. This manual must be approved by the Marine Administration for all types of vessels which are engaged in the carriage of all cargoes, other than liquid or solid bulk cargoes.

Function of the Manual

The purpose of the manual is to cover all aspects of cargo stowage and securing of cargoes, other than bulk cargoes, which are covered by alternative regulations. Cargo Securing Manuals are meant to be uniform in their guidelines on the carriage and stowage of cargoes respective to the individual ship.

Definitions Respective to the Manual

Cargo securing devices – Fixed and portable devices used to secure and support cargo units.

Maximum securing load – A term used to define the allowable load capacity for a device used to secure cargo to a ship – namely the SWL of the device.

Standardised cargo – Cargo for which the ship is provided with an approved securing system based on the specific cargo unit types.

Semi-standard cargo – Cargo for which the ship is provided with a securing system which is capable of securing a limited variety of cargo units such as vehicles or trailers.

Non-standard cargo – Cargo which requires individual stowage and securing arrangements.

Use of the Manual and Its Contents

The guidelines set out by the manual are expected to suit the stability and trim criteria of the vessel and are not meant to infringe the loadline requirements. Neither is the manual trying to replace the principles of good seamanship or experience of recognised stowage and securing practice. It is meant to specify cargo securing methods provided on board the ship, for the type of cargoes that the vessel can be expected to transport. By use of suitable securing points and fittings, the manual should be designed to provide guidance in order to prevent movement of loads which would expect to experience transverse, longitudinal and vertical forces when the vessel is at sea. General advice should also include recommendations on the number, strength and application of securing devices.

SECURING DEVICES

A deck plan of securing devices like pad eyes, eye bolts and bulwark securing points should be provided within the manual. These fittings should note the respective SWL. All fixed securing's should be provided with documentation stating:

1   Name of manufacturer.
2   Designation of use.
3   Material of manufacture.
4   Identification markings.
5   Maximum Securing Load (MSL) (alt: SWL).
6   Results of any non-destructive testing.

For existing ships where non-standardised fixed securing devices are employed, the information regarding the MSL and its location are deemed sufficient.

Portable securing devices such as chains, rods, interlocking fittings, trestles, turnbuckles, etc. would also be expected to have their details documented as above but may have the following additional requirements:

7   Strength test details.
8   Minimum safe operational temperatures.
9   Sketch details of their use.

Regular inspection of securing fitments should be conducted by the ship's Master, but may be periodically taken up by the Marine Administration. Such inspection by ship's personnel or an outside body should be recorded within a 'record of inspections'.

To ensure that securing devices are placed safely and correctly, relevant instruction in the handling of these devices should be undertaken by respective personnel. The manual should contain these instructions and safe handling details within a sub-chapter.

CALCULATIONS OF FORCES (ACTING ON CARGO PARCELS)

The contents of the manual must include tables and/or diagrams showing the accelerations that can be expected in various positions on board the vessel when in adverse sea conditions over a range of GM values. Such content would also include examples of how these forces can be calculated. These examples would utilise cargoes that the vessel is most likely to carry. The values so calculated would be reflective of the forces that could expect to be experienced by the securing devices employed aboard the vessel.

Chief Officers, cargo superintendents, cargo surveyors, supercargoes, etc. should be familiar with the calculation of such forces being experienced by the lashings to ensure that suitable chain sizes or wire lashings engaged are of adequate Safe Working Load, to prevent unwanted movements of the load.

A supplementary section in the manual should be concerned with Ro-Ro vessels and their fixed securing points for vehicles and containers. Portable lashings for such cargoes would be recommended by type, number and SWL.

EXAMPLE: TRACKED VEHICLE (STEELWORK) SECURING ON DECK

(where: 1kN = 100 kgs)

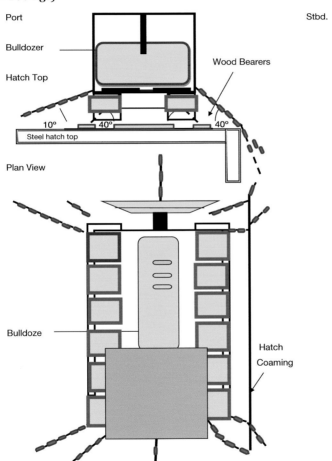

FIGURE 11.7 Example Steelwork Cargo, in the form of a tracked bulldozer shown as being loaded on wood bearers, and secured on a steel hatch cover. Securings by chains both fore and aft and athwartships, to port and starboard. Chains need to be of suitable SWL, to ensure no movement of the load in any extreme weather conditions.

In the interests of good seamanship practice 'Chain lashings' would be preferred for all heavy duty steel manufactured vehicles. However, the following example uses the lesser factor of Steel Wire Ropes inclusive of shackles, bottle screws, and deck eye pads.

Where Ships length (L) = 120 m, Breadth (B) = 20 m, GM = 1.4 m, Speed 15 knots
And Tracked vehicle, Weight = 62 mt. Dimensions 6 x 4 x 4 m.
Stowed in a position at 0.7 L on deck, in low aspect.

Steel Wire Rope B.S. = 125 kN. (MSL (Maximum Securing Load) based on 80% of the B.S. = 100 kN)
(For single one off use)

Calculated Strength (CS) of Securing Devices $CS = \dfrac{MSL}{1.5}$ (safety factor)

Calculated Strength $= \dfrac{100}{1.5}$ = 66.6 kN
_____

Shackles/bottle screws/pad eyes, etc. (as per rigging plan)
Breaking Strength = 180 kN (MSL based on 50% of the B.S. = 90 kN)

Calculated Strength (CS) of securing devices $CS = \dfrac{MSL}{1.5}$ (safety factor)

Calculated Strength $= \dfrac{90}{1.5}$ = 60 kN
_____

Breaking Strength (B.S.) of Flexible Steel Wire Rope (FSWR) 6 x 24 wps, can be found from the following formula:-

$\dfrac{20D^2}{500}$ (where D represents diameter of wire.)

TABLE 11.1 Maximum securing load table for securing equipment

Material	MSL
Shackles, rings, deck eyes, etc.	50% Breaking strength
Web lashings	70% Breaking strength
Fibre rope	33% Breaking strength
Wire rope (single use)	80% Breaking strength
Wire rope (reusable)	30% Breaking strength
Steel band (single use)	70% Breaking strength
Chains	50% Breaking strength

Steel Loads and Tracked Vehicles

Tracked vehicles like mobile cranes, bulldozers and military hardware (tanks) can be loaded to most types of floating transports. The concern for such loads is that they are not only heavy parcels to lift being usually manufactured in steel, but the 'tracks' can cause permanent damage to steel decks and composite decks.

The way to alleviate such ship damage is to have the cargo load secured to a 'Low-Loader', vehicle. This allows the load to be driven on as Ro-Ro cargo without incurring deck damage.

An alternative to drive on is to carry out a heavy lift on process, where the area of loading is covered with robust timber bearers (pit-props or railway sleepers) to keep the deck area clear of steel contact. This also provides a non-skid surface, spreads the load over a greater area and reduces the risk of unwanted movement in a seaway.

In every case steel loads of this nature would be well secured by 'chains'. Chain lashings can be tensioned with bottle screws (US Turnbuckles).

Breaking Strength in tonnes of Stud Link Chain is found from the following formula:

Grade 1 chain	12.5 mm to 120 mm	$\dfrac{20D^2}{600}$
Grade 2 chain	12.5 mm to 120 mm	$\dfrac{30D^2}{600}$
Grade 3 chain	12.5 mm to 120 mm	$\dfrac{43D^2}{600}$

CONTAINER STOWAGE

Chapter 4 of the manual is directly concerned with the stowage and securing of containers and provides information for the Master regarding any cargo stowage situations which may deviate from the general recommendations. It would be expected to show the accelerations that the stowage and securing system is based on, information on the overall mass of a 'container stack', the anticipated wind load, and any permissible reduction in securing methods for stacks of reduced height.

FIGURE 11.8 The container ship *Lircay* (IMO No. 9294824) seen with deck container stack fore and aft of the navigation bridge position. The tiers of the containers are five (5) high above the weather deck. The larger container vessels of today frequently have up to ten or eleven tiers high. The *Lircay* has been renamed *Pohorje*.

CARGO DOCUMENTATION SUMMARY

Note: Container vessels are particularly vulnerable to Parametric Rolling because they are generally constructed with wide beam and a large bow flare. High container stack securing's can experience heavy stresses on stack securing's and may cause tiers to topple causing actual loss of containers.

Ballast management record – The dangers to aquatic life have instigated the need for tighter controls on the movement and in particular the discharge of ballast waters. The ship's geographic position when ballast/de-ballast operations take place, must be recorded. Positions of ballast changes, dates, amount and tank location, must all be entered into the record.

Bill of Lading (B/L) – The consignee's title to the goods which have been shipped or are about to be shipped. The B/L will quantify the goods and reference their condition at the time of shipping and the consignee would expect to receive the goods at the port of discharge in the same good condition as when shipped. In the event that the goods are damaged at receipt or in loading or discharging, the B/L would be endorsed to specify the damage. Such an endorsed B/L would be considered as a dirty or foul B/L, as opposed to a clean B/L which is without endorsement. The Bills of Lading are usually drawn up by the shipping agent and signed by the Master of the vessel.

Cargo manifest – The official listing of all cargo parcels carried on board the vessel. This document is what the Master bases his declaration on when entering port. All cargoes are officially declared on the manifest, which is subject to inspection by customs officers and port security.

Cargo record book – A vessel engaged in the carriage of noxious liquid substances must carry a record of the cargo movements affecting the ship. It would also be expected to carry an MCA approved Procedures and Arrangement Manual, reflecting the operational aspects of the vessel.

Cargo securing manual – A legal requirement for every ship other than those engaged in the carriage of solid or liquid bulk cargoes. The purpose of the manual is to cover all relevant aspects of cargo stowage and securing. Securing devices and methods must meet acceptable criteria for strength, applicable to relevant cargo units, inclusive of containers and Ro-Ro transports. Each manual is prepared in a manner to reflect the individual ship's needs, relevant to the type of cargo parcels it is engaged to carry and customised to each vessel. Each manual is approved by the Marine Authority of the country of registration.

Cargo stowage plan – A charted plan of the vessel's cargo-carrying spaces which illustrates the type, tonnage and description of goods for designated discharge in the various intended ports of call for the voyage. The plan is constructed by the Cargo Officer and is meant to provide an overall illustration of the distribution of the ship's cargo. The plan is copied and despatched to the various ports of discharge prior to the ship's arrival on the berth. It allows relevant cranes to be ordered and stevedore gangs to be employed in advance which subsequently speeds up the time of the vessel lying in port. Although it is considered essential for dry cargo vessels, tankers, bulk carriers and container vessels all carry stowage plans respective to their relevant cargoes. (See pages 71-73 Chapter 3 and 315-316 Chapter 5.)

Certificate of Fitness – Required by every United Kingdom tanker and gas carrier. They are issued by the MCA and are valid for a period not exceeding five years, being subject to initial, annual and intermediate surveys. This certificate cannot be extended.

Charter Party – A formal agreement (contract) between the principal parties to the agreement to hire, rent or lease a ship and is evidence of who the operator of the ship is. Charter Parties are set in three categories: Time Charter, Voyage Charter, or Demise Charter (US:

Bare Boat Charter). Variations of the three categories are drawn up based on the operational requirements of the ship and the intended voyage.

Container Packing Certificate – The packing and unpacking of containers usually takes place at shore terminals or at the address of the shipper or consignee. Prior to loading on board the vessel a valid Container Packing Certificate must be received as evidence that the goods have been packed in such a way as to withstand carriage at sea. It is also a security check that the unit does not contain contraband goods and that the merchandise is as described on the certificate. Container units are now electronically scanned at entry to many shipping terminals.

Document of Authorisation (to carry grain) – As required by SOLAS Chapter VI, a certificate issued by a surveyor of the Marine Authority of the country of loading, following a survey of the ship's cargo holds and its capability to carry grain safely. Unless the ship is in possession of an exemption certificate, the Document of Authorisation would be an official requirement.

It is evidence that the ship is capable of complying with the requirements of the Grain Code.

Document of Compliance (for carrying dangerous goods) – A certificate of compliance issued to a type of ship which is permitted to carry certain categories of dangerous goods in packaged or dry bulk form. Not all ships can carry hazardous cargoes – e.g. passenger vessels are not allowed to carry Class 1, explosives.

Enclosed entry permit – A work permit which is issued prior to entry into an enclosed space compartment. The permit is issued only after all the required safety checks and inspections of the compartment have been made and the relevant precautions have been taken.

Export licences – These are supplied by the shipper as required for certain specific cargoes: computers, foodstuffs, livestock, armaments, etc. The export licence is required by government/state officials for certain types of cargo which are subject to inspection by customs – e.g. armaments, drugs, etc.

Freeboard Verification Form – Displayed showing ships draughts prior to sailing, to ensure the ship is not overloaded and has a safe level of freeboard.

International Pollution Prevention Certificate – For the carriage of noxious liquid substances in bulk.

International Security Certificate – Issued to a vessel by a recognised security organisation, confirming that the ship is compliant with the ISPS code.

Mate's Receipt – A receipt for goods received and delivered on board the vessel. As the name implies, it is signed and issued by the Mate of the ship, i.e. the Chief Officer. It may form the basis for the final Bill of Lading.

Note of Protest – This is where the Master of a ship makes a declaration of 'Protest' under oath before a Notary Public, Magistrate or British Consul. The declaration often affects cargo damaged or suspected damage due to a 'peril of the sea'. The main use of Protest in the UK is to support a cargo owner's claim against his underwriters. The Note of Protest is admissible as evidence before legal tribunals in many countries, but not in the UK unless both parties agree. Masters should note Protest as soon as possible after arrival in port and before 'breaking bulk'. The Master may 'Extend Protest' once the situation has been further assessed and the full extent of damage is revealed.

Oil record book – Current legislation requires oil tanker vessels to carry two (2) oil record books, one for cargo movement and one for fuel movements. Non-tanker type vessels are only required to carry one oil record book. Entries into oil record books should cover any movement of oil in or out of the vessel, including internal transfers between tanks. Each entry in the book should by signed by the ship's Master and another officer.

Register of lifting appliances and cargo handling gear – A record of all the ship's cargo handling equipment, usually retained and updated by the ship's Chief Officer. It contains all the certificates for such items as shackles, blocks, wires, derrick and crane tests, hooks, chains, etc. The register is open to inspection by Port State Control Officers and would be required by the surveyor when carrying out the cargo handling equipment survey.

Rigging plan – A ship's arrangement plan which illustrates the operational aspects of the ship's lifting appliances. Safe Working Loads and maximum permissible outreach limits would expect to be displayed alongside the related positions of cargo stowage compartments.

Stability information booklet – The ship's stability criteria may be in booklet format or in the form of a series of plans or even carried in a combination format of both. Either way the documents are in the control of the ship's Chief Officer and will contain the following:

General particulars of the vessel.
General arrangement plan showing cargo compartments and tank dispositions.
Special notes on the stability and loading procedures.
Hydrostatic particulars.
Metric conversion table.
Capacity plan showing centres of gravity of cargo stowage compartments (to include free surface moment of oil and water tanks).
Notes on the use of free surface moments.
Cross curves of stability (known as KN curves) with examples of their use.
Deadweight scale.
List of ship conditions and typical condition sheets.
Statical stability curve for conditions.
Simplified stability information together with damaged stability criteria.

Transportable Moisture Limit (TML) Certificate – A certificate issued within seven days of measuring the moisture limit of the bulk product to be shipped.

RELEVANT DEFINITIONS FOR COMMERCIAL CARGO ACTIVITIES

Administration – The government of the state whose flag the ship is entitled to fly.

Arrived ship – A ship which is ordered to a port and as soon as she enters the ports limits and is at the charterer's disposal, ready to load cargo, she is considered an 'arrived ship'. The vessel does not have to be at the loading berth to be considered an arrived vessel unless expressly described in the Charter Party.

Average Adjuster – Average is a marine insurance term used where a partial loss is incurred. An Average Adjuster is that person who is an expert in determining loss proportions. (See also: General and Particular Average.)

Average Bond – A signed document which gives an undertaking by the cargo owner that on delivery of his/her goods they agree to pay their proper portion of any General Average charges which may subsequently be declared. The most common bond is a Lloyds Average Bond, which would generate a General Average Deposit Receipt.

Barratry – A wilful act of violence to affect the ship or its cargo. It can also include the wrongful appropriation of the ship or her cargo, or any fraudulent activity carried out by the Master or crew, without the owner's consent, which exposes the ship to confiscation.

Bill of Lading (B/L) – A receipt for goods either received from the shipper ashore, or for goods loaded on board the vessel. It acts as evidence of a contract between the shipper and the carrier and as a document of title. It can also be endorsed to indicate that freight has been paid. Clean 'Bills of Lading' describe a B/L which declares the goods as being free of defect. A B/L that is endorsed to expressly state that the goods are second-hand, used or defective in some way is described as a foul B/L or a dirty B/L.

> Note: Multiple copies of the Bill of Lading may be issued where cargo is shipped by more than one shipper, or where the cargo is consigned to more than one consignee, or where more than one type of grade specification of a cargo is shipped by one shipper.

Cargo insurance – Is normally taken up by the buyer or the seller of the goods. The cover arranged is usually by a voyage policy for a one-off shipment. Regular shipments are generally covered by what is known as open cover.

Cargo manifest – Documented list of all cargo parcels that the ship can expect to transport during the period of the voyage.

Cargo plan – A pictorial plan of the ship's internal cargo compartments showing the stowage arrangement of all ship's cargo parcels. Each parcel of cargo would be identified by its weight and title, together with port of discharge. Each commodity for specific ports is usually colour coded.

Cargo ship – Any ship which is not a passenger ship.

Carriage of Goods by Sea – This is an Act of the UK Parliament which provides legal standing to the Hague-Visby Rules under UK statute law. The carrier has basically three obligations under these rules: a) to ensure that the vessel is seaworthy as far as is reasonably possible; b) to look after any cargo; c) to issue a Bill of Lading if the shipper requires one.

Certificate of Fitness – A Certificate issued by the MCA to every UK tanker or gas-carrying vessel. The certificate is valid for a period of five (5) years and is subject to annual inspection and satisfactory surveys of the vessel. It cannot be extended beyond its period of validation.

Consignee – The party to whom goods are addressed or consigned.

Container – A rectangular, standard sized metal box used for the transport of cargo and freight in modern-day shipping practice. Sizes of containers fall into several categories: 8ft x 8ft, 10, 20, 30, 35 and 40 ft sizes are all engaged in cargo movement by sea, rail and road.

Container vessel – A ship specifically designed to carry containers both below and above decks.

Charter Party – A formal document of agreement to hire, rent or lease a ship or yacht. There are various types of Charter Party: Time Charter for a fixed period of time, Voyage Charter for a specific voyage, Bare Boat (Demise) Charter where the charterer provides the crew, equipment, fuel and stores.

> Note: The Charter Party relates directly to the ship and is evidence of contract. Although there are some commonalities with the Bill of Lading regarding seaworthiness, freight, deviation, etc, the B/L relates to a specific cargo carried by the vessel.

Deadweight tonnage – This is defined as the actual number of tonnes of cargo, bunkers, stores, etc. that can be loaded aboard the vessel to bring the ship down to her loadline marks. The term is used when calculating Hull Insurance Premiums to assess the size of the ship to be insured. It is the difference between the lightweight of the ship and the loaded displacement.

Demurrage – A financial pay-out by the charterer to the owner of the ship for delays in loading and unloading in accord with the terms of the Charter Party. The rate of demurrage is fixed and agreed between the owner and charterer at the time of completing the Charter Party and cannot be altered. Can be taken in contrast with **Despatch money** – opposite to demurrage, where the owner pays reward money to the charterer for completing loading and/or discharging earlier than the expiry of 'lay time'.

Deviation – Any period of time that a vessel makes a departure from her scheduled route without lawful reason or excuse.

Displacement mode – The regime, whether at rest or in motion, where the weight of the craft is fully or predominantly supported by hydrostatic forces.

Document of Compliance (for dangerous goods) – A certificate of compliance issued to a type of ship that is permitted to carry certain categories of hazardous goods

> Note: Not all vessels are permitted to carry all classes of dangerous cargoes.

Export licence – A licence supplied by the shipper as required for specific cargoes such as computers, foodstuffs, livestock, armaments, drugs, etc. The licence is required by state officials, e.g. customs, and cargoes under the licence are generally subjected to inspection in countries importing the goods.

Ferry – A place where boats ply regularly across a river or arm of the sea for the conveyance of goods and/or persons. (Ref: *Encyclopaedia Britannica*)

Flotsam – A term which describes goods thrown or lost overboard which are recoverable by reason of remaining afloat.

Free on board (FoB) – Where the seller delivers goods aboard the ship and where the buyer accepts all of the subsequent charges.

Freeboard Form (FRE 13) – Following the completion of cargo operations in a port, the ship's draughts will be expected to have changed. The draughts fore and aft should be read and the mean draught obtained. It is a legal requirement that before the ship sails outward, the Freeboard Certificate is completed and displayed for the crew to see that the vessel is not overloaded and the minimum bow height is not exceeded.

Freight – The charge made for the carriage of cargo in ships. Advance freight is that amount paid before the delivery of the cargo. Back freight is that amount that is claimed when cargo is refused or cannot be delivered. Lump sum freight is that which is paid for all or part of the capacity of the ship. Dead freight is that which is claimed for space which is not provided by the charter, despite the written agreement.

Freight forwarding agent – An agency firm that specialises in the import and/or export by sea, air, rail or road, working on behalf of the merchant. When handling imports the forwarder is the consignee's agent or receiver of goods. When engaged in exporting the forwarder is the shipper, making a contract of carriage with the carrier.

Freight only Ro-Ro ship – A Roll on, Roll off vessel with accommodation for not more than 12 (driver) passengers.

General Average – A term in marine insurance meaning an expression of joint financial responsibility, involving the ship owners, the owner of the cargo, and the Master and crew. It is invoked when loss or damage occurs due to an Act of God, where the financial loss is shared by all.

Gross Tonnage (GT) – Defined as the measurement of the internal capacity of the ship. The GT value is determined by the formula: $GT = K_1 V$

When $K_1 = 0.2 + 0.02 \log 10V$
and V = total volume of all enclosed spaces measured in cubic metres

Hague-Visby Rules – A set of internationally agreed rules which define the contractual obligations and liabilities of sea carriers and cargo shippers where a Bill of Lading or Sea Waybill is issued.

High Speed Craft – A craft capable of a maximum speed, in metres per second (m/s), equal to or exceeding

$3.7 V^{0.1667}$

Where

V = displacement corresponding to the design waterline (m^3)

Hydrofoil boat – A craft which is supported above the water surface in a non-displacement mode by hydrodynamic forces generated on foils.

International Maritime Dangerous Goods (IMDG) code – A code of conditions and requirements that are required for the carriage of dangerous or hazardous cargoes. It covers the carriage and stowage of incompatible cargoes and details requirements for segregation of commodities, packaging and labelling of goods.

International voyage – A voyage from a country in which the present Convention applies to a port outside such country, or conversely.

Jetsam – The term given to goods which have been thrown or lost overboard from a ship, which are recoverable by being eventually washed ashore or recoverable from shallow waters.

Lagan – Goods which have been thrown overboard from a ship and buoyed, to allow future recovery.

Lay days – Denotes the number of days which are allowed for loading or discharging of cargo, as expressed by the Charter Party. They may be considered as 'reversible' lay days, which allow for both the services of loading and discharge. Charter Parties may express the lay days as working days or in the rate of cargo tonnage to be worked in running days.

Lay time – The time which is available to the charterer to load or discharge the chartered cargo, free of any charge other than the freight charges. Lay time cannot commence until three conditions have been satisfied:

1   that the vessel is an arrived ship (under legal terms)
2   that Notice of Arrival has been tendered (it does not need to have been accepted)
3   that the vessel is in all respects ready to load or discharge.

Lloyds Agents – Persons acting on behalf of Lloyds (Insurance) Underwriters. They carry out cargo surveys and damage surveys with the view to settling small claims. They also pass on relevant information affecting shipping movements, casualties, natural disasters, ship arrests and detainment of ships.

Lloyds of London – An association of insurance underwriters which specialises in maritime insurance cover.

Lloyds (Open Form) Salvage Agreement – An international salvage agreement accepted worldwide. It is often employed where parties cannot agree a mutual contract for a salvage operation and operates on a clause of 'No Cure, No Pay'.

Lloyds Register of Shipping – An organisation comprising ship owners, ship builders and underwriters with the prime function of establishing standards of construction and maintenance of ships. It is the British Classification Society upon which its counterparts in other countries modelled their own classification administration.

Manifest – A detailed list of all the loaded cargo aboard the vessel, usually based on the Bills of Lading. This list contains all cargo information and is one of the documents required when the ship is entered inwards by the Master.

Mate's Receipt – A receipt for goods received on board and part of the ship's tally. It would normally carry identity marks of the cargo it represents. It is issued on board to the shipper and is a prelude to being copied to a Bill of Lading. A copy of the Mate's Receipt would be retained on board the vessel.

Notice of Readiness – A written notification that the ship is in all ways ready to load her cargo. The vessel must be an 'arrived ship' with her holds and loading appliances rigged and ready for use. Any permits, or special requirements, like shifting boards or dunnage required, must be in place and the ship should be ready for loading before the notice can be issued. The notice, prepared in duplicate by the ship's Master, must be administered during normal working hours and served to the charterer.

Open cover insurance – A long-term insurance cover taken out by the buyer or seller of goods to be insured. The cover is usually arranged over a twelve-month period and is generally taken out by regular shippers of goods, as opposed to a one-off shipment. Open cover limits the risk assured in any one vessel and the amount of risk to one location.

Over-carriage – The term relates to carrying cargo past its destination. The cargo is said to be over-carried.

Particular Average – An insurance term to express the loss of a ship and/or cargo.

Passenger – Any fare-paying person on board the ship, other than the Master, officers or crew, or owners and family members. (Exception: on occasions, the Master's wife.)

Passenger car ferry – A passenger or ferry ship which has Ro-Ro access of sufficient dimensions to allow the carriage of Ro-Ro trailers and/or passenger's cars (RoPax).

Passenger return – A list of all passenger numbers carried aboard the vessel. This must be disclosed to immigration and customs officials at the time a ship enters inwards.

Passenger ship – A ship which carries more than twelve passengers.

Passenger Spaces – Those spaces which are provided for the accommodation and use of passengers, excluding baggage, store, provision and mail rooms. (Exception: spaces provided below the margin line for crew accommodation shall be regarded as passenger spaces with regard to regulations 5 and 6, respectively, 'Permeability in passenger ships' and 'Permissible length of compartments in passenger ships'.)

Perils of the Sea – An insurance term which relates to an accident involving collision, stranding, fire or heavy weather. Human errors and mistakes are not considered in this category.

Primage – A lump sum paid to a ship's Master as part of the freight charge. It is meant to cover the care and protection of his cargo. It is now more commonly used as an additional expense cover, where cargo has special transportation problems.

Protest – The action of 'Noting Protest' is made by the ship's Master as soon after entering port as is practical, when he feels litigation is possible following loss or damage which has occurred or when the Charter Party has been violated. Protest is an affidavit made before a proper authority, e.g. British Consul.

Receiver – The party which takes delivery of the goods from the carrier at the port of destination. It could well be the consignee, or a designated party, such as a road transport authority.

Receiver of wreck – A national official of a district, usually appointed by the government's Department of Trade, such as a customs officer, coastguard or Inland Revenue official, to receive reports on shipwrecks and take control of any wreck situation.

Reefer unit – A mobile/vehicle Ro-Ro unit, designed and capable of carrying refrigerated cargoes.

Right of ferry – An exclusive right to convey persons or goods (or both) across a river or arm of the sea and to charge reasonable tolls for the service (goods may include vehicles).

Note: So long as the right is exercised, the proprietor has a corresponding obligation to provide and maintain an adequate service and to restrict tolls to such as are considered reasonable. However, the right must be exercised in such a way as not to interfere with ordinary navigation.

Ro-Ro cargo space – A space not normally subdivided in any way and extending to either a substantial length or the entire length of the ship in which goods (packaged or in bulk) in or on rail or road cars, vehicles (including road or rail tankers), trailers, containers, pallets, demountable tanks or in or on similar stowage units or other receptacles can be loaded and unloaded normally in a horizontal direction.

Ro-Ro passenger ship – A passenger ship with Ro-Ro spaces or special category spaces as defined by SOLAS regulation II-2/3.

Ro-Ro vessel – A vessel providing horizontal means of access and discharge for wheeled, tracked or other mobile cargo.

Salvage Association – The leading salvage organisation in the world, originally established in London. Its main function is to accept instructions to salve ships and cargoes from interested parties such as ship owners, Average Adjusters, from casualties in the marine environment.

Salvage surveyor – Usually highly qualified Marine Engineers or other similar marine personnel engaged in providing impartial surveys to effect recovery of shipboard/cargo interests. Very often involved with heavy lift and project cargo transportation.

Sea Waybill – A document which is often shipped with the goods and has a similar function to that of a Bill of Lading. However, it is not a document of title and cannot be used to transfer goods to a third party. It is employed where there is a trust between the buyer and the seller of the goods and is often used where Bills of Lading tend to lag behind the cargo, especially in the short sea trades.

Shipper – A company or person who contracts with a carrier of goods by sea. Sometimes referred to as the merchant.

Shipper's Export Declaration – A customs form completed by the shippers of goods to a foreign country. It contains relevant data used in compiling trade statistics. May be referred to as a shipper's manifest.

Shipper's Guarantee – An indemnity given to the ship owner in exchange for clean Bills of Lading for goods to which an endorsed Mate's Receipt has been issued.

Shipping Permit – A written authorisation given by the owner of goods to the shipper, permitting delivery of the goods to the loading pier/dock. An alternative name is a 'Shipping Note'.

Ship's Tackle Delivery – An endorsement to a Bill of Lading or to the cargo manifest that the consignee of goods will take delivery of the goods from alongside the ship directly from the ship's derricks or cranes.

Short international voyage – An international voyage in the course of which a ship is not more than 200 miles from a port or place in which the passengers and crew could be placed in safety. Neither the distance between the last port of call in the country in which the voyage begins and the final port of destination nor the return voyage shall exceed 600 miles. The final port of destination is the last port of call in the scheduled voyage at which the ship commences its return voyage to the country in which the voyage began.

Special category space – Any enclosed space above or below the bulkhead deck intended for the carriage of motor vehicles with fuel in their tanks for their own propulsion, into and from which such vehicles can be driven and to which passengers have access.

Standard Shipping Note – A multi-purpose shipping document for non-hazardous goods, which accompanies goods to the freight forwarding agent's depot or to a container depot. It refers to the information about the shipment and designates the receiving authority.

Tally – A recorded count of the number of cargo parcels handled.

Waybill – A non-negotiable receipt which contains terms of contract.

Wharfage – A charge placed on a cargo vessel for use of the wharf, quayside or dock.

PIRACY AND ARMED ROBBERY

Modern day piracy continues to be on-going in specific regions of the world. It is the act of robbery from ships at sea with the intent to steal cargo or other valuable goods. The International Maritime Bureau has established a piracy reporting centre based in Kuala Lumpur, Malaysia. They offer a free service to ships masters and maritime organisations by way of a 24-hour hotline Tel. +60 3 2078 5763. Or e-mail the Piracy Reporting Centre and a helpline Tel, +60 3 2031 0014.

Mariners are currently advised (2022) to be cautious and take precautionary measures when in the following areas:

South East Asia and the Indian Sub-continent
West Africa and the Gulf of Guinea
South and Central America and the Caribbean waters
Eastern Europe/Ukraine

Reported incidents during 2022, have been received from the following countries:

India
Philippines
Indonesia
South Africa
Guyana
Peru
Angola
Haiti
Brazil
Somalia
Nigeria

Coastguards and military authorities continue to try to prevent acts of piracy as and when they occur with the consent of the governments of the countries affected.

Location and communications of incidents become essential in investigations in their attempts to counter acts of piracy.

CHAPTER 12

Passenger Vessels

INTRODUCTION

Passenger vessels are and always have been considered a prestigious sector of the Mercantile Marine. However, it does not sit well as being cargo-related within the binding of a cargo book such as this. The fact is that all passenger vessels have a tendency to take on considerable stores, using a selection of cargo handling methods that may require crews to have some knowledge and experience of lifting gear, mobile units and general stowage.

The early years of passenger ship development saw the cargo/passenger vessel carrying up to 12 passengers on small/medium sized freighters. The trade and the destinations were determined by wherever the commercial cargoes were assigned. The growing ferry markets encroached dramatically into the cargo/passenger routes with the subsequent reduction in designated cargo/passenger traffic, all be it in somewhat of a reduced form.

Cruise ships and liner trades, however, expanded considerably from the 1980s onwards and have developed so much so that excessive berths are now available in the Class 1 cruise trades. The ships themselves vary in range from the smaller 200 to 1,200 passenger range to the larger category carrying in excess of 6,000 passengers. The passenger capacity influences the crew complement considerably. Although the navigation and engineering personnel are generally of a fixed number, the catering and hotel staff is extensive to cater for large passenger numbers.

Theoretically you would not refer to passengers as cargo but in reality the commercial aspects are generated through passenger cruise fees just as freight is generated from cargo parcels. The passenger business is now well established around the world and offers a degree of luxury with a recognised high degree of safety within the maritime environment. The sector boasts high-profile ships operated with multinational crews to all the oceans and waterways of the world and by such reason is rightfully included within this text.

MANAGEMENT OF PASSENGERS

Fare-paying passengers enter the cruise industry through a variety of methods, where travel agencies tend to play a major role. Advertising through the media, press, television,

DOI: 10.4324/9781003407706-12

radio, etc. all attract the general public towards the holiday of a lifetime. Once inside the industry, large numbers of personnel of every age need to be catered for in a manner that would hopefully encourage them to cruise again at a future date.

FIGURE 12.1 The *Queen Mary 2* (IMO No. 9241061) seen moored port side to in Southampton Water.

Boarding Passengers

In this day and age the wellbeing and security of passengers aboard liners and cruise ships has become paramount. Identity and security checks are conducted by means of location address and individuals personal details inclusive of passport data.

These checks are reiterated at various stages of and during the intended voyage, to satisfy immigration, customs officials and local authorities.

It is normal practice for cruise ship passengers to be directed to a holding area close to the ships berth, where documentation, identification and registration take place. The allocation of cabin accommodation is made and boarding passes, usually with photographic ID are issued.

Baggage is cleared and delivered direct to allotted accommodation and individuals are then permitted to board the vessel via gangway security. This logging system provides the total number and identities of those embarking.

Once on board the vessel, a designated Passenger Reception, manned by ships personnel from the Purser's Department, will welcome individuals to the ship and initiate electronic cabin keys with directions to relevant deck and cabin locations. The personal baggage will usually be inside the passenger's accommodation.

With often many thousands of passengers boarding a large vessel it must be anticipated that this is an extremely busy time for the ship's catering department while being an anxious time for would-be passengers. To ease concerns most cruise ships hold a welcome introductory get together over drinks and hors d'oeuvres.

> Note: Attention of all persons involved on and in passenger vessels, that the **MSN 1794,** directs attention to the need to count and record the names, age and gender of all passengers prior to departure of the vessel.

Passenger Reception on Board

Passengers with boarding passes are monitored via gangway access to the on-board reception area. Many passenger companies now provide online boarding passes to would-be passengers The operational service desk on board the ship will then allocate accommodation, issue necessary cabin/suite keys. Most modern ships will usually use computerised electronic pass (key) cards. Location and movement towards passenger accommodations being found from the displayed ship plans found in the reception area and with guidance from ships staff directions.

The organisation and handling of hundreds of passengers during the boarding period is usually well regimented and people will usually find their pre-labelled luggage waiting for them inside the cabin or in the alleyway outside by the cabin. Crew members from the catering department and Pursers department are fully employed in getting passengers well settled in their accommodation.

It is normal practice that on the day of boarding several public deck/spaces or lounges conduct 'welcome aboard' events with light refreshments. Often these will also show video programmes of ongoing activities current to joining the vessel. A Public Address (P/A) system will keep passengers informed of meal times and restaurant allocations and similar notifications inclusive of any emergency announcements.

The reception desk will tend to deal with all the housekeeping needs of passengers and the on board ship interconnected telephone communication system. In most cases separate desk areas are designated to excursions ashore, foreign currency exchange, and general booking arrangements. Social activities aboard the vessel are usually conducted through the office of a separate cruise director.

Well-being of Passengers

Passenger vessels in the main provide not only transport but a level of enjoyment for travellers while on board. The entertainment includes live shows, numerous bars, restaurants, cinemas, casino, dance venues, educational lecturers, sports and swimming pool facilities, even an ice rink on some of the larger ships. The whole entertainment package is built into the cruising model, to ensure on-board passengers have a rewarding vacation or holiday period. Generally the larger the ship, the greater the number of entertainment and features are available.

A large number of ancillary workers (often contracted) such as performing artists, lecturers, hairdressers, etc. are engaged on board to ensure a wide range of facilities for the fare-paying customer. Some of these are regularly changed between company vessels to provide a varied schedule of entertainment across a company's fleet.

FIGURE 12.2 The *Star Princess* (IMO No. 9192363), seen in front of a glacier in the Northern waters of Canada and Alaska, involved in scenic cruising.

The ships usually have a large complement of catering personnel with numerous chefs for speciality foods. Bedroom stewards, bar staff, laundry personnel, security persons, as well as extensive administration personnel are encompassed within the housekeeping operations of the Purser's domain. Additional personnel for navigation, engineering, communications and medical staff are considered more as the ship's nautical, operational people.

It is worthy of note that the passenger complement when added to the crew numbers gives rise to a large number of personnel on a single floating platform. The larger ship could well be carrying in excess of 7,000 souls and it is this figure which dictates the need for adequate life-saving and fire-fighting resources to be placed on board and ready for immediate use in case of emergency.

PUBLIC ROOMS AND PASSENGER FACILITIES

Passenger accommodation (cabin blocks) tend to include numerous public spaces and compartments by way of lounges, bars, cafes, theatres usually spread over several decks.

The number, size and quality of public spaces will vary from ship to ship, but most carry a central passenger area spread over several decks. Substantial lounge areas, shops or similar other communal areas. Such larger capacity spaces like theatres and cinemas can be doubled as use for emergency muster stations where large numbers of persons can be brought together for organised emergency drills and control.

FIGURE 12.3 Upper deck swimming pool and jacuzzi situated at the aft end of the cruise ship *Explorer of the Seas* (IMO No. 9161728) operated by Royal Caribbean International.

THE LARGEST PASSENGER SHIP

The *Wonder of the Seas* is currently the largest passenger vessel afloat carrying 6,988 guests, with a crew of 2,300. It has 15 decks and a wide range of entertainments.

Facilities include 20 restaurants, a sports hall and ice rink, a central park area with living plants. It operates with numerous bars and cafes and has its own shopping mall, a solarium, spa and salon, swimming pools with water slides, climbing wall, nursery and child play areas. Three theatres, live show facilities as well as cinemas together with shops, bars and cafes.

The ship is operated by Royal Caribbean at a cruising speed of 22 knots. It is fitted with stabilising fins and all up to date manoeuvring aids. It has a crew of 2,300 persons in deck, engineering and catering departments.

The ship is 362 m long with a beam of 64 m fitted with luxurious balconied accommodation. It is equipped with sufficient survival craft and tenders to satisfy the Safe Manning Certificate issued by the Marine Authority,

Margins of Safety – Passenger Ships

The reputation of the cruise and passenger sector of the industry relies totally on its safety record, a record that cannot afford to be lost under any circumstances. Similar to the civil aviation industry which emphasises the safest way to travel, it is not absent from the occasional mishap. But operators the world over strive to ensure that the basic principle of the safety of life at sea is paramount and always held in the highest esteem.

Ships carrying passengers must comply annually with a seaworthiness certificate as stipulated by the regulations. Such standards encompass a second line of defence for every person on board with adequate allocation of survival craft places for every soul on board. This is supported extensively by additional life-saving appliances in the form of lifeboats, life rafts, lifejackets, lifebuoys, evacuation systems and comprehensive fire-fighting arrangements.

Research and development has moved alongside the twenty-first century in the form of enhanced construction methods for all passenger vessels. Satellite communications are now a standard feature appropriate for all sea areas, together with Life Saving Appliances (LSA) to meet the personnel numbers as stipulated by the Safe Manning Certificate. The ships must operate under a Safety Management Certificate with the Document of Compliance (copy) in order to meet the conditions and regulations of the International Safety Management (ISM) system, operational for all ships worldwide as stipulated by the International Maritime Organization (IMO).

Confidence is established throughout the industry showing experienced seafarers participating in emergency drills for Abandon Ship, Fire, and damage control limitations. Boat drills are held weekly within the passenger ship sector of the industry. Passengers go through a boat muster drill and don lifejackets prior to commencing a voyage.

FIGURE 12.4 The passenger vessel *Queen Mary 2* engages in a lifeboat launching drill while secured port side to, in Southampton. Partially enclosed boats are seen turned out and operating on the surface.

FIGURE 12.5 Emergency muster positions are usually identified specifically inside the rear of cabin doors with the emergency muster point localities. Signage in alleyways and around the vessel will highlight typical assembly stations.

Gangway Access

The ships are fully equipped with safe access facilities, usually from optional disembarkation deck positions. Many are fitted with upper and lower shell doors to accommodate alternative gangway arrangements. Shoreside gangway facilities are also an option at most of the main passenger ship destinations.

These large vessels tend not to be fitted with conventional 'accommodation ladders' but favour well-fenced portable companionways to ensure passenger safety.

FIGURE 12.6 The passenger gangway arrangement from the *Sea Princess* (IMO No. 9150913) rigged from a side shell door to the quayside. Both inboard and outboard ends of the gangway are attended by security watchmen. A swipe on-board, swipe off-board system is usually in place so that the ship knows total persons on board at all times.

FIGURE 12.7 Gangway arrangements vary from ship to ship and port to port, usually because of changing tidal levels. The gangway shown is for the Royal Caribbean ship *Explorer of the Seas* (IMO No. 9161728) deployed with a security/safety fenced barrier area on the quayside.

FIGURE 12.8 Twin short gangways are deployed from low shell doors set into the side of the Cruise vessel *Norwegian Epic* (IMO No. 9410569), rigged with safety gangway nets and quayside barriers with security watchman in place.

FIGURE 12.9 *Queen Victoria* (IMO No. 9320556) passenger liner seen on arrival entering Southampton Water. Small craft are in the habit of welcoming a Queen's arrival.

FIGURE 12.10 A passenger ship engages in close proximity to a seaward edge of a glacier while engaged in scenic cruising in the Northern waters of North America. Such activity in and around the ice regions not only requires a precision ship handling technique but also reliable machinery and manoeuvring equipment.

Operational Activities

FIGURE 12.11 The passenger vessel *Silver Wind* (IMO No. 8903935) seen port side to alongside the passenger terminal in Barcelona, Spain while taking bunkers from an oil barge on the starboard side.

FIGURE 12.12 The passenger ship *Norwegian Epic* (IMO No. 9410569) seen loading stores and gas bottles into the forward stores by means of an electro- hydraulic gantry hoist.

Ship Handling and Manoeuvring (Large vessels)

FIGURE 12.13 Large Class 1A passenger vessel *Liberty of the Seas* (IMO No. 9330030) operated by Royal Caribbean Shipping Company, seen berthing with tug assistance at fore and after ends.

FIGURE 12.14 The passenger vessel *Amsterdam* (IMO No. 9853890) seen in dry dock fitted with twin 'pod' propulsion units either side of a centre line skeg. A centre line stern anchor is also featured. Such ships are usually built with a high level of manoeuvring aids like stabilisers, bow thrust units and CPP propellers. The Master/pilot will tend to have 'joystick' fingertip bridge control of all ship movements, particularly useful to ensure ease of docking and berthing activities.

FIGURE 12.15 The passenger cruise vessel *Crystal Symphony* (IMO No. 9066667) seen under way and making way with escort tug, in the River Mersey, Liverpool.

FIGURE 12.16 *Carnaval Pride* (IMO No. 9223954) lying port side to, at the Admiralty Dockyard passenger berth in Bermuda.

HIGH SPEED CRAFT (HSC)

High-speed craft are now a well-established section of the industry and deeply embedded into the ferry and passenger markets. They operate under the High Speed Craft code and could have as easily been included in the ferry chapter of this text. The fact that they are extensively involved with the carriage of passengers on the high seas and waters connected thereto, justifies placing them in this section.

FIGURE 12.17 High-speed hydrofoil passenger-only *Flying Dolphin 17* (IMO No. 8331467) operational in the non-displacement mode in the Mediterranean Sea in the vicinity of the Greek Islands.

HSC CATEGORIES

The IMO, HSC code was introduced in 1994 and had mandatory implementation in 1996. Under the auspices of the code High Speed Craft were placed into one of three categories:

Category 'A' Craft

Defined as any high-speed passenger craft carrying not more than 450 passengers, operating on a route where it has been demonstrated to the satisfaction of the flag or port state that there is a high probability that in the event of an evacuation at any point of the route, all passengers and crew can be rescued safely with the least of:

1 The time to prevent persons in survival craft from exposure causing hypothermia in the worst intended conditions;

2 The time appropriate with respect to environmental conditions and geographical features of the route; or

3 Four hours.

Category 'B' Craft

Defined as any high-speed passenger craft other than a Category 'A' craft, with machinery and safety systems arranged such that in the event of damage, disabling any essential machinery and safety systems in one compartment, the craft retains the capability to navigate safely.

Cargo Craft Class

Defined as any high-speed craft other than a passenger craft and which is capable of maintaining the main functions and safety systems of unaffected spaces, after damage in any one compartment on board.

MAXIMUM SPEED FORMULA

Speed must be equal to or exceed 3.7 x the displacement corresponding to the design waterline in metres cubed (m³), raised to the power of 0.1667 (metres per sec).

Applicable to most types of craft, corresponds to a volumetric Froude number greater than 0.45.

CERTIFICATION AND SURVEY

Safety Certificate

All high-speed craft, following an initial survey to ensure that the vessel complies with the HSC code, will be issued with a Safety Certificate (period of validity for five years with an annual endorsement, c. 2000). This will specify the passenger capacity and the necessary radio communication and Life Saving Appliances that are applicable. It may also state operational limitations and it would also contain details of shore-based as well as onboard capabilities regarding maintenance and support facilities.

Permit to Operate High-speed Craft

A Permit to Operate will specify the name of the operator, the geographic area/routes and base port(s) that the vessel can work to. It will reiterate the passenger numbers and specify the manning scale and provide a maximum range from a place of refuge. Any other operational restrictions including the worst intended condition to allow continued operations would also be itemised. The permit has a period of validity similar to the Safety Certificate.

Note: Chapter 18 of the current HSC code details the many other conditions that must be met prior to the issue of the permit, inclusive of crew training and qualifications, shore based facilities, loading limitations, emergency procedures, navigation equipment and maintenance schedules, etc.

Certificate Issue

Both the Safety Certificate and the Permit to Operate are issued by the marine administration of that country where the vessel/craft is registered, following consultation with the port state where it is intended for operation.

ADDITIONAL DOCUMENTATION CARRIED BY HSC

All high-speed craft would expect to carry a craft operating manual and a specific route manual providing details of the vessel and route or routes that the vessel is engaged. Additionally, a service and maintenance manual together with a training manual would also be a requirement to satisfy the HSC code.

Operational documents would include statutory navigation publications including charts, as with any other type of vessel, and a stability information booklet. Any specialist equipment not incorporated into the craft operating manual would be accompanied by its own instructional operator's manual.

CONTENT OF THE CRAFT OPERATING MANUAL

The following is based on Chapter 18 of the IMO International Code of Safety for High Speed Craft.

The manual will contain at least the following information:

1 General particulars of the craft.
2 A general description of the craft and its equipment.
3 Procedures for checking the integrity of buoyancy compartments.
4 Details of the requirements of Chapter 2 – Buoyancy, Stability & Subdivision. Likely to be of a direct practical use to the crew in the event of emergency, i.e. stability and damage stability information.
5 Damage control procedures.
6 A description of and the operation of machinery systems.
7 A description of and the operation of auxiliary systems.
8 A description of and operational details of all remote control and warning systems.
9 A description and operational detail of electrical equipment.
10 Loading procedures together with limitations, inclusive of the maximum operational weight, centre of gravity and distribution of loads.
11 A description of the operation of fire detection and extinguishing equipment.
12 Diagram presentations indicating the structural fire protection arrangement.
13 A description and operation of radio and navigational equipment on board.

14 Information with regard to the handling and control and the performance of the craft, as determined in accordance with Chapter 17 of the HSC code.

15 The maximum permissible towing speeds and towing loads where applicable.

16 Dry-docking procedures or lifting procedures with limitations.

The manual should also provide information for emergency procedures including evacuation procedure in the worst anticipated conditions together with limiting machinery parameters required for safe operation.

Note: Machinery data should take into account Failure Mode & Effects Analysis (FMEA) as reflected by Annex 4 of the HSC code.

QUALIFICATION FOR A PERMIT TO OPERATE

(Ref. Chapter 18 of the HSC code)

The administration will issue a 'Permit to Operate' a high-speed craft when it is satisfied that the operator has made adequate provision for general safety throughout all aspects of the vessel's operation. It will revoke the permit if such provisions are not maintained to a satisfactory degree. It would be specifically concerned with the following:

1 The suitability of the craft for the intended service having due regard for the safety limitations and information contained in the Route Operational Manual.

2 The suitability of the operating conditions in the Route Operational Manual.

3 The arrangements to obtain weather information, on the basis of which the commencement of a voyage may be authorised.

4 Provisions in the area of operation of a base port fitted with facilities in accordance with Chapter 18.1.4. of the HSC code.

5 The designation of the person responsible for decisions to cancel or delay a particular voyage – for example, in the light of bad weather.

6 That a sufficient complement is in place to operate the craft, deploying and manning of survival craft, supervision of passengers, vehicles or cargo in both normal routine and emergency conditions as defined by the Permit to Operate. The crew complement should be that two officers are on duty in the operating compartment when the craft is underway.

7 That crew qualifications and training, inclusive of competence in relation to the particular type of craft and service intended, together with instructions, are in accord with safe operational procedures.

8 Restrictions with regard to working hours and crew rosters to ensure adequate rest periods to prevent fatigue.

9 That training of crew in craft operations and emergency procedures takes place.

10 That crew competence is maintained with regard to emergency procedures.

11 That adequate safety arrangements and safety compliance taking place at terminals is considered appropriate.

12 That traffic control arrangements are suitable and comply with any existing traffic control procedures.

13 Restrictions in use of position fixing to operations at night or in restricted visibility inclusive of radar and/or other electronic navigation aids.

14 Additional equipment which may be required due to specific characteristics of the intended service – e.g. night operations (night vision equipment).

15 Adequate communication systems between craft to base port, coast radio station, emergency services and other ships. To include radio frequencies to guard.

16 Suitable records to be kept to enable the administration to verify the following:

That the craft is operated within specified parameters.

That emergency and safety drills are observed.

The working hours of crew operating on board.

The numbers of passengers carried on board.

Compliance of any law to which the craft is subject.

Craft operations.

Maintenance of the craft and its equipment to approved schedules.

17 Arrangements to ensure that equipment is maintained in compliance with the Administration's requirements, and to ensure co-ordination of information as to the serviceability of the craft and its equipment – the arrangements to be conducted through the operator's organisation.

18 The existence and use of adequate instructions regarding:

   a) Loading the craft so that the weight and centre of gravity limitations are observable and that cargo is adequately secured when necessary.

   b) Adequate provisions for fuel reserve.

   c) Action in the event of reasonable foreseeable emergencies.

19 The provision of contingency plans by operators including all land-based activities for each foreseeable emergency scenario. The plans should provide operating crews with information regarding Search & Rescue (SAR) authorities and local Administrations and organisations which would complement tasks undertaken by crew.

20 The Administration should determine the maximum allowable distance from a base port or place of refuge after assessing the provision made under Chapter 18.1.3. of the HSC code.

MISCELLANEOUS FACTS ON HIGH-SPEED CRAFT

The High Speed Craft (HSC) code was derived from the previous Code of Safety for Dynamically Supported Craft adopted by the IMO in 1977. The original HSC code was

FIGURE 12.18 *Speed Runner 1* (IMO No. 9141871). High speed RoPax ferry seen alongside in the port of Piraeus, Greece.

adopted by IMO in 1994. A revised HSC code was submitted to the Maritime Safety Committee and has been adopted to apply from 2002 for vessels of such category built after such date.

The first HSC operation was established by Sea Containers on the Dover/Calais route, employing 'Seacats' in 1991.

The world's first commercial seagoing hydrofoil was the *Freccia de Sole* (1956). She was built by the Italian shipbuilder, Rodriquez Cantieri.

Hoverspeed, one of the main operators of hovercraft, ceased operations with hovercraft in the English Channel in September 2000. Competition from catamarans and the Channel Tunnel were stated in the press as being behind the decision to stop sailings. Hoverspeed recommenced operations in March 2001 with high-speed 'Seacats' operating the Channel Ports inclusive of the Dover/Calais route.

The world's longest commercial diesel-powered trimaran is the 127 m length *Benchijigua Express*. The vessel carries 1,350 passengers and 337 vehicles and is built with an aluminium construction by Austal Ships, Australia. She is currently operated by the Fred Olsen shipping company between Spain/Canary Islands. The vessel is of 8,973 tonnes and operates at up to 42 knots. Similar craft have since been built and operate with the US military.

Annual Summary Notices to Mariners – The attention of mariners is drawn to Notice No.23 of the annual summary which highlights the routes and types of high-speed craft operating around the shores of the United Kingdom.

The High Speed Craft Safety Certificate – Issued for a period specified by the Administration, which should not exceed five years.

'Dead ship' test – In order to establish the craft motions and the direction of lying to wind and waves for the purpose of determining the conditions of a craft evacuation, the craft should be stopped and all main machinery shut down for sufficient time that the craft's heading relative to wind and waves is stabilised. This test should be carried out at an opportune time to establish 'dead ship' behaviour, under a variety of wind and sea conditions.

WiG Craft – Wing in Ground (effect). These are a comparatively new concept which operate at very high speeds. In addition to the normal navigation lights they must also exhibit a high intensity all round flashing red light when landing and taking off.

HSC (lead authority in the UK) – The contact authority in the United Kingdom for related HSC matters is: Passenger Ship and High Speed Craft Branch, MSPPIA, The Maritime and Coastguard Agency, Spring Place, 105 Commercial Road, Southampton SO15 1EG.

AUTHOR'S COMMENT

Catamarans and trimarans are now commonplace in the high-speed sector of the industry and the US military have also acquired similar hulls to carry out amphibious landings. The design tends to accommodate a mix of vehicles and up to 500 passengers depending on size. Speed in operation varies but 40 knots+ would not be considered exceptional in good weather conditions.

CHAPTER 13

Offshore Trades

INTRODUCTION

The commercial aspects of the marine industry are easily recognised among the passenger ships, container vessels and tanker markets. What is not as easily seen is the numerous work boats and inshore traffic movements that integrate with shipping as a whole. From pilot boats and dredgers to the fuel-supplying 'bunker barges', the industry is saturated with numerous working craft that are essential to maintaining cargo movement.

The coastal waters of the maritime nations offer added complications to those deep-sea vessels on route towards the commercial ports. The inshore waterways and the offshore regions are becoming more and more threatening to the mariner, with development regions for offshore oil and gas and the designated wind farm areas posing greater navigational hazards.

Oil and gas construction sites and energy operational sites continue to survive with active support from stand-by and supply vessels, coastal traffic, shuttle tankers, barge activity, towing operations, pilot craft, customs/immigration launches and maintenance craft.

Specialist operations such as civil engineering projects, as well as the recovery of natural energy resources, are continuing through virtually all of the major oceans and seas of the world. They need every type of support vessel in order to sustain their positions in the offshore environment. Their activities deliver oil cargoes to the tankers while survey vessels are keeping the channels navigable. All are engaged in a role to ensure cargo supplies are not interrupted and an improving way of life is achieved for all.

> With the revealing aspects of global warming our way of life would seem to be under constant threat. Energy in all its forms is demanding a greater price in new technology, to endeavour to clean the planet not just from plastic and all other forms of pollution but from obsolete platforms and installations to return surfaces back to their original status.

DOI: 10.4324/9781003407706-13

OFFSHORE ACTIVITY

Maritime activity to recover the earth's resources of oil and gas have been around for at least 100 years, with developing nations keen to maximise energy sources within their own borders. The more recent advances with wind and tidal power being exploited has seen the expansion of wind farms populated by windmill turbines together with tidal turbine vessels being positioned off coastlines. Just as oil and gas installations encroached into the shipping lanes of the Arabian Gulf, Gulf of Mexico and the North Sea, so have wind farms expanded off our coastlines.

All these sites, whether for oil, gas, or wind-generated energy, all require trained personnel to go offshore, not only for development constructions, but also for recovery work and continued maintenance. They are continuously serviced and supplied by numerous smaller support craft and maintain offshore sites in operational conditions.

Helicopter pilots and mariners are the essential transport links which sustains this major industrial sector of the Mercantile Marine. At the same time, however, potential hazards are presented to the shipping fraternity as the offshore business directly affects safe navigation. Although such obstacles can be overcome by effective 'passage planning', it is in everyone's interest to ensure a correct lookout is maintained while vessels are on passage.

Energy is expensive and it is not going to get any cheaper. To obtain the level of energy we need in our society, we will be dependent on labour both on shore and offshore. Skilled manpower is a major element of any industry, but the skills offshore are somewhat more specialised. It is a fact of life that many shoreside characteristics do not readily lend to seagoing activity.

The control of supply vessels in often poor weather conditions is not the most comfortable of tasks; controlling the same deck in gale and storm conditions, can only be described as horrendous. The work can be restrictive, not only by the weather but also by geographic conditions which may prevent seamanlike practice. Restriction on anchor use because of sub-sea pipelines may limit use of equipment, while manpower is never in a fail-safe mode.

The industry is supplemented by many alternative craft besides deep-sea vessels and virtually all of them are engaged in coastal and offshore waters. It is a high-risk area because of geography, weather and high traffic density, but it is sustained by the experience and the seamanlike practice of the men within its perimeters.

OFFSHORE SUPPLY AND SUPPORT VESSELS

FIGURE 13.1 The new build *Pacific Retriever* (IMO No. 9236810, built 2002) anchor handling and offshore support vessel seen on trials testing and checking operational systems.

FIGURE 13.2 A 15 tonne 'Stevpris Anchor' being recovered to the aft open deck of an anchor handling vessel in the offshore regions of the North Sea.

OIL AND GAS – OFFSHORE ASSOCIATION

Legislation requires that offshore installations must have a second line of defence for those persons so engaged offshore, in the event of an emergency incident. Stand-by vessels are therefore a requirement as per international regulations. The operators need to keep the installations working and so they employ stand-by/supply vessels to fulfil the needs of the regulations and to sustain rig operations. The emergency role of the stand-by vessel is one of defence to provide a secondary safe platform for operational personnel. Where the installation is engaged in drilling operations, supplies of drilling pipelines becomes a regular cargo for offshore supply vessels, just as fresh water and stores become essential to sustain personnel in the field, so to speak. The offshore supply vessel caters to all the needs of the installation.

FIGURE 13.3 The ex-*Toisa Mariner* (IMO No. 7623916) renamed *Excellent Diver*, lies at anchor. Offshore support/supply vessel with heavy duty crane over the aft deck and prominent heli-pad. It is a multi-purpose offshore vessel capable of safety standby, offshore supply and diving support duties.

FIGURE 13.4 Ariel view of an offshore complex installation. Each platform is linked by covered companionways and has its own heli-pad landing facility. The drill tower is seen in the top right hand corner. All sections have local cranage to receive relevant stores from offshore supply craft.

Image reproduced by kind permission of Witherby Publishing Group.

DEVELOPMENT AND CONSTRUCTION OFFSHORE

Offshore activities like the building of installations and positioning of drilling rigs, has now come to include the de-commissioning of installations, all falling inside the perimeter of what we know as the offshore industry. Offshore supply vessels are regularly engaged in carrying not only day-to-day cargoes but frequently heavy lift parcels. Large project cargoes like submerged rig jackets and topside modules are continually active. Semi-submersible vessels are being regularly employed in the offshore construction aspects of the industry and can expect to be fully occupied during any de-commissioning activities.

The handling of such heavy lifts and project cargoes requires the seamanship skills of an experienced labour force, usually mariners with the experience gained from other cargo sectors of the shipping industry.

FIGURE 13.5 The *Nordica* offshore support vessel seen in operations in ice conditions of the Baltic Sea in 2003. Built as a multi-purpose ice-breaker and platform supply vessel. Also capable of being employed in offshore construction duties.

FLOATING SHEER LEGS IN OFFSHORE CONSTRUCTION

FIGURE 13.6 *Asian Hercules II* (IMO No. 8639297) a crane barge, floating sheerleg platform of the Asian Lift Pte Ltd, engages in a modular build up, to an offshore steel jacket of a developing offshore installation.

Most sheer leg platforms and floating cranes are motorised, with their own bunker space. The modern platforms may also have bow/stern thrust units to provide defined positioning. They usually operate with limited standard crew and engage a specific rigging gang for operational needs. Many sheer leg platforms have the facility to extend the outreach by fitting a jib extension. Extending the outreach is a useful option but can reduce maximum load capacity, when engaged. Such heavy lifting platforms having such flexibility are often engaged with project cargoes in and around offshore oil and gas fields in the early stages of construction and development of the energy fields.

OFFSHORE ZONES – NAVIGATIONAL DIFFICULTIES TO THE MARINER

FIGURE 13.7 A 'Jack-Up' drilling rig is seen in close position to the installation in the Norwegian 'Ecofisk' complex. Drilling Tower block, accommodation block and high flame gas burner are visible amongst extensive cranage of the installation.

It is highly likely that any vessel passing near or through an offshore region will encounter erroneous traffic pertinent to the operational sector. Fairways and safety exclusion zones around installations must be expected, as well as a high level of working boat activity.

Vessels engaged in towing operations, anchor handling or survey work would not be considered unusual sights. The fact that these vessels could all be operating in a restricted in ability to manoeuvre mode, would usually place the responsibility of having to give way on through traffic.

Also these areas are actively engaged with helicopter activity, both to rigs and to stand-by/support vessels, which virtually all have helicopter landing facilities. As such, any vessel engaged in launching or recovering aircraft would be expected to display restricted in ability to manoeuvre signals.

Such areas as the Gulf of Mexico, the North Sea, the Irish Sea and the Arabian Gulf all have extensive sub-sea pipeline operations ongoing, which could restrict anchor use by through shipping. It would be an expected issue to ensure that ships' Passage Plans contain relevant contingencies in the event of on-board technical problems.

The working ships of the area may well be engaged in loading/discharging from floating storage units, SBMs or other isolated units. Where these are encountered, safe passing distances at reduced speed may become a necessity and could still encroach on the movements of additional through traffic.

The energy cargoes are sought all over the world by the large oil/gas companies but their activities engage not just the tanker vessels and their subsidiaries, but directly affect all other sectors of the marine industry.

The hazards of the offshore industries are realised and it comes down to the watch-keeping abilities of the mariner to keep personnel inside and outside its perimeter safe within the environment.

CIVIL ENGINEERING INTO THE MARINE ENVIRONMENT

The use of heavy lift vessels and sheer leg platforms are extensively involved with project cargoes. These are frequently found to encroach into the more general areas of shipping. This is especially noticeable in port developments, bridge construction as well as establishing offshore sites for oil, gas and wind turbines.

Many operations dictate the need for tug and lifting services in order to complete complex engineering tasks. The transport of specialist equipment to out of the way sites, often in extreme weather conditions, underlines the nature of hazardous duties.

> Example: Ecofisk Oil & Gas production complex.
> This was Norway's first producing offshore oil field discovered in 1969 and
> started production in 1971. It has recently had its operational licence extended to 2048.
> It is found in a position 200 nm South West of Stavanger and is engaged in oil and gas production. Running an oil pipeline into Teesside, England, and a NORPIPE natural gas line to Emden, Germany.

OFFSHORE OPERATIONS

An offshore installation is defined as a man-made fixed or floating platform used for offshore activities, inclusive of drilling for subsea natural resources.

Various types of installation from production platforms, semi-submersibles, tension platforms, jack-up rigs and drilling ships make up the offshore sector. Associated operations will also use FPSO's Floating Production, Storage and Offloading units. The geographical position of these units is constantly changing by either being towed or using azipod power sources. Once established in a desired drilling position, they may be anchored within an anchor pattern of several anchors or held by Dynamic Positioning (DP).

By the very nature of their work servicing, supply and safety regulations are followed by the use of the Offshore Stand-By vessels (OSVs).

FIGURE 13.8 A semi-submersible moves her position with towed assistance from three stand-by/ offshore support vessels. Seen passing a fixed North Sea oil installation.

OFFSHORE ANCHOR WORK

Depending on the type of installation, it will depend on its positioning and holding system. Many rigs like jack-ups may be employed for test drills and have limited time in any one position. Others may be positioned with an anchor pattern of several anchors being deployed in diverse directions to resist prevailing weather and tidal directions.

Anchor handling vessels will carry out and lay anchors together with position indicating and recovery buoys. The task of laying an anchor pattern can be time consuming and involving several vessels at any one time. Other marine traffic are advised to give a wide berth if on route through, or near too, such an activity. The safety zones around an installation is set at 500 metres but at least 2 miles of sea room would normally be expected.

FIGURE 13.9 An AHV *Maersk Achiever* (IMO No. 9245902) in close proximity to the installation prior to using the chaser method to recover an anchor.

ANCHOR RECOVERY

Recovery of anchors can be achieved by the 'chaser' method or by recovering the 'Anchor Buoy' followed by an anchor pendant. The Anchor Handling Vessel (AHV), first landing the buoy, then heaving on the anchor pendant to recover the anchor to the stern roller position of the AHV. Both methods are popular and well used within the offshore environment.

FIGURE 13.10 Offshore anchor buoys employed to locate and recover anchors to the deck of anchor handling vessels. Offshore installations (non-permanent) are often positioned amidst a pattern of anchors. Depending on size and nature of work, 8, 10 or 12 anchors would not be considered abnormal. Each anchor would be identified and marked by a recovery buoy and suitable riser.

TUGS AND TOWING MOVEMENTS

FIGURE 13.11 The jack-up installation 'Magellan', seen under tow by rig support and anchor handling vessels.

FIGURE 13.12 Designated ice-breaker engaged in towing a coastal vessel in Baltic ice conditions. Most of the countries around the Baltic Sea offer ice-breaker services to keep water access free to shipping.

PORT AND HARBOUR TRAFFIC

Tug Operations

FIGURE 13.13 The Liverpool docking tug *Brocklbank* (IMO No. 642048) seen operating off Albert Dock Liverpool. The large cargo vessels, container, bulk carriers, tankers and passenger vessels generally all engage tugs for berthing.

FIGURE 13.14 The deck shows the additional option to the use of the towing hook by way of the towing winch and use of a towing spring hawser. The winch can also be engaged to control the gob rope when deployed with a towing hawser.

ADDITIONAL OFFSHORE AND COASTAL TRAFFIC

Pilot Boats

FIGURE 13.15 Pilot boats operate for all the major ports of the world, transporting the marine pilot to and from the navigation bridge of virtually all shipping to the area.

FIGURE 13.16 Small high-speed pilot boat, seen underway and making way operating at high speed.

AUTHOR'S COMMENT

The reader of this cargo book might very well question the input of this final chapter on offshore trades. As the author I questioned the inclusion myself. Given the fact that the off-shore sector has experienced massive growth around the globe, I felt it would be wrong to ignore the numerous cargoes shipped out to installations, the numerous project cargoes and heavy lift parcels which continue to be moved over and through our marine environment.

To remain a successful industry many additional craft play important roles in the shipping of cargoes, tugs, dredgers, security vessels and pilot boats to name a few. The tug might not carry a cargo as we recognise it in commercial shipping, but when that tug tows a jack-up rig into position, that rig is that tugs cargo. It might be towed instead of being carried, but the task has the same principle of being charged with a duty of care, under a charter party.

We are all inside the perimeter of the maritime environment, whether on a tanker or a passenger cruise vessel. Each aware of the hazards that are inherent with the bulk carrier and the tug boat alike. The weather is common to us all, just as the cargo container has become common to the Ro-Ro, box boats and to the offshore supply sector.

I feel all sectors have a vested interest in cargo movement and that this fact justifies my inclusion of this chapter within this text.

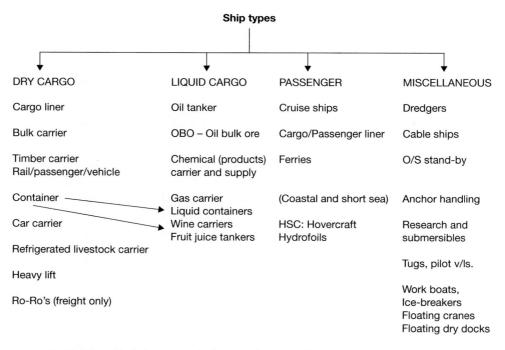

Ship types

DRY CARGO	LIQUID CARGO	PASSENGER	MISCELLANEOUS
Cargo liner	Oil tanker	Cruise ships	Dredgers
Bulk carrier	OBO – Oil bulk ore	Cargo/Passenger liner	Cable ships
Timber carrier Rail/passenger/vehicle	Chemical (products) carrier and supply	Ferries	O/S stand-by
Container	Gas carrier Liquid containers	(Coastal and short sea)	Anchor handling
Car carrier	Wine carriers Fruit juice tankers	HSC: Hovercraft Hydrofoils	Research and submersibles
Refrigerated livestock carrier			Tugs, pilot v/ls.
Heavy lift			
Ro-Ro's (freight only)			Work boats, Ice-breakers Floating cranes Floating dry docks

FIGURE 13.17 Specialist fishing vessels, factory ships, auxiliaries for warships, light vessels, sail trainers, harbour craft, tenders and bunker barges, oil recovery vessels, emergency support and salvage craft do not easily fall into the above ship type categories. However, they still operate on the peripheral of cargo movement.

Miscellaneous Cargo Information

INTRODUCTION

A text of this size cannot hope to cover every commodity or every situation that ships' officers encounter. Cargoes are varied and can be carried in many alternative forms. The following brief notes are meant to portray the fundamentals that go along with and support the various chapters and specifics expressed throughout this work.

FREEBOARD FORM (REVISED)

Following the completion of cargo operations in a port, the ship's draughts will expect to have changed. The draughts fore and aft should be read and the mean draught obtained. It is a legal requirement that before the ship sails outward, the 'Freeboard Certificate' is completed and displayed for the crew to see that the vessel is not overloaded and the minimum bow height is not exceeded.

FUMIGATION

The marine environment and several types of cargo often require pest control in one form or another. Fumigation of cargo spaces and sometimes cargoes themselves need treatment for reasons of health and safety.

Fumigation is defined as the act of introducing a toxic chemical into an enclosed space in such a manner that it disperses quickly and acts on the largest pest in its gaseous or vapour state.

There are several methods of introducing the toxic substance depending on the nature of the cargo. Short or long probes can be used, or a combination of both, or the chemical can be trenched in. In every case the compartment must be sealed prior to injection. Once injected into a hold, the space should remain sealed with hatch lids battened down.

Infestations of insects, rodents or similar species can be effectively controlled with these toxic substances. Examples include Hydrogen Phosphide, Methyl Isocyanine, or Methyl Bromide (restricted use).

Maritime & Coastguard Agency	DRAUGHT OF WATER AND FREEBOARD NOTICE	MSF 2004 / REV 0219 / VERS.1

Ship .... **PORT OF REGISTRY** ....

GROSS TONNAGE ....

(1) Summer freeboard .... millimetres corresponding to a mean draught of ....millimetres
(2) Winter freeboard .... millimetres corresponding to a mean draught of ....millimetres
(3) Tropical freeboard .... millimetres corresponding to a mean draught of ....millimetres
(4) Winter North Atlantic freeboard..................... millimetres corresponding to a mean draught of ....millimetres
(5) Allowance for fresh water for all freeboards other than Timber freeboards ....millimetres
(6) Timber Summer freeboard .... millimetres corresponding to a mean draught of ....millimetres
(7) Timber Winter freeboard ....millimetres corresponding to a mean draught of ....millimetres
(8) Timber Tropical freeboard .... millimetres corresponding to a mean draught of ....millimetres
(9) Timber Winter North Atlantic freeboard .... millimetres corresponding to a mean draught of ....millimetres
(10) Allowance for fresh water for Timber freeboards ....millimetres

NOTES:
1. The particulars to be given above of freeboards and allowances for fresh water to be taken from the load line certificate currently in force in respect of the ship.
2. All freeboards given on the load line certificate must be stated.
3. The mean draught to be given above is the mean of the draughts which would be shown on the scales of measurement on the stem and on the stern post of the ship if it were so loaded that the upper edge of the load line on each side of the ship appropriate to the particular freeboard were on the surface of the water.
4. Where the draught is shown on the scales of measurement on the stem and on the stern post of the ship in feet the mean draught must be given in millimetres.

		PARTICULARS OF LOADING						
1	2	3	4	5	6	7	8	9
Date	Place	ACTUAL DRAUGHT			MEAN FREEBOARD		SIGNATURE OF MASTER AND AN OFFICER	
		Forward	Aft	Mean	Actual (see notes 1 and 2 below)	Corrected (see note 3 below)	Master	An Officer

NOTES:
1. The actual mean freeboard (column 6) is the mean of the freeboards on each side of the ship when the ship is loaded and ready to leave.
2. If the actual mean freeboard is less than the appropriate minimum saltwater freeboard as shown on the load line certificate there must be entered in Column 7 the corrected freeboard arrived at after making any allowances for density of water, rubbish to be discharged overboard and fuel, water and stores to be consumed on any stretch of river or inland water, being allowances duly entered in the ship's official log book.
3. If the actual mean freeboard is greater than the appropriate salt water freeboard, Column 7 need not be filled in.

FORMERLY FRE 13 **AP.01**

The use of Hydrogen Phosphide may be used in ships in-transit during a voyage to reduce idle time. However, if carried out it must be issued with a Certificate of In-Transit Fumigation. It must be done at the discretion of the Master and any spaces treated must be inspected and checked for leaks. The vessel must also carry gas-detection equipment, medical treatment recommendations and at least four respiratory devices. The use of fumigation activity would normally be carried out in port or at anchor, after crew have been disembarked. The ship would then be ventilated prior to sailing.

Additional information on this subject can be obtained from the International Maritime Fumigation Organisation (IMFO), http://.imfo.com, IMFO Code of Practice.

OIL FILTERING AND MONITORING EQUIPMENT (OIL WATER SEPARATORS)

Every ship of 400 GT and above, but less than 10,000 GT, shall be fitted with oil filtering equipment which complies with an approved design in accordance with the specifications for such equipment as set out in the recommendations on International Performance and test specifications for oily water separating equipment and oil content meters.

Every ship 10,000 GT and above shall be provided with:

1 Oil filtering equipment (as stated).
2 Oil content measuring equipment fitted with a 15 ppm alarm device and with arrangements for automatically stopping any discharge of oily mixture when the oil content in the effluent exceeds 15 ppm.

OIL RECORD BOOKS

Current legislation requires oil tanker vessels to carry two (2) oil record books, one for cargo movement and one for fuel movements. Non-tanker type vessels are only required to carry one oil record book.

Entries into oil record books should cover any movement of oil in or out of the vessel, including internal transfers between tanks. Each entry in the book should be signed by the ship's Master and another officer.

VAPOUR RECOVERY SYSTEMS

More use is now being made of vapour recovery systems in order to reduce emissions of volatile organic compounds (VOCs) in ships and offshore operations. Where cargo tanks are loaded with crude, hydrocarbon gas emissions are vented to atmosphere with an estimated loss of between 100 to 300 tonnes from each loading. A recovery system in place is meant to be economical and the following three types are now widely used:

1   Absorption of non-methane volatile organic compounds (NMVOCs) into the crude being loaded.

2   Liquefaction and the storage of NMVOCs to be discharged at a later time, or used as on-board fuel.

3   Sequential transfers of tank atmospheres during the cargo operations of loading and discharging.

Liquefied VOC emissions can be used to fuel the VOC recovery plant as well as similar steam-driven onboard systems. It can also be used as a blanket gas in cargo tanks instead of inert combustion gases. Further use can be achieved by discharging it ashore for use as fuel or further refining.

NOTE OF PROTEST

Masters would 'Note Protest' for any of the following cargo reasons:

1   Whether damage has been caused or is suspected of being caused to cargo.

2   Whenever the ship has encountered heavy weather which may have resulted in damage to cargo.

3   Where cargo is known to have been damaged through a peril of the sea.

4   Where cargo is shipped in such a state that it is likely to deteriorate during the voyage. (Bills of Lading should be also endorsed in this case.)

5   Where a serious breach of the Charter Party has occurred, by the charterer or his agent – e.g. refusing to load cargo, delaying the loading, loading improper cargo.

6   In all cases of General Average being declared.

7   When consignees fail to take delivery of cargo and pay due freight in accord with the terms of the Charter Party or Bills of Lading.

Note of Protest is a declaration by the ship's Master made under oath to a notary public or other officer, such as a British Consul, having the authority to administer oaths regarding loss, damage or delay to cargo, due to circumstances beyond his control. Note of Protest should be made when necessary as soon after arrival in port and always within 24 hours after arrival before 'breaking bulk'. The ship must also be an arrived ship.

The right to extend protest can also be made when the full extent of any damage is not fully known. A Master could be barred from claiming the cargo's contribution in a General Average action if he failed to note protest with 24 hours of arrival. He is also obliged to inform the consignee that he has noted protest.

In the UK, Note of Protest cannot be accepted as evidence but in other countries it can be admitted into a legal tribunal. The main reason for making protest is usually to support a cargo owner's claim against the underwriters.

Protest is made through a 'Proper Officer' i.e. British Consul or Magistrate, on arrival at the next port. In the case of cargo protests, before breaking bulk.

The Master should reserve the right to extend the declared Protest.

Example: MARINE NOTE OF PROTEST

On this ....day of ....20**...., personally appeared and presented himself before me................... (insert name of master), Master of the......... (insert name of vessel) of.............(insert name of Port of Registry, official number.......... and...... Net Tons, which sailed from.............. on or about the .... (insert date) in ballast/ with cargo of .... (Insert brief description of cargo if any) bound for................ (insert destination port), and arrived at port of ....

...............on (insert port name & date) and declare that during the voyage having experienced a collision with the......................... (insert name of colliding vessel) at about..........hrs on... ... (insert date of collision) in .... (insert general location of collision e.g. "in Black Sea") and fearing damage to the vessel's hull, rigging, machinery and/or cargo, he hereby Notes his Protest against all claim, losses, damage etc., reserving the right to extend the same at time and place convenient.

Sworn and sign before me........................ Captain (insert name of Master)

At....................... Notary Public.
Notes _____

Similar Note of Protest would be applicable for other situations like 'Stranding' where the word collision would be changed for stranding or fire relevant to the circumstances.

Self Examiner – Questions and Recommended Answers

CARGO RELATED – EXAMPLE CALCULATIONS AND QUESTIONS

Abbreviations used in the following example calculations:

A	represents Aft
AP	Aft Perpendicular
cm	centimetres
CoT	Change of Trim
d	distance
DW	Dock Water
DWA	Dock Water Allowance
F	Forward
FP	Forward Perpendicular
FWA	Fresh Water Allowance
G	Position of the ship's Centre of Gravity
GG_1	Movement distance of the ship's Centre of Gravity
GM	metacentric height
KG	measured distance between the Keel and the C of G of the vessel
KM	measured distance between the Keel and the Metacentre
L	Length of ship
l	A proportionate length of the ship's length
M	Metacentre
m	metres
MCTC	Moment to Change Trim 1 cm.
mm	millimetres
RD	Relative Density
Stbd	Starboard
SW	Salt Water
Tan	Tangent
TPC	Tonnes Per Centimetre
W	displacement of vessel
w	added or discharged weight

CARGO WORK – STABILITY EXAMPLES

Example 1

A vessel of 5,870 tonnes displacement has a load draught of 5.4 metres with a TPC = 11. Calculate the load draught of the vessel if she is working cargo in fresh water.

$$FWA = \frac{W}{4 \times TPC} = \frac{5870}{4 \times 11} = 133.4 \text{ mm}$$
$$= 0.133 \text{ m}$$

Load Draught in Fresh Water = 5.40 + 0.133 = **5.533 m**

Example 2

A vessel has a load draught in salt water of 6.4 m. Calculate the maximum load draught in dock water of Relative Density 1.010. The ship's FWA is 75 mm.

$$DWA = FWA = \frac{1025 - \text{Density of Water}}{25}$$
$$= 75 \times \frac{1025 - 1010}{25}$$
$$= 75 \times \frac{15}{25}$$
$$= 45 \text{ mm} \quad = 0.045 \text{ m}$$

Maximum Draught $= 6.40 + 0.045$
$= $ **6.445 metres**.

> Note: Dock Water Allowance (DWA) is the amount the vessel may legally submerge her disc (Plimsoll Line) when loading in a dock water of less density than that of sea water.

Example 3

A vessel of 10,000 tonnes displacement with a KG of 7.0 m loads 100 tonnes of cargo goods at KG 12 m. Calculate the new KG of the vessel by taking moments about the keel.

Weight	KG	Moment
10,000	7.0	70,000
+ 100	12.0	+ 1,200
10,100		71,200

$$\text{Final KG} = \frac{Total\ Moments}{Total\ Weights} = \frac{71,200}{10,100}$$

$$= 7.0495\text{m}$$

Example 4

A vessel of 12,000 tonnes displacement has a KG of 7.8 m and a KM of 8.6 m. She then loads the following parcels of cargo:

250 tonnes at KG of 11.0 m
100 tonnes at KG of 7.0 m
50 tonnes at KG of 3.0 m

Calculate the vessel's final GM after completion of loading.

Weight	KG	Moment
12,000	7.8	93,600
250	11.0	+ 2,750
100	7.0	+ 700
50	3.0	+ 150
12,400		97,200

$$\text{Final KG} = \frac{Total\ Moments}{Total\ Weights} = \frac{97,200}{12,400} = 7.839\ \text{m}$$

$$\text{GM} = \text{KM} - \text{KG} = 8.6 - 7.83 = \textbf{0.761 metres}$$

Example 5

A vessel of 7,500 tonnes displacement with KG of 6.0 m and KM of 6.8 m is expected to load timber on deck in a position of KG 11.0 m.

Calculate the maximum weight of timber that can be loaded in order to arrive at the destination with a GM of 0.2 m if an allowance of 15% increase in weight is anticipated with water absorption by the deck cargo.

Let the weight of cargo to be loaded = w

Weight	KG	Moment
7,500	6.0	45,000
1.15w	11.0	12.65w

7,500 + 1.15w   45,000 + 12.65w

Final KG = 6.8 – 0.2 = 6.6

But

$$\text{Final KG} = \frac{\text{Total Moments}}{\text{Total Weights}} = \frac{45,000 + 12.65w}{7,500 + 1.15w}$$

$$\therefore \ 6.6 = \frac{45,000 + 12.65w}{7,500 + 1.15w}$$

$$\therefore \ 49,500 + 7.59w = 45,000 + 12.65$$

$$4,500 = 5.06w$$

$$w = \frac{4500}{5.06}$$

$$w = \textbf{889 tonnes of timber to load}$$

Example 6

A ship of 10,000 tonnes displacement is to load a heavy lift of 100 tonnes with a KG of 3.0 m by means of the ship's heavy derrick. The head of the derrick is 24 m above the keel. The ship's KM was 7.0 m, with a KG of 6.2 m before loading. The load is to be stowed on the ship at a KG of 6.0 m.

Calculate (i)   the minimum GM experienced
          (ii) the final GM.
When

(i) $GG_1 = \dfrac{w \times d}{W}$

$$= \frac{100 \times (24 - 3)}{10,000}$$

$$= 0.21 \text{ m}$$

$\therefore$ KG = 6.2 + 0.21 = 6.41
       GM = 7.0 – 6.41 = **0.59**

(ii) $GG_1 = \dfrac{100 \times (6-3)}{10{,}000}$

$= 0.03$ m

$\therefore KG = 6.2 + 0.03 = 6.23$ m

$\therefore GM = 7.0 - 6.23 = 0.77$ m

Example 7

Note: When the centre of gravity of a vessel moves off centre, an upsetting lever is produced which causes the vessel to list until G and M are in the same vertical line. The angle of heel due to G being off centre is found by the formula:

$$\text{Tan } \varnothing = \frac{\text{Transverse GG1}}{\text{GM}} \text{ but } \text{GG1} = \frac{w \times d}{W}$$

$$\therefore \text{Tan } \varnothing = \frac{w \times d}{W \times GM} = \frac{\text{Listing Moment}}{W \times GM}$$

A vessel with 4,000 tonnes displacement which is initially upright moves a 12 tonne weight 7 metres transversely across the deck. The ship's GM with the weight on board is 0.3 m. Calculate the resulting list.

$$\text{Tan } \varnothing = \frac{w \times d}{W \times GM} = \frac{12 \times 7}{4{,}000 \times 0.3}$$

$\varnothing = 4°.0$

Example 8

A vessel of 11,000 tonnes initial displacement loads a 50 tonne weight by a floating crane in a position 12 m to port off the ship's centre line. Assume that the KG and KM remain constant and that the vessel is upright prior to loading. Calculate the angle of list if the ship's GM is currently 0.25 m.

$$\text{Tan } \varnothing = \frac{w \times d}{W \times GM} = \frac{50 \times 12}{11{,}050 \times 0.25}$$

$\therefore \varnothing = 12°.25$ to port

Example 9

A vessel of 10,000 tonnes displacement with an initial GM of 0.2 m conducts the following cargo operations:

Loads 50 tonnes 4 m to starboard of the centre line.
Loads 70 tonnes 5 m to port of the centre line.

Discharges 90 tonnes 3 m to starboard of the centre line.
Shifts 40 tonnes 6 m to starboard.

Assuming KG and KM remain constant, determine the final list.

Weight	Distance off Centre	List Moment	
		Port	Stbd
10,000	–	–	–
+50	4 m Starboard		200
+70	5 m Port	350	
–90 (discharge)	3 m Starboard	270	
+40 (transferred)	6 m Starboard		240
10,030		620	440

$$\frac{440}{180}(\text{port})$$

$$\therefore \text{Tan} \, \emptyset = \frac{180}{10,030 \times 0.2}$$

$$\emptyset = 5°.13 \text{ to port}$$

Example 10

A vessel of 160 m length loads 40 tonnes in a position 60 m from the AP. Calculate the final draughts if:

The initial draughts are Fwd. 5.0 m Aft 6.0 m.
The Longitude Centre of Flotation is 70 m forward of the AP.
The ship's TPC is 20, and MCTC is 100.

$$\text{Sinkage due to load} = \frac{\text{Weight loaded}}{\text{TPC}} = \frac{40}{20} = 2 \text{ cms}$$

$$\text{CoT} = \frac{w \times d}{\text{MCTC}} = \frac{40 \times 10}{100} = 4 \text{ cms (by stern)}$$

$$\text{CoT Aft due to CoT} = \frac{1}{L} \times \text{CoT} = \frac{70}{160} \times 4 = 1.75 \text{ cms}$$

$$\text{CoT Forward} = 4 - 1.75 = 2.25 \text{ cms}$$

	Forward	Aft
Initial Draughts	5.0	6.0
Sinkage	0.02	0.02
	5.02	6.02
CoT	−0.0225	+0.0175
Final draughts	4.9975 F	6.0375 A

GENERAL CARGO QUESTIONS AND ANSWERS

Example 1

Calculate by how many millimetres a ship may submerge her loadline when she is currently loading in dock water of relative density 1.013, if the vessel has a FWA of 190 mm.

$$\text{Dock Water Allowance} = \text{FWA} \times \left(\frac{1.025 - \text{water density number}}{1.025 - 1.000}\right)$$

$$= 190 \times \left(\frac{1.025 - 1.013}{25}\right)$$

$$= 190 \times \left(\frac{0.012}{0.025}\right)$$

$$= 91 \text{ mm}$$

Summer loadline may be submerged by 91 mm

Example 2

A rectangular tank of 9 metres length and 6 metres breadth has a depth from the ullage plug of 11 metres. How many tonnes of oil of a relative density of 0.83 does the tank contain when the ullage is 350 mm?

Depth of tank	= 11.0 m
Ullage	= 0.35 m
Depth of Oil	= 10.65 m
Therefore the volume of oil	= 9 × 6 × 10.65 m³
	= 575.1 m³
Weight of Oil	= Volume × Density
	= 575.1 × 0.83
	= 477.33 tonnes

Example 3

What is the smallest purchase that could be used to lift a 5 tonne weight with flexible steel wire rope having a SWL of the wire equal to 3.125 tonnes?

$$S \times P = W + \left(\frac{nW}{10}\right)$$

Assuming the purchase is to be used to dis-advantage P = n.

$$\text{Then } 3.125 \times P = 5 + \left(\frac{5 \times P}{10}\right)$$

$$2.625P = 5$$

$$P = 2$$

Therefore a Gun Tackle is the minimum purchase to use for this lift.

Example 4

Calculate the maximum number of tonnes which can still be loaded into a vessel whose TPC = 19, and FWA = 190 mm. Her loaded salt water freeboard is 2,310 mm and her present freeboards are 2,420 (starboard) and 2,404 mm (port), in water of Relative Density 1.009.

Present freeboards 2,420 mm Stbd

2,404 mm Port

True mean freeboard = 2,412mm

$$DWA = 190 \times \frac{(1.025 - 1.009)}{1.025 - 1.000}$$

$$= 121.6 \text{ mm}$$

Corresponding salt freeboard

2,412 + 121.6	= 2,533.6 mm
Permitted freeboard	= 2,310 mm
Sinkage allowed	= 223.6 mm
[TPC = 19, TPmm	= 1.9] × 1.9
Cargo to Load	**= 424.84 tonnes**

Example 5

A cargo tank with an area of 75 m² is being filled from a pipe of 200 mm in diameter. The ullage is now 1.6 m. Calculate how much longer the filling valve must be left open to obtain an ullage of 800 mm, if the average rate of flow through the pipe is 1.75 m/s.

Difference in ullages	= 1,600 mm – 800 mm	= 800 mm (0.8 m)
Volume of liquid to load	= 75 m² × 0.8 m	= 60 m³
Area of pipe	= π r²	= 3.1416 × 100 × 100
	= 31,416 mm²	
	$= \dfrac{31,416}{1,000 \times 1,000 \, \text{m}^2}$	

Volume of liquid loaded per second

$$= \frac{31,416}{1,000 \times 1,000} \times 1.75 \text{ m}^3$$

Time to load 60 m³
$$= \frac{60 \times 1,000 \times 1,000}{31,416 \times 1.75} \text{ seconds}$$

$$= 1,091 \text{ seconds or } 18.18 \text{ minutes}$$

The valve should be left open for a further 18 minutes.

Example 6

A ship is 140 metres long and displaces 10,000 tonnes is floating at draughts 6.5 m Forward and 7.7 m Aft. The vessel is scheduled to enter a canal where the maximum draught allowed is 7.2 m. Calculate the minimum amount of cargo to discharge from a compartment which is 30 metres forward of the Aft Perpendicular. The ship's TPC = 16, MCTC = 180, and the Centre of Flotation is amidships.

Let 'w' represent the cargo to discharge.

Change in draught aft = 7.7 − 7.2 = 0.5m (50 cm)
Change of draught aft = Rise + Change due to Change of Trim

Therefore 50

$$= \frac{w}{TCP} + \frac{1}{2} \times \frac{(w \times d)}{MCTC}$$

$$= \frac{w}{16} + \frac{1}{2} \times \frac{w \times (70-30)}{180}$$

$$= \frac{360w + (16 \times 40)}{16 \times 360}$$

50 × 16 × 360   = 369w + (16 × 40w)

288,000   = 360w + 640w

w $= \frac{288,000}{1,000}$

w   = 288 tonnes represents the minimum cargo to discharge

Example 7

Q: What do you understand by the term 'Loadicator' and what information would you obtain from it? (*Most ships now use customised software with the shipboard computer*)

A: A Loadicator is the term given to a cargo loading computer, which is configured to suit the ship's loading programmes. The loadicator will provide the Cargo Officer with the following information once the weight distribution is entered into the programme:
   Distribution of weights or cargo units in the ship's compartments.
   The status of relevant tank weights and commodities.
   The seagoing shear force and bending moment conditions.
   The stability aspect with values for KG and GM.
   Ballast distribution and quantity would also be available.
   The loadicator is often linked to a shoreside monitor to allow data transmission on unit weights for cargo distribution and special stowage requirements. Particularly relevant to Roll on, Roll off vessels engaged on fast turn round, short voyage trades.

Example 8

Q: How would you load a bulk carrier with iron ore?

A: Ensure that the hold is clean and that bilge suctions are tested to satisfaction prior to commencing loading. Draw up a pre-load plan and a ballast/de-ballast plan calculating the stress factors affecting the ship throughout the proposed loading programme. Assess final bending moment and shear forces over the vessels length.

The maximum angle of heel would also be calculated for a potential shift in the cargo volume, bearing in mind that a moisture content is present in the cargo.

The loading rates for the cargo would commence slowly and gradually increase. Fast rates of loading can cause serious damage by generating rapid stress values throughout the ship's length.

The important aspect is that iron ore is a dense cargo and heavy. The cargo compartments would only be about ¼ full. The Chief Officer would calculate the stability based on the load draughts. Condition formats for the bending moment and shear force affecting the loaded condition would be drawn up. (Stowage factor Iron Ore 0.34/0.50.)

Example 9

Q: What are the concerns for the Master of a container vessel carrying containers stacked on deck, engaged on the North Atlantic trade in winter?

A: The Master would be concerned with his Chief Officer about the positive stability of the vessel, bearing in mind that the possibility of encountering sub-freezing air temperatures on this trade route at this time is likely. Such conditions could lead to ice accretion and added weight from icing of the container stack could detrimentally affect the stability of the vessel.

Masters would monitor all weather reports and consider re-routing farther south to warmer latitudes if practical. A reduction in speed could also effectively reduce the rate of ice accretion occurring on the vessel. Where possible the crew should be ordered to make their best endeavours to remove ice formations if safe to do so.

Additionally the Master would maintain a course heading that would reduce the roll motion of the vessel in a seaway, in order to prevent development of parametric rolling.

Example 10

Q: When working as a Cargo Officer aboard an oil tanker, how would you keep the tanks outside the flammable limit?

A: The introduction of inert gas into any tank containing hydrocarbon gas/air mixture will decrease the flammable range until a point is reached where the LFL and the UFL coincide. This point corresponds to an oxygen content approximately 11% at which no hydrocarbon gas/air mixture can burn.

Note: Additional reference should be made to the Flammability Composition diagram found in ISGOTT.

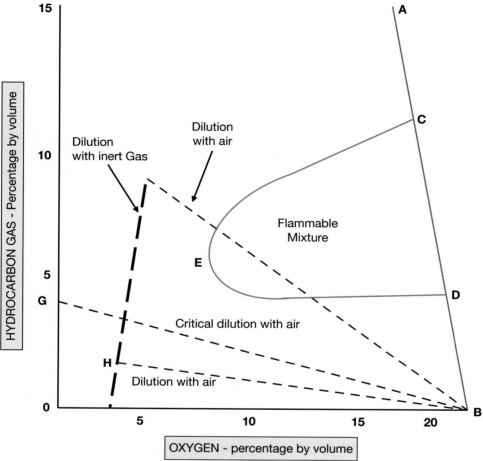

Key:
C represents the UFL,
D represents the LFL Area,
CED is the flammable mixture area.

Example 11

Q: What and when is 'lateral drag' evident and what can the cargo officer do to reduce the effects?

A: Lateral drag is associated with heavy lifts causing the vessel to heel over as the weight is taken up by the ship's derrick/crane. It can occur during loading or discharging of the load and is effectively a sideways movement of the load as the vessel returns to the upright. If unprepared for, the lateral movement of the load can be violent as the ship rolls against the angle of list.

The effects of lateral drag can be reduced by retaining the line of plumb of the derrick head above the point of landing. This can be achieved by 'coming back' on the topping lift and cargo hoist runner, quickly. This action tends to reduce movement of the load when discharging. If loading the weight, a steady slow lifting operation should be carried out.

Example 12

Q: When about to make a heavy lift by means of the ship's heavy derrick, how can the vessel's stability condition be improved so that positive stability is retained throughout the loading period?

A: The concern with loading a heavy weight is that the C of G of the weight effectively acts from the head of the derrick. The GM of the ship should be increased by filling the double bottom tanks before the lift is made. This will increase the GM value.

Additionally, eliminate any free surface moments in tanks, as this will also act to reduce the GM value.

Example 13

Q: How can the risk of a grain cargo shifting be reduced?

A: Grain should be loaded in accord with the Grain Regulations and the risk of shifting of the cargo can be reduced by:

a   Fitting of temporary longitudinal subdivisions (shifting boards).
b   Use of bagged cargo in a saucer formation on top of the bulk stow.
c   Bundling in bulk.

Example 14

Q: How would you describe the 'stowage factor' of a commodity?

A: The stowage factor can be defined as that volume that is occupied by a unit weight of cargo and is usually expressed in cubic metres per tonne (m^3/tonne).

$$\text{By example: Stowage Factor} = \frac{\text{Volume of Space}}{\text{Tonnage}}$$

e.g. How much cotton at a S/F of 2.0 m^3/tonnes could be loaded into a tween deck space of 200 m^3

$$\text{Tonnage} = \text{Volume of Space} = \frac{\text{Volume of Space}}{\text{S}/\text{F}} = \frac{200}{2}$$

$$= 100 \text{ tonnes cotton.}$$

Example 15

Q: When loading drop trailers and mobile units aboard a Roll on, Roll off ferry, why is it essential that the vessel is kept in the upright position?

A: Roll on, Roll off ferries load their mobile units via vehicle ramps either at the bow or more often through the stern door. These ramps are lowered onto link spans which provide the landing connection between ship and shore. If the vessel develops a list, the ramps become angled to the flat shore connection and prevent the movement of vehicles to and from the ship's garage spaces.

Most modern ferries will have automatic stabilising tank systems to counter any overload to port or starboard, so keeping the vessel always in the upright position and vehicle ramps flush on the shore or the link span.

> Note: Over-reliability on tank stabilisers should be avoided and safe practice is always to load and discharge in an even manner to avoid any one side ever becoming adversely affected by localised tonnage.

Example 16

Q: When would it be considered appropriate to carry out a draught survey?

A: The purpose of a draught survey being conducted would usually be:

i.   To ascertain any bending along the length of the vessel, usually after the loading of a bulk cargo.

ii. To determine the exact displacement in order to calculate the total weight of cargo loaded.

Example 17

Q: A vessel is scheduled to load sacks of mail. How would these be loaded on a general cargo vessel if they are loose and not in a container and what precautions would a prudent Chief Officer take?

A: Mails are classed as a specialised cargo and as such would be given lock-up stow. The bags would be tallied in and tallied out at the ports of loading and discharge, respectively. Watchmen or responsible ship's officers would monitor the movement of the mails probably being loaded by means of cargo nets or cargo boxes.

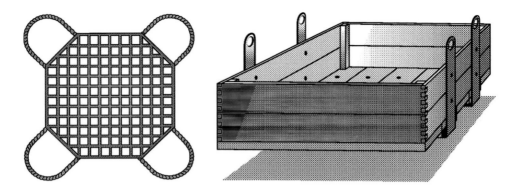

Example 18

Q: What cargo information would the Master of a bulk carrier pass to the loading terminal when expecting to berth, to take a full cargo of coal?

A: In addition to passing the ship's particulars, a pre-loading plan of cargo stowage by hatch together with the hatch loading order and respective quantities on each pour, assuming that the vessel has sufficient information to prepare such a plan. Confirmation that holds are in a state of readiness to load would be given.

The provisional arrival and departure draughts together with details of the ship's own cargo handling gear and respective capacities of same, if appropriate. Details of the ballast capacity and the time required to de-ballast.

Additional ship-keeping details reflecting the gangway position, the number of moorings, etc. would also be included as standard information.

Example 19

Q: For what purpose would a Cargo Officer use the Load Density Plan?

A: The ship's Chief Officer would use the Load Density Plan to check the capacity of cargo compartments to ascertain the volume of the space and consider the point loading factor to ensure that the deck strength is adequate to accommodate the intended cargo tonnage to be stowed in the space. Particularly useful with heavy lifts, where a concentrated weight over a small area may exceed the tonnage per square metre.

Example 20

Q: What is contained in the Register of Lifting Appliances and Cargo Handling Gear?

A: The register is kept up to date by the ship's Chief Officer and contains all the certificates for the lifting appliances, wires, shackles, hooks, chains, etc. used aboard the vessel for cargo operations.

Codes and Conventions Affecting Cargo Work Operations (under Merchant Shipping Act 1995)

Code of Safe Working Practice for Merchant Seamen

IMO International Code for the Construction and Equipment of Ships Carrying Dangerous Chemicals in Bulk (IBC Code)
IMO International Maritime Solid Bulk Cargoes Code (IMSBC code)
IMO Code of Safe Working Practice for Cargo Stowage
IMO Grain Regulations (International Code for the Safe Carriage of Grain)
IMO ISPS Code
IMDG Code (Hazardous Cargoes)
Inert Gas Code
MARPOL Convention and subsequent amendments
Merchant Shipping Regulations for Control of Noxious Liquid Substances in Bulk
Merchant Shipping (Load Lines) (Deck Cargo) Regulations
SOLAS'74 Convention (and subsequent amendments)
ICS Tanker Safety Guide (Chemicals)
International Safety Management Code

ADDITIONAL REFERENCES – STATUTORY INSTRUMENTS

SI 1509: 1997 The Merchant Shipping (Cargo Ship Construction) Regulations 1997
SI 1644: 1999 The Merchant Shipping (Additional Safety Measures for Bulk Carriers)
SI 336: 1999 The Merchant Shipping – Safety. MS (Carriage of Cargoes) Regulations
SI 929: 2004 Gas Carrier Amendment Regulations
SI 930: 2004 Dangerous or Noxious Liquid Substances in Bulk: Amendment regulations
SI 2183: 2006 PUWER Provision and Use of Work Equipment Regulations

CURRENT 'M' NOTICES (MGNS AND MSNS)

Relevant MGNs

36 Documentation of Compliance for ships carrying dangerous goods in packaged or dry bulk form

37 The M.S. Regulations 1977 (dangerous goods and marine pollutants)

60 (M) CSWP for Solid Bulk Cargoes (BC Code) 1996, Amendment to the Carriage of Coal Cargoes

107 (M) Loading Manual for bulk carriers

108 (M) Hull stress monitoring systems with 'Bulk Cargoes'

144 (M) The M.S. Regulations (Additional Safety Measures for Bulk Carriers)

146 The Carriage of Packaged Cargo and Cargo Units (Cargo securing manual)

198 (M) Bulk cargo handling terminals

223 (M) Carriage of bulk products in 'Deep Tanks'

282 (M) Carriage of dangerous goods on offshore supply vessels

283 (M) Back loading of contaminated bulk liquids

284 (M & F) Ships carrying fumigated bulk cargoes

332 (M & F) Lifting operations and lifting equipment

340 IMDG code. Cargoes carried in Cargo Transport Units

341 Roll on Roll off ships vehicle decks: Accidents and access

418 Roll on Roll off ships. Stowage and securing of vehicles

440 Measures to counter Piracy, armed Robber

511 (M) Solid Bulk Amend 1. 02–13 to IMSBC

512 (M) Solid bulk cargo: Cargoes not listed

513 (M) Solid bulk cargo: controlling moisture content

514 (M) Solid bulk cargo: iron ore fines

516 (M) IMO codes and guidelines for offshore vessels

531 (M) Cargo stowage and securing

534 (M & F) Cargo Safety: verification of gross mass of containers

545 Transportation of Dangerous Goods

552 Cargo safe stowage of specialised vehicles

576 Guidance for undertaking fumigation operations alongside

579 Load Line convention regulations 2018

603 Oil mixtures for Offshore Installations

619 LOLER and PUWER regulations 2006

621 Ro/Ro Guidance for stowing and securing vehicle

659 Enclosed space entry

Relevant MSNs

1196 Marine Pollution Manual on oil pollution

1231 Cargo handling operations – Offshore Supply vessels

1458 Offshore support vessels – guidance

1589 Merchant Shipping Regulations: Noxious Liquid Substances

1643 Prevention of oil pollution
1669 Ships carrying Dangerous Goods – Fire safety measures
1671 Cargo Ship Construction schedule
1671 Amendment 1, to Construction Regulations
1671 Amendment 2, merchant ship regulations 1999
1704 Carriage of firearms (use for animals)
1706 The Carriage military and commercial explosives
1706 Amendment 1, Military and Commercial explosives
1717 IGC code amendments
1718 Safe use of pesticides in ships
1790 Stability requirements for ro-ro passenger ships
1790 Amendment 1: Stability on Roll on-Roll off ships
1794 Amendment 1: Counting and registration or persons on Passenger Ships
1799 Rabies: Carriage of animals on ships
1852 Requirements for tanks and taking Dangerous Goods by sea
1906 Carriage of Dangerous Goods and Marine pollutants, amendments
1908 The Merchant Shipping (Control and Management of ships ballast water and sediments) Regulations 2022

BIBLIOGRAPHY

Activities in Port Areas: IMO.
Cargo Access Equipment: published by Clarke Chapman.
Cargo Securing Manual.
Cargo Stowage and Securing: A Guide to Good Practice: by Charles Bliault.
Code of Safe Practice for Cargo Stowage and Securing: IMO.
Code of Safe Practice for Ships Carrying Timber Deck Cargoes: IMO.
Design and Operation of Ships' Derrick Rigs: British Standards Institute.
Hatch Cover Inspections: Nautical Institute, by W. Vervloesem.
International Maritime Dangerous Goods code: IMO.
Irradiated Nuclear Fuels Code (INF Code): IMO.
International Maritime Solid Bulk Cargoes Code (IMSBC code).
International Safety Guide for Oil Tankers and Terminals (ISGOTT): published by Witherby.
Lashing and Securing of Deck Cargoes: Nautical Institute, by Capt. J. R. Knott, BA, FNI.
Marine Pollution & Climate Change – CRC Press Taylor Francis Group.
Recommendations on the Safe Transport of Dangerous Goods and Related Activities in Port
Thomas's Stowage: published by Brown Son and Ferguson.

Additional Relevant Texts by the Author:

Seamanship Techniques (5th edn): published by Routledge.
The Command Companion (to Seamanship Technique): published by Routledge.
Marine Emergencies for Masters & Mates (2014): published by Routledge.

Marine Heavy Lift and Rigging Operations (2nd Edn): published by Brown Son & Ferguson.
Navigation for Masters (4th Edn): published by Witherby.

Relevant Digital Contacts/websites affecting cargoes

Fumigation http://www.imfo.com, code of practice
IMB Piracy Reporting Centre E-mail: piracy@icc-ccs.org/imbkl@icc-ccs.org
Regulations http://www.legislation.gov.uk/
Radioactive dangerousgoods@mcqa.gov.uk
Website www.gov.uk/government/organisation/maritime-and-coastguard-agency

Commodity Index and Stowage Factors

This commodity index lists numerous cargo types and product varieties, but it is not exhaustive. Not all items are detailed within the main text and users are advised to reference shippers' documentation as well as other associated works.

Limited detail on commodities is enclosed together with the stowage factor, where appropriate, and the relevant page number if considered within the main body of this work.

	Commodity name and details	Stowage factor (m^3/tonne)	Page reference
Acetone	In drums, see IMDG code	2.35–2.5	
Acids	Highly corrosive. IMDG code. Handling precautions required.	–	355
Agricultural machinery	In crates	1.39–2.23	
Ammonia	LPG cargo carriage. IMDG code reference	–	171, 216–246, 348
Ammunition	Magazine stowage. (Dynamite) Dangerous goods IMDG code.	Various	345–347
Anaesthetics	May need temperature controls. Special lock-up stowage as drugs.	–	214
Anchor Cable	Flaked flat athwartships in holds and generally overstowed.	Variable width Cable diameter	
Apples	Temperature control carriage. Cartons	2.37–2.65	261
Apricots	Dried fruit\nFresh	1.39–1.45\n2.56–2.78	87
Asbestos in cases	See IMDG code	1.53–1.67	
Asphalt	Contains drying oils and liable to spontaneous combustion. Should be properly dry.	1.39	

	Commodity name and details	Stowage factor (m³/tonne)	Page reference
Bacon	Cool stowage	1.73–1.84	261
Bale goods	Various, e.g. cotton – 700 lbs per bale	3.62–3.76	80
Bagged goods	Various commodities	2.0–2.23	78/79
Bananas	Mostly in cartons at a carriage temperature of 12/13°C	3.63–3.90	262
Barbed wire	In reels	1.56–1.67	
Barley	Grain Regulations. Bulk Bagged	1.36–1.50 1.45–1.67	151
Barrels	Stowed on side bung up, e.g. castor oil (also in drums) codfish (salted) colza oil (also in drums) creosote (also in drums or bulk) fish oil (also in bulk) Glucose Tung oil	1.73–1.78 1.67–1.73 1.90–2.00 1.67–1.73 1.67–1.87 1.62–167 1.28–1.34 1.78–1.81	82/83
Bean cargoes	In bags, e.g. cocoa, coffee, soya, etc. average S/F Some products also shipped in bulk	2.17	77
Beef	(Chilled) cartons Frozen (Chilled) boneless	1.53–3.76 2.37–2.79 1.67–1.74	260
Beer	Bottled in cartons or in casks Empty casks/drums Cartons Casks Bottled	1.39–1.84 1.50–1.56 1.95–2.09 1.36–9.75	83
Bitumen	Inflammable shipped – In solid or liquid form. Will taint other cargoes. (RD 1.00–1.10) Barrels Drums Casks	 1.25–1.35 1.28–1.39 1.53	212
Bone meal	Stow clear of edible goods. Bags Bones in bulk	1.11–1.25 2.23	78
Bricks	In crates	0.70	
Bulk cargoes	Various commodities	Varies between 0.31–2.81	141–169

Commodity name and details		Stowage factor (m³/tonne)	Page reference
Bullion, bank notes, stamps, etc.	Valuable cargoes Tally in and tally out, lock up stow	Various	338
Butane	LPG cargo carriage		227
Butter	Dairy product in cases	1.45–1.50	260
Cable	Stowed on reels which must be secured against movement	Size variant	86
Canned fruits	Cartons	1.67	
Carbon Black	Bags often on pallets. Very dirty cargo may shift Protect other cargoes, see IMDG code.	1.67	
Cars	Designated car carrier. Individual cars may also be shipped in containers or in open stow. 400 mm required between car units Crated motor cycles	4.18–8.3	295–301
Carpets	Valuable cargo in bales	3.34	80
Case goods	Various	2.79–3.34	84 / 85
Casks	Various commodities, e.g. China Clay Chutney Copper Sulphate (highly corrosive) Glue (liquid) Ginger Stearin (natural fat)	1.23–1.34 1.06–1.11 1.23–1.28 3.07–3.34 1.58–1.81 1.73–1.78	82 / 83
Cattle meal cake	Bags or bulk	1.95–2.09	
Caustic Soda	IMDG code reference. Drums	0.95	
Cement	Different specific gravities. Unitised Bags Drums Bulk	0.72–0.79 0.65–0.70 0.98–1.11 0.61–0.64	78
Cheese	Temperature sensitive Crates Cartons Cases	1.48–1.62 1.00–1.34 1.20–1.25	216
Chemicals	Various IMDG code, bulk chemical code	–	220–229, 348/349
China ware/ porcelain	Various packages careful handling – usual for container shipment	3.34–5.57	

	Commodity name and details	Stowage factor $(m^3/tonne)$	Page reference
Cinnamon	Highly scented stow away from other cargoes		91
	Bundles	3.62–3.90	
	Cases	2.79	
Cloves	May damage by moisture. Ventilate well and stow away from all other goods. Liable to damage.		91
	Chests	3.07–3.21	
	Bales	3.07–3.34	
	Bags	3.38–3.42	
Coal	See IMDG code. Bulk requires surface ventilation Stowage factor variants depending on country of origin	1.25–1.35	160–163
Cocoa	Beans in bags	2.0–2.15	78
Coconut oil	Bulk, deep tank carriage	1.06	92
Coffee	Beans in bags	1.81–2.09	78
Coir	(Coconut fibre) in bales	2.79	81
Coke	Bulk. Absorbs 20% of its weight in moisture if carried as deck cargo.	1.95–2.79	162
Concentrates Bulk	May need shifting boards	Varies on commodity	143
	Average	50–0.56	
	e.g. Copper concentrates	39–0.50	
	Zinc concentrates	0.56–0.61	
Condensed milk	Cases	1.25–1.28	
Confectionery	In cases or cartons	2.34	
Containers	Generally, goods stowed in containers are under the same conditions as open stow		304–337
Copper	Ingots, ore, coils or concentrates Coils	0.84	86
Copra	Stow away from edible foods		79, 166
	Highly infested with copra 'bugs' Troublesome to humans, bulk (hold) avoid steelwork contact (tween deck)	1.95	
		2 2.09–2.15 T/D	
	Bags (hold)	2. 2.09–2.37 T/D	
Corn	Grain Regulations apply, bulk or in bags	1.25–1.41 1.39–1.53	151

	Commodity name and details	Stowage factor $(m^3/tonne)$	Page reference
Cotton	Waste. Liable to spontaneous combustion shipped in bales. Cotton goods in cartons. Note: Cotton seed classed as grain, under IMO.	3.90–4.46	80
Crude oil	Tanker cargo	(S.G. 0.8/0.9)	171
Dairy products	Various. Usually shipped in cartons or cases e.g. eggs, butter, cheese, etc.	Varies with commodity	254–261
Diesel oil	(S.G. 0.6/0.9)		85, 212
Dried blood	In bags	1.11–1.67	79
Dyes	May be powder, liquid or in paste form. See IMDG code. May cause staining.	Varies on package type	85
Earthenware	Mixed parcels Pipes Crates Cases Unpacked	 1.48–1.67 2.79–3.34 1.81–1.95 5.57	
Eggs	In boxes liquid form. Frozen cases and packs	2.93–3.48 1.11–1.25	261
Elephants	On deck. Fully grown animals weigh up to 3 tonnes. Allow for 120 litres of water and 280 kg of food per day. Bills of Lading should be endorsed to show that the ship is not responsible for mortality during passage. (see Livestock)		267
Esparto grass	(Fibre in bales) liable to spontaneous combustion	3.62–4.74	81
Ethyl acetate	Inflammable liquid, drums	1.50–1.78	226
Ethyl chloride	Inflammable liquid, drums	3.62	226
Ethylene	Fully refrigerated		226 / 227, 229
Explosives	Dangerous Goods. Ammunition, dynamite and fireworks. See IMDG code – may require magazine stowage depending on type.		345–348
Fertilisers	In bags or bulk	1.39–1.67	226
Fibres	In bales	2.79–3.34	81

	Commodity name and details	Stowage factor (m^3/tonne)	Page reference
Fish	(Frozen) Boxes or cartons		261
	−18° to −15°C	2. 2.50	
	little danger of taint.		
	Shellfish, crates/cartons	2.2.28	
	Crustaceans, crates/cartons	2.2.34	
Fishmeal	Liable to spontaneous combustion	1.73–1.81	77–79
	Bags must be dry and well dunnaged to provide adequate ventilation		
	Bulk fishmeal may be in pellet or powder form	1.34	
	Space must be full to avoid shifting		
Fish oil	May be shipped in bulk or tins in cases.		
	Bulk	1.09	
	Cased tins	1.39–1.48	
Flour	Bags. Keep off steelwork.	1.39–1.59	79
Formic acid	Corrosive. See IMDG code.		
Fruit	Green – clean spaces with mechanical ventilation (extractor fans) Cases or		87, 261
	cartons	2.37–2.65	
	Dried, cases	1.95–2.09	
	Cartons	1.42	
Fuel oil	(RD 0.92/0.99)		212
Furniture	Large packing cases	1.1–2.2	
Garlic in bags	Strong smelling	2.65	87
Gases	(Compressed) in approved cylinders		228–243
Gas oil	(RD 0.84/0.87)		212
Gasoline	Cases, drums or bulk. Highly inflammable.	1.39–1.4	212
Ginger	Preserved in syrup, wet cargo: Casks	1.58–1.81	
	Cases	2.95–2.09	
Glass	Crates stowed end on and supported	1.26–1.53	
Glue	Various methods of carriage. Bales see IMDG code,	4.18–5.57	
	Drums	3.34	
	Cases	1.81–2.09	
	Casks	3.07–3.34	

	Commodity name and details	Stowage factor (m³/tonne)	Page reference
Grain	Bags or bulk. Grain Regulations apply. Bags Bulk	 1.67–1.81 1.45–1.67	151–160
Grapes	Must have cool ventilation Cases/cartons	3.29–4.18	87
Grass seed	Bags	1.39–4.18	
Guano	Must not be carried with foodstuffs. Bulk or bags. Bags Bulk	 1. 1.17–1.23 1. 1.11	166
Gunpowder	See IMDG code (Class 1 Explosives)		345–352
Hay/straw	In bales	3.34–4.46	
Hides	Shipped in dry or wet condition. Casks, barrels, bales or loose. May be on pallets. Strongsmelling – ventilate. Loose (dry) Loose (wet) Barrels Bags (wet) Bags (dry) Bundles	 2.79–4.18 1.95 1.53 1.81–1.95 2.09–2.23 1.39–1.67	95
Ingots	Copper, lead, etc. Aluminium Lead Tin Zinc Copper loose	 0.50–0.64 0.28–0.33 0.22–028 0.22–0.33 0.28–0.33	86
Iron	(Pig) bulk Galvanised sheet Galvanised coils Ore bulk	0.28–0.33 0.56 0.84 0.33–0.42	41, 141, 163
Jute in bales	(High fire risk)	1.81–1.87	
Kerosene	Cases Drums	1.39–1.45 1.73–1.78	212
Lamb	Carcases Chilled or frozen carriage – cartons	4.18 1.81	260
Lard	Liable to melt with heat In in cases or pails Oil in drums	 1.53–1.61 1.67–1.78	

	Commodity name and details	Stowage factor (m³/tonne)	Page reference
Latex	Bulk – deep tank stow or drums	1.03 1.38–1.53	94
Leather	Rolls or bales may be valuable Bales Rolls	 1.95–2.79 5.57	
Logs	Different wood types stow at various stowage factors because of differing material densities Teak Mahogany sq. logs	 2.23–2.37 0.75–0.84	249, 252 / 253
Lubricating oil	Usually in cases or 50 gallon drums. Drums may be deck stowed. (RD 0.85/0.95)	1.48–1.62	355
Machinery	Sometimes cased	1.12–1.53	
Mail	Lock-up stow Parcels	2.79–4.18 3.34	39, 346 / 347
Maize	Grain Regulations apply. Bulk Bags	1.25–1.41 1.39–1.53	151
Meats	Chilled or frozen. Cases/cartons Mutton – frozen Meat meal in bags	1.81–2.23 2.92–3.06 2.23–2.37	260
Melons in crates	Adequate ventilation	2.79–3.34	87, 261
Molasses	(RD 1.20/1.45) Bulk Drums	0.74 1.1.39–1.67	213
Nitrates	In either bags or bulk. See IMDG code.	1.11	166
Nuts	In bags or bulk. Cool, dry stowage. S/F varies on type.	1.95	166
Oakum	In bales Pressed bales	2.51–2.79 1.1.95–2.09	81
Oats	Liable to heat. Grain Regulations apply. Bulk Bags	 1.67–1.94 1.81–2.06	151
Offal	Frozen	2.32–2.37	260
Oil (palm)	Heating required – Bulk Barrels	1.09 1.62–1.67	92

	Commodity name and details	Stowage factor $(m^3/tonne)$	Page reference
Oil cake	In bags See IMDG code.	1.53–1.95 Varies	
Olives	In kegs or drums	1.90–1.95	
Olive oil	Barrels, drums or bulk	1.67–1.73	
Onions	Good ventilation May taint. Cases and crates 20 bags per ton	 2.23–2.29 2.37–2.51	87
Oranges	Cases or cartons Tainting damage possible	1.67–1.81	261
Ores	Various types of varying densities (in bulk or as stated otherwise) Iron Zinc Bismuth (in bags) Chrome Aluminium Manganese (Galena) Lead	 33–0.42 56–0.67 0.84 0.34 0.84–0.92 0.47–0.50 0.36–0.39	164/165
Paint	In drums	0.50–0.56	91 86/87
Paper	Keep dry and careful handling Reels Bales Rolls	 1.20–2.65 1.1.3–1.8 1.67 & 1.81	
Peaches	In cartons. Refrigeration.	3.78	261
Pears	Fruit cases or cartons	2.05–2.96	261
Pepper/spices	In bags	2.06–2.51	91
Personal effects	Usually in crates	2.83	338
Phosphates	In bulk: Granular Rock	1.12 0.92–0.98	166
Pig iron	Bulk. Angle of repose 36°	0.30	163
Pipes	Bundles	1.67	185–187
Pit Props	In bundles	6.41–7.25	249
Plums	In cartons	2.34–2.41	261
Potatoes	Bags Crates or cartons	1.53–1,81 1.62–1.90	79

	Commodity name and details	Stowage factor $(m^3/tonne)$	Page reference
Poultry	Crates or cartons. Deep frozen	1.67–2.23	260, 265
Prunes	In cases or bags (dried fruit)	1.39–1.45	87
Pulses	In bulk	0.47	151
Radioactive materials	Stow away from crew. See IMDG code.	Varies	344–346
Rags	In bales	1.53–2.09	
Railway iron	As rails	0.36–0.42	164
Rice	In bags. Liable to heat and sweat and susceptible to strong odours. Must be kept dry. Paddy rice White rice	1.81–1.95 1.39–1.45	79/80, 157
Rope	In coils	2.23–2.78	
Rubber	In block or crepe form. Cases Bales Sheet Crepe	1.90–1.95 1.81–1.87 1.67 3.3.34	81
Rum	See spirits		83
Rye	Grain Regulations apply. Requires extensive trimming. Bulk Bags	1.39 1.53	151
Salt	In bags or bulk. Bags Bulk	1.06–1.11 0.98–1.11	79, 166
Salt rock	Granules. Angle of repose 30°. Bulk	0.98–1.06	
Sand	In bulk	0.53–0.56	
Sanitary ware	In cases/crates	4.18	
Seeds	Stowage factor varies with product	1.28–3.76	151
Sheep dip	In drums	1.25–1.53	85
Soda ash	In bags (Treat as dirty cargo)	1.11–1.25	79
Soya bean	Bulk or bags. Bulk Bags (from US)	1.23–1.28 1.59–1.62	

	Commodity name and details	Stowage factor (m³/tonne)	Page reference
Spirits	In cartons. Inflammable. Special lock-up stow, highly pilferable. Also carried in bulk tank containers.	1.67–1.81	83
Steel work	Heavy cargo		74 –76,
	Bars	0.33–0.45	163–166
	Billets	0.28–0.39	
	Castings	1.12–1.39	
	Plates	0.28–0.33	
	Pig iron	0.28–0.33	
Scrap		Various	149, 164
Steel coils	May weigh up to 20 tonnes. Also as pipes, castings and plant machinery.	Various	74–76
Sugar	Dry sugar and green, (raw wet sugar)		79, 166
	Dry sugar in bulk	1.11–1.25	
	Dry sugar in bags	1.28–1.34	
	Green sugar in bags	1.11–1.17	
Sulphur	Shipped in bulk. Fire, dangerous cargo, see IMDG code.	0.84–0.89	166
Tallow	Deep tank stow with heating	1.67–1.78	92
Tea	Chests: Delicate cargo and must be stowed away from odorous commodities	2.79–3.07	91
Tiles	Crates	0.98–1.39	
Timber	Carried in many forms and as deck cargo. Danger from absorption when on deck.	Various	248–254
Tin	See Ingots		86
Tin plate	Bulk packs	0.28–0.39	
Tobacco	In cases	2.23–3.34	81
Tomatoes	In crates and boxes	1.95–2.09	
Tyres		4.18–4.87	
Vegetable oils	Oil or fat from plants shipped in drums or deep tanks	1.67	93/94
Vehicles	See Cars. Ro-Ro and car carriers.		297–301
Whale oil	Drums	2.09	213/214
	Bulk	1.14	

	Commodity name and details	Stowage factor (m³/tonne)	Page reference
Wheat	Bagged or bulk. Grain Regulations apply Bulk Bags	 1.18–1.34 1.34–1.50	151
Whiskey	Bottled in cartons (see Spirits) S/F 1.67. Also in bulk in container tanks.		83
Wild animals	Livestock	–	267 / 268
Wine	Cases (bottled). Also now in bulk tanker vessels. Cases	 1.67–1.95	62,82– 84, 214
Wood pulp	Liable to damage by moisture. Shipped in bales.	1.25–1.39	81, 248 254
Wool	In bales will vary depending on country of origin. Average.	0.48	81

The reader should note that with the dominance of containers, many packaging systems have changed and may be obsolete. However, produce packed into containers tends to generally follow the normal standards regarded as necessary for the safe carriage of commodities as General Cargo procedures dictated.

General Index

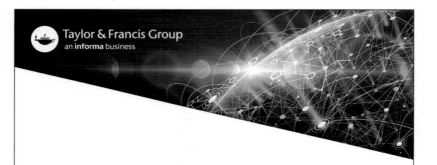

For Product Safety Concerns and Information please contact our
EU representative GPSR@taylorandfrancis.com Taylor & Francis
Verlag GmbH, Kaufingerstraße 24, 80331 München, Germany